International Praise for
O: The Intimate History of the Orgasm

"You might be forgiven for plunging straight into this hot pink tome with unbridled enthusiasm. After all, it promises nothing less than a ride through the development and quirks of that most sought after and occasionally elusive sensation. . . . Gratifyingly, it delivers. . . . The book will have you guffawing on the tube, but it's a double-edged sword; a text which amuses but also illuminates." —*The Observer* (London)

"One of the best books on human sexuality that I've come across . . . an excellent and very down-to-earth history of human sexuality . . . It also takes a very critical but constructive look at sexuality in our western Christian society and the damage that has been done by religious guilt and repression. . . . If I had any control over the school syllabus, I would make this book required reading for sixteen- to eighteen-year-olds."
—Bishop Pat Buckley, *News of the World* (London)

"[Margolis] knows more than you dreamt possible about climaxes. . . . [He] discusses bigger themes, such as how the power of the orgasm, the relentless human drive towards it, has led to a history of religious and political suppression." —*The Times* (London)

"What this rather amazing book is about [is] how our knowledge of, and attitudes to, sex change dramatically with every generation. . . . A serious piece of work. Everything ever written about sex, ever, seems to be referred to in Margolis's 403 pages . . . [though] his style is fluent and light." —*The Daily News* (New Zealand)

O: THE INTIMATE HISTORY OF THE ORGASM

Also by the same author

The Hot House People
Cleese Encounters
The Big Yin
Lenny Henry
Bernard Manning
Uri Geller: Magician or Mystic
A Brief History of Tomorrow

O: THE INTIMATE HISTORY OF THE ORGASM

Jonathan Margolis

GROVE PRESS
New York

First published in the United Kingdom in 2004 by Century,
The Random House Group Limited, London, England

Printed in the United States of America

All quotations are reproduced with the kind permission of the authors.
All works are fully cited in the bibliography.

Every effort has been made to trace all original copyright holders, but if any have been inadvertently overlooked, the publishers will be pleased to make any necessary changes to future printings.

FIRST GROVE PRESS PAPERBACK EDITION

Library of Congress Cataloging-in-Publication Data

Margolis, Jonathan.
O : the intimate history of the orgasm / Jonathan Margolis.
p. cm.
Originally published: London : Century, 2003.
ISBN 0-8021-4216-8 (pbk.)
Orgasm—History. 2. Sex—History. I. Title: Intimate history of the orgasm. II. Title.
HQ12.M346 2004
306.7—dc22 2004052387

Grove Press
an imprint of Grove/Atlantic, Inc.
841 Broadway
New York, NY 10003

05 06 07 08 09 10 9 8 7 6 5 4 3 2 1

Contents

Preface

My grandmother once told my mom that in sixty years of marriage, she and my grandfather never saw each other naked. I never saw my mother naked.

My children, by contrast, are so accustomed to their parents walking around in the nude that they have been known to remind my wife and me to 'put some clothes on' when they have friends to stay.

It goes without saying that the late twentieth century was a time of considerable sexual liberalism. I prefer not to imagine what kind of sex life my grandparents had. I suspect it would have been, like Thomas Hobbes' description of primitive life, '... nasty, brutish and short.'

Yet we fall into the trap of generational smugness if we imagine that our particular time or our culture invented sex.

More than a hundred million acts of sexual intercourse take place every day, according to the World Health Organisation. Men and women have practiced procreative sexual intercourse for approximately a hundred thousand years. A back of the envelope calculation suggests, then, allowing for expanding world population since 98,000 BC, that human beings have had sex some 1,200 trillion times to date.

It cannot, surely, have been bad *every* time.

Sexual history, as a British psychotherapist, Brett Kahr, has put it, is 'a minefield of progression and regression.' Some of the greatest eras of sexual freedom are far in our distant past; some of the most repressed times are within living memory.

What follows is a history not so much of sex, but of sexual *pleasure*, of sex as the culmination of a bonding process between two people or as a recreational activity that involves reproduction only as an optional by-product.

The orgasm is the ultimate point of such sex. It is what we hope to attain. As any soccer fan will confirm, there is enjoyment to be had from a game that ends in a zero-zero score, but a great match requires goals. And many of the greatest matches, for real connoisseurs, have been high-scoring draws, with equal satisfaction for both sides.

Yet orgasm has a highly paradoxical role within sex. My grandparents, I expect, will have been extremely vague about what an orgasm was. My grandfather was a corporal in the First World War trenches, so he will have had a rough idea that it had something to do with 'coming off' as ejaculation was then known.

I am confident that my grandmother, however, will have died with only the fuzziest notion of what an orgasm was or where or how it was possible to acquire one.

Even growing up in the 1960s, I was fantastically innocent by today's standards. It was the custom then in wholesome boys' books like *The Hardy Boys* series for characters to do almost anything rather than 'say' their dialogue. Frank and Joe and friends like Chet and Biff, would asseverate, smile, chuckle, declare, expostulate, muse, mutter and grin their lines, but most commonly, they would 'ejaculate' them. (As in: 'What!' ejaculated Mr. Hardy. 'That's ridiculous! Why would you steal his cane?')

I was vaguely aware that this was a somewhat awkward usage, but I could never understand, until I was long beyond *The Hardy Boys*, why my parents sniggered at it.

And I wasn't alone in such innocence. *Sex in History,* a notably progressive 1953 book by a renowned author of the day, Gordon Rattray Taylor, does not contain the word orgasm. Even its polite alternative, climax, only appears twice.

Yet aside from the need to breathe and eat, the pursuit of

orgasm has been one of the strongest single determinants of human behaviour throughout history.

Every bit as profoundly as the craving for love (more so for much of humanity) the unquenchable desire for orgasm has made and destroyed marriages and dynasties, inspired poetry, drama and novels, destroyed people's health through sexual diseases and fired up a world-wide, cross-cultural sex industry. Sex, rather than money, is the most happiness-inducing factor in modern Americans' lives, according to economics professor David G. Blanchflower of Dartmouth College, New Hampshire, co-author of *Money, Sex and Happiness: An Empirical Study*, a 2004 survey of 16,000 people for the National Bureau of Economic Research. Blanchflower's study quoted previous work in which 1,000 employed people had been asked to rate nineteen activities for happiness; sex came top, commuting, bottom.

In objective terms, however big and overpowering and breathtaking they seem at the time, however exciting and passionate the build up, however warm and satisfying the prized 'afterglow', both the male and the female orgasm are actually relatively minor events.

Even when sexual partners are skilled, compatible and having sex on a frequent, regular basis, a combination that all research suggests is a rare thing, the capacity for orgasm as a life-sustaining mechanism, hardly rates with digestion or vision.

Readers will need to be aware throughout what follows that 'sexology' is an inexact science, in which researchers notoriously are at odds with one another, relying as they do on a somewhat rickety mixture of laboratory work, questionnaires, and, not infrequently, mere hunches. But even if some of the research I present in the pages that follow seems to be contradictory, there is a near consensus that, physically, an orgasm for either sex is barely more than a flash in the pants.

Averaging out for both men and women, they last just around ten seconds apiece. With the mean frequency of intercourse standing at once to twice a week, most individuals will experience a mere twenty seconds of orgasm a week, a minute or so a month, or a total of twelve ecstatic minutes a year.

Over a typical, verging on the optimistic, sexually active lifetime of fifty years, then, we can expect to enjoy some ten hours of orgasm, twenty or thirty for the most avid of masturbators.

Given the time we spend thinking about, worrying over, preparing for and analysing sex and our performances of it, ten or even thirty hours of 'product' in a lifetime seems a very low return for all the effort.

But whether through sex or masturbation, orgasm's serotonin rush and momentary muscular relaxation comprise the most potent and popular drug we have.

And even if we lack the addiction, we are assaulted daily by a barrage of social, cultural and media pressure to acquire it. An English religious commentator, Malcolm Muggeridge, observed in the 1960s that 'the orgasm has replaced the Cross as the focus of longing and the image of fulfilment'.

That was a decade before almost every women's (and, latterly, men's) magazine in the world made it mandatory to include a sensational, monthly feature on orgasm.

Despite its iconic importance in virtually every culture and country in the world, however, no definitive account has ever been written of the orgasm – of its little-known, largely secret history, its biology, anthropology, psychology, technology, and sociology, its cultural role and its literature.

That, aside from the fact that I really rather like orgasms, but could never quite work out why, is the reason I have written this.

The Human Sexuality Collection at Cornell University, the Magnus Hirschfeld Archive for Sexology at the Humboldt University of Berlin and the libraries of RELATE's Herbert Gray College in Rugby, England, the Women's Sexual Health Clinic at the University of Boston Medical School, the Department of Medical Sexology at the University of Utrecht, the Henry Koch Institute in Berlin, the Kinsey Institute in

Indiana, the Institute of Psychology and Sexology in St. Petersburg and the Center for Sexual Health in North Carolina, have all been of great assistance in the research for this book.

My thanks additionally for all their help and wisdom to: Gabrielle Johnson, the late Professor Marcello Truzzi, Dr. Marc Demarest, Rabbi Shmuley Boteach, Mike Robotham, Mike McCarthy, Clare Kidd, Wahida Ashiq, Tracey Cox, Matthew Norman, Bryony Coleman, Jemima Harrison, Jon Laine, Owen Scurfield, Ruth Margolis, Claire Ockwell, Lucy McDowell, Jason Williams, Lisa da Souza, Dr. Marvin Krims, Fiona Wentworth-Shields, Hannah Shepherd, Jon Gower, Julia Cole, Dr. Robert Douglas-Fairhurst, Professor Chiara Simonelli and not forgetting, for obvious reasons, Sue Margolis.

Jonathan Margolis, 2004

1

The Sexiest Primate

'Your children are not your children. They are the sons and daughters of life's longing for itself'

Kahlil Gibran, 1883–1931

The first act of sexual intercourse, the earliest example of two like creatures coming into intimate contact for the purpose of combining their DNA to create a new creature, probably took place about one and a half billion years BC, at a location deep in the primordial oceans. The lovers and parents to-be are thought to have been a type of single-celled blob called *eukaryotes.*

Sexual intercourse subsequently became the normal method of reproduction for practically all animals. Compared to asexual reproduction – or 'cloning', as practised by bacteria, shrimps and stick insects – sex between a male and a female is a far better way of improving the genetic stock of a species and ensuring the long-term benefits of natural selection.

However, the first sexual act by which two like creatures sought intimate contact expressly to give one another physical and emotional pleasure, in an explicit and mutually understood spirit of social, political, intellectual and economic equality and regardless of whether or not they succeeded in reproducing their DNA, may well not have taken place until some time in the twentieth century AD, most likely at a location in Western Europe or North America.

The paramours on this occasion, needless to say, were of the genus *Homo*, species *sapiens*, a distant and highly adapted descendant of *eukaryotes*. Many million examples of these *Homo sapiens* have since refined their sexual behaviour and begun to enjoy as a joint, democratic pleasure the powerful orgasmic spasm exclusive to their species, and differentiated again between males and females whose orgasms are so different, yet so similar.

So while sex is nothing new or particularly unique to humans, orgasm – in the sense of the pleasurable sensation enjoyed by the two sexes outside a reproductive context and sought in a pre-meditated, practised way – is both. On the evolutionary scale, *Homo sapiens* is a global newcomer and the orgasm is a complex, sophisticated phenomenon unique to these strange, new, bipedal creatures. A few isolated species and sub-species aside, non-humans do not share our studied pleasure in orgasm. Even in the modern era, most *Homo sapiens* who have begun to appreciate this subtle, tricky, fickle but deeply moving neurological reward for the drudgery of reproduction have yet fully to exploit its delights. And by extrapolating Western surveys, which repeatedly report on the lack of sexual fulfilment suffered by a large proportion of sexually active people, we can reliably surmise that for the majority of humankind, satisfactory exploitation of the capacity for orgasm remains an unfulfilled ambition, a rigorously proscribed societal taboo – or a pleasure of which they are simply unaware.

The paradoxes and inconsistencies of orgasm make it a phenomenon to rival quantum mechanics in its fickleness. One indication of the orgasm's immaturity in the scheme of things is that the anatomical machinery designed, or so it appears, for male/female pairings to enjoy the orgasmic spasm simultaneously and thereby promote the worthy cause of a couple's mutual happiness and spiritual bonding, often works creakily, if at all.

Men are prone to have orgasms too easily, while women tend not to have them easily enough. The existence of prostitution by women for men in every society, but the reverse only in a tiny minority of Western cities, suggests additionally, and eloquently so, that men are also more physically dependent on frequent orgasm than women – dependent, that is, in the crude, urgent, mechanical 'offloading' sense that only they perhaps know. As a famous London madam, Cynthia Payne, once succinctly put it: 'Men are all right as long as they're de-spunked regularly. If not, they're a bleeding nuisance.' Masturbation too tends to be quicker and less of a production number for men than for women – although, so far as we can tell without the benefit of anyone who has masturbated as both a man and a women, it seems to be a rather less satisfying activity for males.

Rarer and the result of a more refined longing as they are, however, women's orgasms, with their satisfying multiple muscular contractions, are an infinitely bigger and more expansive experience than the sensation men have when they ejaculate – a fleeting feeling not dissimilar, when the emotion is stripped out of it, to common-or-garden urination from an overfull bladder, a sneeze or an urgently needed bowel movement. The most prosaic analogy to be heard from a woman along such lines is that having one's ears syringed is not unlike a very small orgasm.

There are more fundamental inconsistencies between the two genders' orgasms, too. One such apparent mismatch in heterosexual intercourse is that men's orgasms are practically essential for reproduction to take place, whereas women's do not have any obvious function other than to be pleasurable. A woman is designed to conceive after intercourse regardless of her sexual response during it.

There are important grey areas here that need to be clarified early on in an account such as this. One is the question of whether male orgasm is a straight synonym for ejaculation. Ejaculation in men is the physiological expulsion of seminal fluid, whereas orgasm is the 'climax', the peak of sexual pleasure. Orgasm and ejaculation generally coincide, but they have been acknowledged for many thousands of years to be distinct processes that can occur independently. Some semen may be emitted before the male has even become very sexually aroused. And most men are familiar with the 'dry' orgasm that can result from a number of sexual dysfunctions, as well as be consciously cultivated, most famously by 'Tantric sex' practitioners, in an attempt to preserve sexual stamina and erection. The muscular pulse of orgasm proper, however, serves to aid conception a little by pushing sperm on its way along the eight to thirteen-centimetre-long vaginal passage.

The female orgasm is not completely divorced from conception, either. In fact, women may, according to one veteran British sex resercher, Dr. Robin Baker, retain slightly more sperm after orgasm than in climax-less sex, and while orgasm is occurring may even draw the sperm up through the cervix and into the uterus. This is a marginal effect, however. Orgasm is functionally unnecessary for successful conception.

Yet for billions of women, neither Baker's contention that orgasm aids conception, nor the model of female orgasm as a pure pleasure finer than anything males will ever know, means a great deal. This is for the very good reason that orgasmic pleasure remains elusive for much or all of their life. Female orgasm even in today's supposedly more knowing world is an all-too-rare thing, and there is little reason to suppose the situation has ever been any better. Lionel Tiger, Charles Darwin Professor of Anthropology at Rutgers University, states a truth for all times when he declares in his 1992 book *The Pursuit of Pleasure* that, 'The gross national pleasure is far lower than it need be.' It is ironic that, mechanically speaking, in terms of reliability of orgasm, male homosexual intercourse and its female parallel 'work' rather better than the reproductive (or pseudo-reproductive) heterosexual variety.

If it is not all, then, an agglomeration of curious design flaws, the best that can be said of the human orgasm is that it is a work in progress. But 'progress' implies a history, and how can we know anything of the human orgasm's history? Does it even have a history?

To answer the second question first, we can be reasonably confident from the study of surviving isolated primitive societies, many of which have language and customs relating to both male and female orgasm, that it has existed for the 100,000 years or so that humans have been 'civilized'. And if we take an overview spanning from those days to now, we can tell that, far from being some fleeting neurological phenomenon like blinking, orgasm has consistently been of disproportionate importance to the way people have evolved both as organisms and in societies.

But has the orgasm *changed;* has it improved or deteriorated, in any progressive or regressive way, down the millennia? History is above all a narrative and without evidence of a traceable journey what follows might as well be an account of constipation through the ages – even though this might, on reflection, not be such a fatuous idea; many great people, from Martin Luther to Mao Ze Dong, suffered from the affliction and may well have owed their temperament and actions in part to its miseries.

But the human sexual climax is more important than constipation. The orgasm's has been a long intellectual journey, during which whole civilisations for long periods in their history have advanced, then backtracked, then advanced again. If orgasm were merely humankind's profoundest pleasure, it would be a matter of some importance – especially given the tortured relationship various cultures at different times have had with the curiously controversial idea of enjoyment.

But the orgasm's existence, as we shall see in this and subsequent chapters, has been influential in more than just the history of human physical gratification. Orgasm has been central, principally, in defining how both men and women and same-sex partners form and maintain couples. This in turn has been crucial to directing the way the human family has developed, to determining important facets of how we live together in broader communities under religious and legal constraints, and even to shaping, via the institutions of marriage and subsequent property inheritance by children of a sexual union, how we distribute land and material goods.

The pursuit of orgasm has, indeed, been one of the most powerful impulses we have. Its iconic importance has been manifest in every culture and country in the world – never more so than today. And a large proportion of the world's literature and art has been preoccupied with the endless, appetitive, unquenchable craving for orgasm; the sexual compulsion, of which orgasm is the goal, has routinely made and destroyed marriages, and occasionally dynasties.

There is a good argument that testosterone, the chemical catalyst for desire in both sexes, has been the most influential compound in human history. Bill Clinton's predilection for oral sex in the Oval Office was only the latest chapter in a long dirty book. Today, additionally, in an era when it is no longer widely taboo, the quest for orgasm has become even more of a business proposition than it was when the oldest profession was the only profession. Orgasm has become the universal yearning that underpins industries ranging from fashion to film to pornography to pharmaceuticals. Britney Spears even released a song in 2003, Touch of my Hand, expressly about masturbating. It is not merely for sensationalist impact that the zoologist Desmond Morris called *Homo sapiens* 'the sexiest primate alive' – and he did so thirty years before that key date in the history of orgasm, March 1998, when the US Food and Drug Administration approved sildenafil citrate (marketed as Viagra), the first oral pill to treat male impotence.

But we have been sexy in different ways at different times. Evidence that human sexuality is a completely different thing when you compare, say, Ancient Rome, Renaissance Florence and 1980s San Francisco, implies that the orgasm is effectively a cultural artefact, and that the sexual urge is shaped as much by society as by hormones. As the scientist and philosopher Jacob Bronowski wrote in 1969, in his book *The Ascent of Man*: 'Sex was invented as a biological instrument by (say) the green algae. But as an instrument in the ascent of man which is basic to his cultural evolution, it was invented by man himself.'

Unlike when palaeontologists find a bone or archaeologists an arrowhead, anthropologists and historians have no access to the cultural artefact of orgasm, to people's actual experience of it. We have contact only with the edifice of text and artwork surrounding the artefact, which is not quite the same thing; a Princeton University scholar, Professor Lawrence Stone, has explained that even when data does survive on historical love-making, it is highly selective. Few historical letters or diaries allude directly to sex, and those that do – like some of the earthy British seventeenth- and eighteenth-century diaries of men like Pepys and Boswell – give an entirely male perspective. Assessing women's experience of sex in these circumstances is extremely difficult. We, the modern and post-modern Western cultures, have accordingly 'created' the orgasm in the same very real sense that we have 'created' radio from naturally existing but disorganised electromagnetic waves.

But the orgasm is, additionally, the principal example of the extraordinary human genius for intellectualising and making a pleasure for its own sake of natural phenomena that happen to be necessities of life. From the need to eat and the resultant discovery of cooking food, we developed gastronomy. From the need to communicate and the resultant evolution of language, we developed poetry. From the requirement to keep warm and the resultant clothing, we developed fashion. From the need to keep fit for hunting and fighting, we developed sport. And from the need to reproduce, we have honed the by-product of our reproductive act, the phenomenon of orgasm, into a leisure pursuit which we follow for the sheer enjoyment of it. Even medical science, with fatal diseases still unconquered, has found time to concern itself with differentiating between pleasure and reproduction.

Not all the evidence for orgasm being a cultural construct is to be found in literature or art, however, nor even in the peculiar moral blanket with which myriad cultures and religions have attempted down the years to smother orgasm and try to suppress or kill it off altogether. The most telling way in which we have built a cultural superstructure on the foundations nature gave our species is to be seen in the manner in which we have succeeded, through contraception, in separating the natural coincidence between sexual climax and babies.

It is remarkable that today heterosexuals barely think about squalling, puking, doubly incontinent infants when they have sex. Sex is primarily seen as being about romance, glamour, pleasure, good living, happiness – almost anything other than nappies, sleepless nights and teething rings. To the straight couple having wild sex in the dunes or simpering at one another in an expensive restaurant, it might even seem 'unnatural' and strange that what they are doing has the slightest connection with baby production. This interest in pure eroticism, in sex as a leisure pursuit, is fundamentally human – a primary symptom of civilisation in the way it presupposes that basic survival needs have been taken care of, and that there is now time and energy to spare for fun for fun's sake.

Professor Richard Dawkins of Oxford University, one of the world's leading thinkers and writers on evolutionary biology, points out that contraception appears to overturn the most fundamental Darwinian dictates by offering the pleasure of sex without the reproduction. He suggests that the explanation for this is that the human brain has evolved its own, advanced, ameliorated spin on survival; in sex-for-pleasure, the brain is seeking and experiencing pleasure as another method of aiding survival. The civilised activity of sex for fun may well go back further than we imagine, too. According to a book by Jeannette Parisot, *Johnny Come Lately: A Short History of the Condom*, a fresco in the Dordogne, France, dating from 10–15,000 BC, provides the (literally) sketchy first record of a sheath being used for sex.

Human culture meanwhile, rather than evolution, dictates even the times when we indulge in the pleasure of sex. Oestrogen and testosterone levels are at their highest at dawn, yet the most common time for lovemaking in modern Western civilisations is 11 p.m., between the smoochy dinner date and the need to get to sleep so as to be up for work in the morning.

So what about the more difficult question of the physical feeling of orgasm, and how we might assess whether that too has a history? This is bound to be a tricky question, given that even sexually aware men and women have never quite managed to impart to the opposite sex what their version of orgasm feels like. Greek myth, informed as it was by the experience of Greek mortals, had it that a man called Tiresius was privileged to spend seven years as a woman, and then invited to Mount Olympus to be debriefed by Zeus on his experiences. Tiresius's principal conclusion was remarkable; after taking seven years to ponder on the huge number of differences between the sexes, he summed up his observations in one line. Women, he informed Zeus, enjoy sex more than men. For being the bearer of this unwelcome message for men, Tiresius was blinded.

In the modern world, transsexuals who have undergone surgery or hormone therapy offer us a hint, possibly, of what Tiresius might have experienced had he actually existed. Even without surgery, an FTM (female-to-male transsexual) can possess both a penis produced by testosterone acting to enlarge the clitoris, and the vagina he was born with. Several people with these characteristics have reported on Internet message-boards a sensation that they describe as a simultaneous male and female orgasm in the same body — two distinct and 'definitely different' sensations. When pressed to differentiate between the sensation these 'competing' orgasms offered, one respondent reported that the only difference he could describe was that while the penis contracted from base to tip, the vagina did the reverse, contracting inward, from outside to inside.

As for gauging what were the sexual feelings experienced by prehistoric men and women, let alone their simian grand-parents, we are obviously confined to informed guesswork. It is perfectly plausible that the greatest sexual difference between prehistoric men and women was the same as, according to

Irma Kurtz, it is today: when it comes to sex, Kurtz wrote in a 1995 book, *Irma Kurtz's Ultimate Problem Solver*, the male's greatest fear is of failure, while the female's is of not being loved.

But we are not completely without biological evidence on which to base our conjecture about what sex was like thousands of generations ago. Desmond Morris contends that man's basic sexual qualities come from his 'fruit-picking, forest ape ancestors', and according to the American anthropologist Helen Fisher, too, the physical facility for orgasm had already evolved before our ancestors came down from the trees.

Of course, we have no preserved soft tissue from prehistoric times to prove conclusively that pleasure has always been sought through sexual intercourse. But it is highly unlikely that the clitoris, the only organ in the human being that exists purely for pleasure, with no known anatomical role, has suddenly developed in the brief five million years of human evolution as a response to our growing social and intellectual sophistication.

An outstanding piece of biological evidence for prehistoric man and woman having been rather erotic beasts is measurable with a ruler. It is the simple fact that the human penis is enormous in proportion to the rest of a man's body, dwarfing by far even that of the gorilla, whose organ is a puny two inches erect. Only the barnacle, improbably, has a larger penis in relation to its body size. (Owing to the barnacle's somewhat sedentary lifestyle, its penis has to be capable of searching the area around it to find a receptive female. It also throws its penis away once a year and grows a new one.)

Not only do human males have gigantic sexual organs, but they flaunt them, too. Gorillas might argue that they do not need such a huge penis because they show off by means of their body size. But human men demonstrably announce their sexual power by displaying their penis – or, if one might lapse for a moment into barroom Freudianism, a symbol thereof, such as a gun or a car with an extravagantly long hood and a name such as *Testarossa*, a real name, chosen without apparent irony by Ferrari. Additionally, any man who has felt under his fig leaf can attest to feeling as if he has as many highly tuned nerve endings down there as lucky women with their clitorises, even though anatomists claim these have 8,000 nerves – twice as many as the penis. This is not to forget the labia, which for 10 percent of women have even more nerve endings than the clitoris. The clitoris is also understood today to be bigger than was once thought; it seems to have two previously undetected 'arms' extending some nine centimetres back into the body and up into the groin.

Another key exhibit in the case of whether prehistoric humans had the ability and desire to enjoy sex for its own sake is to be found in the existence of a substance called oxytocin, or 'hormone of love' as it has been called. Oxytocin is a neurotransmitter synthesised by the hypothalamus or 'master gland' at the base of the brain and stored in the posterior pituitary, from which it pulses out when required, which is during sexual activity and later in childbirth. At the end of a pregnancy, oxytocin stimulates uterine contractions and milk production. Immediately after a woman has given birth, it also prompts the desire to nuzzle and protect infants. This effect of the hormone, its tendency to prompt sensuality, helps us understand its role in promoting the pleasurable feelings of sex.

Oxytocin, which we have no reason to assume is anything new in the body's inventory of catalytic chemicals, induces feelings of love and altruism, warmth, calm, bonding, tenderness and togetherness, of satisfaction during bodily contact, sexual arousal and sexual fulfilment. And it is during orgasm in both men and women that oxytocin floods through our bloodstream more notably than at any other time besides the peak of childbirth.

As with most other bodily substances, oxytocin has more than one job. It makes us feel warm, content and at one with our partner. It also coordinates the sperm ejection reflex with the male orgasm, and, it is believed, the corresponding sperm reception reflex that operates in the female orgasm. As part and parcel of inducing an altered state of consciousness, oxytocin released by female orgasm helps women lie still for a while after orgasm. This crucially increases the likelihood of conception – as well as making it probable that women will seek further coitus because they enjoyed it so much the last time.

Oxytocin is, then, far more than a side effect of orgasm, or a necessary component in the formula that produces successful reproduction. It is what makes orgasm nature's sugar coating to disguise the bitter pill of reproduction, the chemical basis for our capacity and longing for romantic attachment. It is the molecule that for 100,000 years or more has made us want to have sex face-to-face, adoring one another, and to live in permanent, monogamous couples – the latter otherwise done only by one species of ape, the bonobo, an endangered chimpanzee existing in small numbers in the Congo and believed by some naturalists to be the closest primate to humankind. Albatrosses, swans, a handful of crustaceans and a rare New Zealand songbird called the hihi also 'mate for life' – but not for remotely 'romantic' reasons.

2

Coming, Coming, Gone

'But did thee feel the earth move?'
Ernest Hemingway, *For Whom The Bell Tolls*

What actually happens to men and women when they reach orgasm, as the process initiated in nervous and psychogenic centres translates itself into the vascular and muscular?

The spasm lasts a few seconds to a minute at the most, but is accompanied by intense physiological activity. Genitals swell with blood, the pulse races, muscles contract involuntarily. Some people's mouths open. Others' faces contort. Many women's toes curl. In men, big toes often stiffen as their little toes twist. Both partners' feet may arch and shake. Sweat typically surfaces on both participants' brow, the heart pumps frantically, and breathing becomes fast and shallow. Both partners' nostrils may flare and seem to heat the air as it surges through them. With climax, each partner is clenched by contractions at consistent 0.8-second intervals. The human sexual summit is a paroxysm of pleasure. A warm glow envelops the waist and chest. The toes relax.

The emotions, too, generally go into a seismic convulsion. For attempting, or pretending to attempt, to add to the species, both parties have received their reward. A mist of goodwill, wellbeing and lazy relaxation temporarily obscures reality. Both men and women may laugh or cry, or become uncommonly ticklish, although all these reactions are less common for men on the basis that they tend to show their feelings less anyway. Both sexes may experience a burst of creative thought since orgasm produces a near lightning storm in the right, creative-thinking, side of the brain. Biological duty fulfilled, there normally follows a lengthy period of exhaustion, rest, and frequently sleep.

Orgasm is so powerful an emotional and cultural icon that it can even be experienced in some cases by people who strictly speaking cannot 'feel it'. Descriptions of orgasm by male and female participants in one study of people with spinal cord injuries transcended accounts of mere ejaculation and muscular contractions. Their focus was on 'warmth', 'tingling', 'energy releasing' and 'energies merging'. The essence of the orgasmic experience, it would seem, survives even sensory disconnection of the genitals from the brain. This happens because the brain is not solely responsible for sexual arousal; the spinal cord is also important. Independent activity in the spinal cord also explains why disabled men can often have erections and father children, although normally cannot feel sexual sensations.

Descriptions of how the outer symptoms of orgasm appear to the other party in a sexual coupling are, curiously, more commonly found than subjective reports of sex as it feels to the participant him or herself. The fact that this is so is an interesting commentary on the bipartisan nature of orgasm, as opposed to the essentially lonely nature of, say, drug taking. People will always describe their own feelings on taking LSD, but focus on their partner's outward appearance when they orgasm, largely because (they hope) it was their unique blend of charm and skill that created the orgasm their partner is enjoying.

There is, then, bound to be a slight element of bragging in a description of female orgasm such as that of the fifth-century Greek author of erotic romance Achilles Tatius in his *Adventures of Leucippe and Cleitophon*: 'When the sensations named for Aphrodite are mounting to their peak, a woman goes frantic with pleasure, she kisses with mouth wide open and thrashes about like a mad woman'.

As for personal, experiential accounts of the sensations they themselves feel when they orgasm, men's are, tellingly, much rarer and more laconic than women's – an indication, perhaps, of the accuracy of a clutch of folk sayings found the world over, all to the effect that the male orgasm was invented by God, but the female's was the work of the Devil. It is notable that in the eighteenth-century pornographic novel *Fanny Hill*, although it is written by a man, John Cleland, all the (highly fanciful) descriptions of orgasm are of the female variety. Additionally, the most obvious physicality of the male orgasm, the rush of liquid through the penis, practically never features in men's descriptions of their orgasmic experience. The question is often raised of whether either gender's orgasm, although particularly the male's, can adequately be described using language.

The American novelist, Jonathan Franzen, in his celebrated 2001 work *The Corrections*, appears uncharacteristically lost for words to cover the moments between a lyrical description of one of his male characters entering his wife and the gratifying aftermath of his orgasm: '. . . with a locomotive as long and hard and heavy as an O-gauge model railroad engine, he tunnelled up into the wet and gently corrugated recesses that even after twenty years of travelling through them still felt unexplored . . . he no longer felt depressed, he felt euphoric . . . She rose and dipped like a top on a tiny point of contact, her entire, sexual being almost weightless on the moistened tip of his middle finger. He spent himself gloriously. Spent and spent and spent.'

A British writer, Toby Young, gives a poignant autobiographical description of the male experience of post-orgasmic delight in a 2001 book, *How To Lose Friends And Alienate People*. Young's on-off girlfriend, Caroline, whom he loves profoundly, finally accepts his proposal of marriage and in celebration, the couple sleep together for the first time in a long while.

'Suddenly, everything seemed to shrink in size, as if I was travelling away from the scene at a hundred miles an hour. Except it wasn't a spatial sensation, not a linear movement. It was as if the gravity that kept my emotions in check had disappeared. It was like being in the swell of the sea, but not quite. Above all, there was the feeling of being outside time, what Freud called "the sensation of 'eternity'". It was like touching something with a part of myself I wasn't normally aware of. I felt as if I'd made contact with the very essence of the universe.'

A 2002 book called *The Joy Of Writing Sex*, by Elizabeth Benedict, a US author and fiction-writing teacher, is notable for the absence among a plethora of fine literary sexual writing of a solitary instance of a writer describing the male experience of orgasm. The finest female description is from Toni Morrison's novel *The Bluest Eye*, in a passage where the character Pauline describes sex with her husband Cholly:

> I pretend to wake up, and turn to him, but not opening my legs. I want him to open them for me. He does, and I be soft and wet where his fingers are strong and hard. I be softer than I ever been before. All my strength is in his hand. My brain curls up like wilted leaves . . . I know he wants me to come first. But I can't. Not until he does. Not until I feel him living me. Just me. Sinking into me. Not until I know that my flesh is all that be on his mind. That he couldn't stop if he had to. That he would die rather than take his thing out of me. Of me. Not until he has let go of all he has, and give it to me. To me. To me. When he does, I feel a power. I be strong, I be pretty, I be young. And then I wait. He shivers and tosses his head. Now I be strong enough, pretty enough, and young enough to let him make me come. I take my fingers out of his and put my hands on his behind. My legs drop back onto the bed. I don't make no noise, because the chil'ren might hear. I begin

> to feel those little bits of colour floating up into me – deep in me. That streak of green from the june-bug light, the purple from the berries trickling along my thighs, Mama's lemonade yellow runs sweet in me. Then I feel like I'm laughing between my legs, and the laughing gets all mixed up with the colours, and I'm afraid I'll come and afraid I won't. But I know I will. And I do. And it be rainbow all inside. And it lasts and lasts and lasts.

What is most interesting to the student of the orgasm about Benedict's collection of examples of what she regards as good sexual literature is that in the rare cases where a male author has gone the distance in attempting to provide a proper description of orgasm, it is not to explain what his own climax feels like, but what his female partner's orgasm feels like to him. John Casey, in his novel *Spartina*, provides a good example of this:

> He turned his head so his cheek was flat against her. He could feel her muscles moving softly – her coming was more in her mind still; when she got closer she would become a single band of muscle, like a fish – all of her would move at once, flickering and curving, unified from jaw to tail . . .

The most common description of the physical sensation of orgasm, male and female, in both literature and everyday conversation is, paradoxically, that it is 'difficult to describe', or 'indescribable'. W.C. Fields expressed this paradox when he observed: 'There may be some things that are better than sex, and there may be some things that are worse. But there is nothing exactly like it.' We frequently follow our dumbfoundedness when it comes to describing sex by comparing it to things we have almost certainly never experienced – volcanoes erupting, cannons firing, 'paradise', an explosion, the earth indeed (with due respect to Ernest Hemingway) moving. Germaine Greer once likened orgasm to childbirth; she is childless. It was Greer, however, who was on hand to rap another writer, Sean Thomas, over the knuckles when he described male lust in *Cosmopolitan* magazine. 'Male lust is like a great river crashing down to the sea – put an obstacle in its path and it will merely find another route,' said Thomas.

Greer commented acidly: 'It is one of the commonplaces of pornography grossly to exaggerate the volume of ejaculation. Male sexuality is more like a sluggish trickle meandering across a delta, dissipating its force in trillions of channels; twentieth-century men are like De Sade's jaded aristocrats, so sated with sexual imagery that they must behold ever more bizarre and extravagant displays before they can achieve potency.'

As for more experiential descriptions of orgasm: an expulsion of tension, a total release, a big shiver, a glorious sneeze, a spasm, a fluttering, a pulsating, a flash, a surge and a rush are among the less prosaic. Kenneth Mah at the psychology department of McGill University in Montreal, Canada, collected descriptions of orgasm for his PhD dissertation. He offered participants a rich variety of adjectives to choose from in describing their orgasms, among them, 'pulsating', 'erupting', 'quivering' and 'rapturous'. But among his respondents, the distinctly dowdy 'powerful' 'intense' and 'pleasurable' were by far the most popular.

In a 1998 article in the Indian magazine *The Week*, writer Stanley Thomas garnered a unique description of female orgasm, best known in India by the Hindi word *nasha*. 'People describe orgasm in different ways,' Thomas wrote, 'and they call it by different names, climax in English, *sukun* in Urdu, *trupti* in Tamil. Describing it is the hard part, because there are so many descriptions. One woman describes the feeling as that of a hot chocolate egg breaking inside her.

'Others liken it to a hiccup, a ripple or a peaceful sigh. Orgasm,' concluded Thomas, quoting Dr Prakash Kothari, head of sexual medicine at the King Edward Memorial Hospital in Mumbai, 'is like a sneeze. If a woman has had one she will know about it. Otherwise any attempt to explain it will be like explaining a rainbow to a blind man.'

Some impressive descriptions of orgasm were collected in 1999 from members of the public by a Montreal-based Web magazine, *www.QueenDom.com* – not, as the name may suggest, a gay publication, but part of a respected and very popular psychological testing site, Plumeus Inc. QueenDom is aimed principally at educated Western women from 25–45 years old.

Among the best female descriptions were these:

> It begins in my vaginal region and moves through my whole body, and I have an uncontrollable urge to moan or scream. I also twitch a lot, experiencing muscle spasms all over and I feel really warm 'down there'.

> I feel like I've been hit by lightning or touched an electrical socket, but after about ten seconds, I relax quickly and the tension flowing out of my body is such a relief.

> I first feel my toes and head tingle, then two waves in my body swell, race towards each other then crash. That is orgasm.

Indisputably male descriptions were rarer, which is interesting considering that men are presumed to be so fixated on acquiring orgasms. This was one man's lyrical account of his, however:

> I hit an egoless space and see a rush of rainbow colours, sometimes little electric swirly things in my vision. I feel white-gold explosions in my prostate and testicles, spreading up my penis, a soft golden feeling all over my body.

Most interesting, though, were the descriptions that are quite difficult to place. Could anyone be absolutely sure which sex provided these reports? They *seem* to be female, because we are accustomed to women being more poetic about their feelings, – but it might be foolish to place bets on which is which.

> It feels like electricity running through every vein of my body.
>
> I see lights, feel like exploding, hear acutely, heat rises from my feet and a tingle rushes up my legs through my thighs, up my torso and throughout my upper body. A feeling of euphoria releases in my abdomen just before what I call 'falling off the mountain'. I can hardly move for the ecstasy I experience.
>
> I see a bright light and hear a roar like the ocean.
>
> I feel like I am going to explode, and then a great relief – often like I am floating down through the air. My body often trembles and shudders. I see stars.

Such academic research as there has been suggests that in fact orgasm may be more similar than we imagine for men and women. In 1976, researchers E.B. Vance, and N.N. Wagner took written descriptions of the sensation of orgasm from 24 male and 24 female psychology students at an American university. They carefully removed from each account any reference to specific body parts and presented the results to a group of 70 gynaecologists, urologists, psychiatrists and psychologists. Vance and Wagner's conclusion, as reported in *Archives of Sexual Behavior*, was that guesswork was the only way possible to establish which descriptions were by men and which by women.

Whether Vance and Wagner's was a rogue finding or not, there is a significant difference in the range of stimuli capable of setting each gender on the path to orgasm. While a number of 'romantic' stimuli can expedite orgasm in women, one in particular, which seems only very distantly related to sex, can nonetheless spark orgasmic feelings in men. This is aggression.

Aggression and male orgasm are closely connected; nerve tissue in the brain associated with aggression is reportedly so closely intertwined with that carrying sexual messages that it is difficult to separate the two. Fulfilling the sexual urge and the aggressive urge can become confused for men. In *Pumping Iron*, the 1977 documentary on Arnold Schwarzenegger, the performer now known as the Governor of California admits that flexing his biceps is as satisfying for him as ejaculating. Furthermore in a wide variety of men, the compulsion to dominate, humiliate and subdue through sex can, additionally, override the desire for orgasm.

The murderer Jack Henry Abbott, who was briefly lionised by literary New York after Norman Mailer championed his writing from jail, once described his sexual thrill on stabbing a man through the heart: 'You can feel his life trembling through the knife in your hand. You must *masturbate* to the violence taking place inside your mind.'

The seemingly inextricable correlation between sex and violent aggression is no better demonstrated than in the behaviour of conquering armies. Not the first such instance by any means, but by far the best documented, was the Soviet Red Army's advance toward Berlin in 1945. By historian Antony Beevor's calculation, the Red Army's 2.5 million men were responsible for the ferocious gang and individual rape throughout eastern Prussia of two million German women. The key question so far as the subject of this book is concerned is, was the raping Soviet soldiers' primary motivation the satisfaction of sexual desire through easily obtained orgasm? Or the gratifying, temporary dousing via orgasm of anger-fuelled aggression? The answer surely has to be the latter, just as in ordinary criminal cases of rape, aggression, with a desire for domination and control, fuelled by a childhood pattern of exposure to violence plus a history, real or imagined, of sexual rejection and inadequacy, are clearly the most important elements for the rapist – rather than some sudden, urgent desire to behave in a sensual manner.

In the case of the Red Army in 1945, Beevor believes, sexual opportunism, enhanced by the feeling among the Russians that they were already effectively dead men with nothing to lose, also played a part.

There is also a small, new revisionist movement afoot in evolutionary psychology arguing that rape, whatever its immediate motivation, is at root no less than an evolutionary strategy. The Greeks often portrayed the god Zeus raping women, and some scholars have argued that rape was the Ancient Greek man's 'right of domination'. Any libidinous male politician or CEO, furthermore, even if he is ugly, will confirm that aggression in males tends to beget a surfeit of sexual partners. This applies even though the more 'red-blooded' type of male is typically a selfish and incompetent lover, often, as we commonly read in anecdotal accounts of alpha-male men in possession of great political power, something close to a licensed rapist.

We have no such anecdotal evidence of sexual proclivities in the case of, to name but one powerful political figure, the late President Mobutu of Zaire. But most people will draw their own conclusions on learning that the dictator's official name, Mobutu Sese Seko Kuku Ngbendu Wa Za Banga, translates as 'the cock who goes from hen to hen knowing no fatigue'. When we look at a dictator like Benito Mussolini, however, we start to get a distinct picture of super-priapic behaviour, poor sexual technique and huge political power going hand in hand.

A New Life, the 2003 biography by Nicholas Farrell of the Italian Fascist leader, reveals that Mussolini managed to have sex with a different woman every day for fourteen years, more than 5,000 in total from September 1929 to the collapse of his regime in July 1943: yet, far from being a successor to Casanova, he was in such a hurry to relieve himself sexually that he rarely bothered to remove his trousers. Mussolini was nicknamed by in-the-know Italians, wrote Farrell, 'Phallus in Chief'. He would spot beauties in the street from his red Alfa

Romeo, have them stopped and checked out by the police, then summon them to the Palazzo Venezia in Rome. There, in his marble-floored office, he would rip off a woman's clothes and have urgent, rough sex with her in seconds – on a stone window seat, against a wall or on the carpet. One woman reported his sexual technique as consisting of squeezing her breasts as if they were 'rubber automobile horns'. After he ejaculated, he would throw the woman her underwear and rarely see her again; one complained of being dismissed without so much as the offer of 'coffee, liqueur or even a piece of cake'.

But rather than this nerdish sexual technique making him a joke among Italian women, even if a very bad one, the little man seemed, bizarrely, to have enormous sex appeal. Thousands of letters would arrive every day from women begging him to have sex with them. A schoolteacher in Piedmont wrote requesting Mussolini to exercise the medieval *droit de seigneur* on her wedding night. To this day, Farrell reveals, dozens of people a day pay homage at Mussolini's tomb, which is adorned with the black riding boots he was executed in, a black shirt and a flask containing a piece of his brain. 'Many of these visitors are young women who perform a Roman salute with yearning in their eyes,' he discovered.

Although no dictator, President John F. Kennedy seems to have been not unlike Mussolini, in both his extremely limited sexual ability and his aggressive sexuality in dealings with women. According to an account by Frank Sinatra's valet and confidant George Jacobs, who was also very close to JFK, the President's lover Marilyn Monroe readily admitted that Kennedy was useless in bed. She said he suffered from premature ejaculation. However, says Jacobs in his book *Mr S: The Last Word on Frank Sinatra*, 'She tried to take it as a positive, evidence of how she drove him out of control. "Jesus, George, he's got a *country* to run. He doesn't have time for the mushy stuff." Frank, on the other hand, made time for the mushy stuff. "He's the best," Marilyn frequently swooned. "Nobody compares to him. And *I* should know."'

When he was running for president, Kennedy also told Jacobs, who used to give him massages, a story about Marlene Dietrich masturbating him when he was a boy: '"Can you imagine what that was like for a Goddam teenager?" . . . By the time I rolled him over, Jack had become aroused. He turned beet red, but didn't ask me to stop or stop talking. "We better get you laid, Jack," I said, "You darn' well better," he replied.'

Could rape which is, after all, as much an extreme case of pursuance of male orgasm as an issue of control and dominance, be nature's way of efficiently spreading available genetic material as far and wide as possible, as a minority of evolutionary psychologists are now arguing? In humans, conception through non-consensual sex or rape is, sadly, all too common. But, while female orgasm may only play a minor part in conception, it has long been thought (albeit more from anecdotal evidence than hard research) that sex perceived by the female as loving, shared, relaxed and mutually pleasurable is more conducive to successful conception than joyless sex forced by a disagreeable male.

This view may, sadly, be fallacious. Recent research by Jon and Tiffany Gottschall, of St Lawrence University in New York, seemed to show that an act of rape may be more than twice as likely to make a woman pregnant as an act of consensual sex. They found that of 405 women raped between the ages of 12 and 45, some eight percent became pregnant when contraception was factored out of the calculation. They compared this with a separate study which found the proportion of women in a similar age group who got pregnant whilst unprotected from a one-night stand or other one-off act of consensual sex was just over three per cent. The Gottschalls believed one explanation may be that women feel more attractive and sexy when ovulating, and unconsciously give off signals that rapists pick up.

Critics of the Gottschall study mentioned have pointed out

that rape cannot be a very efficient evolutionary strategy when only 38 percent of the conceptions studied led to a live birth, with abortion or miscarriage accounting for nearly two-thirds. The odds, in other words, that a rapist will successfully father a child from a single attack are still less than 1 in 100.

The idea that rape is an attempt at reproductive rather than belligerent behaviour was first mooted in a 1999 book, *A Natural History of Rape*, by a biologist, Randy Thornhill of the University of New Mexico at Albuquerque, and Craig T. Palmer of the Anthropology Department at the University of Colorado, and built on the Gottschall findings to construct a theory – or polemic, more properly – that rape is best understood as a behavioural adaptation moulded by sexual selection, a viable alternative to regular courtship. Thornhill and Palmer suggested that all young men be educated frankly about their supposedly genetic desire to rape.

The greatest problem (among many) with the Thornhill and Palmer idea is guessing how it might explain rape of children and elderly women – not to mention rape of men. It is not easy to see how violent paedophilia, for instance, can be construed as even the distant cousin of an evolutionary strategy. Nevertheless, their theory has been praised by the eminent likes of Steven Pinker, the Massachusetts Institute of Technology Professor of Psychology.

As regards the aggression/sex relationship, it must also be remembered that anger can be a two-sided coin. For some people, both women and men, aggression can cause such strong emotional feelings against a partner as to inhibit orgasm; in others, meanwhile, it can provide positive stimulation. This is why for some couples, fighting is a form of foreplay, and leads often to satisfying, mutually orgasmic sex. There have even been reported instances or couples seeking therapy to end their fighting, and, duly cured, as a result, in the absence of any other erotic stimulus, promptly ending their relationship; as soon as the well of emotional energy dried up, the fighting stopped – but so did the sexual attraction.

The sex/aggression connection greatly intrigues anthropologist Lionel Tiger, because in male masturbation, he sees an easy, natural way in which excess-to-requirements belligerence in young males can be calmed – in a cheaper and safer way, one might add, than drugs like Ritalin and tranquillisers, which are routinely prescribed for such antisocial behaviour. The fact that successive societies have so strenuously outlawed such masturbation fascinates and perplexes Tiger. In the introduction to *The Pursuit of Pleasure*, he writes: '. . . while male sexual intercourse usually causes an increase in bodily testosterone, which is associated with assertion and aggression, masturbation leaves the level of this influential substance unchanged. It may actually reduce the tension and sense of frustration adolescent males often experience.' Tiger might have added that women who are accomplished masturbators are also familiar on a daily basis – hourly in times of stress – with a similar tension and aggression-dousing effect from, you might say, a quick twiddle.

There is a special orgasmic consideration too in the sex/aggression nexus, which is sadomasochism. Why do sadomasochists need to give and experience pain to enjoy the pleasure of orgasm? Tiger in *The Pursuit of Pleasure* suggests it could be that they were beaten as children by parents whose attention they craved, and thus associate love with pain; or, he speculates, they are people who view their lives as worthless and find in hurting others some relief to this sense of valuelessness. One gay US writer, Edmund White, argues that the sadists he knows are gentle pacific types outside the sexual arena. He concludes in his 1981 book *States of Desire* that: 'S&M sex may merely be a more frank expression of the dynamics underlying *all* sex; perhaps gay liberation has merely given the leather boys permission to make manifest what is latent in everyone.'

A still more extreme manifestation of the link men may feel between orgasm and aggression is provided by professional fighters. Boxers, in particular, seem prepared to keep fighting through any pain, any degradation, to have one more one chance at the championship belt. It was former *Sports Illustrated* boxing writer Mark Kram who suspected this masochistic behaviour may have sexual connotations for some fighters – that they outwardly abjure pain while secretly warming to it. 'Old trainers used to tell me that they had known fighters who got hit so much that it became [so] pleasurable, they even ejaculated,' he wrote.

Despite pervasive cultural myths that only males set in train the procedure that leads to the more conventional, intercourse-related orgasm, the probability is that either the man or the woman will have initiated the above scenario. The modern assumption that men have a greater sex drive while women need to be coaxed and often coerced into sex may be faulty. One of the confusing factors in this is the wide difference in the ages at which men and women reach their peak of sexual desire and ability: nineteen in men, close to forty in women. The idea that women are intrinsically less interested in sexual pleasure may also be self-fulfilling – that they absorb societal notions of what their sexual desire *ought* to be.

It is also arguable that anatomy (with a bit of help from evolution, as we shall see) has conspired to make sexual delight a little more elusive for women. The concealed position of the vagina and clitoris is in stark contrast to the more obvious and convenient placement of the penis. A clitoris can be aroused without its owner, if she is not attuned to its moods, knowing it; it is less easy to ignore an erection. Men are, as a result, the more likely to masturbate and become aware of the possibilities of sexual pleasure.

And aware of those possibilities – obsessed with them, no less – men demonstrably are. It is hard to exaggerate the male sex's single-mindedness when it comes to ejaculating as often as possible in as short a time as feasible. After the comedian Bob Hope's death in 2003, a story emerged of his and his professional partner Bing Crosby's insatiable appetite for sex.

According to the London *Sunday Times*, Crosby, while recovering in hospital from an appendectomy, spoke of the unusually attentive post-operative care he was receiving in the form of one nurse who, he said, 'gave me the greatest blow-job of my life.' The next day, according to the story, Hope checked himself into the hospital, pleading 'shattered nerves'. He got the same room, the same nurse – and the same unorthodox treatment.

The whole prostitution industry, of course, is predicated on the ease with which men can reach orgasm – and the urgency with which they seek it, especially for enjoyment in emotionally undemanding conditions. A rare, stark and alarming quantification of the scale of demand for easy sex by men was made by an undercover police operation in northern England in the 1970s. A serial killer of prostitutes – the Yorkshire Ripper – was at large and officers were logging car registrations of regular 'punters'. Over 21,000 men were identified as prostitute users in two relatively small northern cities, Leeds and Bradford – this in spite of it being well known locally that men's car numbers were being tracked, a circumstance that that might have cut the numbers significantly.

The desire for ejaculatory release by brain damaged and mentally ill men suggests how elemental the need for orgasm is in the human male. When all else is stripped away, the desire to masturbate for comfort sometimes seems to remain. This syndrome of compulsive, neurotic masturbation was movingly portrayed by the novelist John Irving in *The World According to Garp*. The hero's father, a brain damaged US Air Force sergeant whose vocabulary has been reduced to one word – 'Garp' – does only one thing incessantly, which is to masturbate. Garp's sexually reticent mother, Jenny, 'harnesses' one of Garp's readily available erections to impregnate herself and produce a son, T.S. Garp.

A broadly equal distribution of desire for sex is, nonetheless, borne out by ethnographic research. People in 72 of 93 societies studied in the 1970s believed that both sexes have an approximately equal sex drive, and that either is equally likely to begin sexual advances. In many societies, the Maoris of New Zealand being one, women more commonly initiate sex than men. It is implausible, given the brief five million years *Homo sapiens* has been around, that equality of sex drive was not also the case for our Flintstone ancestors – who, even though they did not actually live in caves, we prefer to believe did because we tend to find their belongings preserved mainly within such temporary, natural shelters.

It is also highly likely that the outline script for orgasm that follows was similar or identical in many particulars for early human beings. There simply has not been time for it to be otherwise. It is the all-important overlay of subsequent cultural development that has refined (and sometimes coarsened) the *dance à deux* that leads to orgasm, and has given this sublime bodily function its mystique and its history.

The all-important pre-copulatory phase, which barely exists in even the highest primates and is another sign of our stellar sexiness, can begin without sexual contact of any overt kind. A shared meal is a popular starting point for sex in human beings; it is interesting that oxytocin, the hormone of coupling and togetherness, flows almost as easily during an enjoyable dinner, thanks to the variety of sensual and intellectual pleasures on offer in 'intimate dining', as during sex and subsequent orgasm.

A great deal of the forgotten sense, touch, dominates the next stage of the precopulatory ballet. Again, this need not be overtly sexual; fingers brush backs of hands or entwine over the table, hands are held tenderly. 'At times we focus in sex upon the most minute motions,' the late Harvard philosopher Robert Nozick wrote, 'the most delicate brushing of a hair, the slow progress of the fingertips or nails or tongue across the skin, the slightest change or pause at a point. We linger in such moments and await what will come next. Our acuity is sharpest here, no change in pressure or motion or angle is too slight to notice. And it is exciting to know another is attuned to your sensations as keenly as you are.'

The skin is the largest organ and as, Lionel Tiger has written, skin 'is not only an envelope containing a person; it is also a means of communication'. Pre-sex touching continues to involve many body parts other than the genitals; experienced lovers will take care to avoid making direct contact with the vagina or penis or even the crotch area in the very early stages of coitus. Touch is such an erotic sense that pre-sex caressing can be an inadvertent shortcut to orgasm. Some people of both sexes can climax from a simple stroking of the back or feet.

Hand-to-face and face-to-face contact ups the ante, while kissing, in the cultures where it exists and is not regarded as disgusting (as it is by the Thonga people in Mozambique and some Finnish Laplanders) or verging on cannibalistic (as the Chinese once believed it to be), raises the temperature several notches further. Hands and fingers, or alternatively feet and toes, may venture towards the genital area, the inner thighs and the woman's breasts, teasingly at first, a little more earnestly as affirmative signals flash back and forth between the putative lovers. The voice, encouraging, cajoling, affectionately teasing, also plays an important part in this 'foreplay' stage.

Facial expressions, too, are key to building and maintaining the heavy sexual atmosphere required for intercourse to occur. Talking can be a double-edged sword in sex, however, for it can bring about a premature end to lovemaking before it has even started. Both parties pre-sex are in a delicate state of high sensitivity and excitement – two rather incompatible conditions. A careless word or form of words by either – but more usually by the male, who tends to be mentally and physically further along the track towards orgasm than the female at this stage, and hence abandoning subtlety by the minute – can ruin things, especially for the female, for whom delicate mental scene-setting is often more important than for the male.

Backstage, you might say, major bodily changes are taking place during this often protracted time of sexual arousal, as it typically segues from a flirtatious incline that might take several hours to develop, into a brief but steeper ascent to intercourse itself. There will also classically be a change of location now, from a standing area such as the street or dance floor, or an upright seat, to something more horizontal – a sofa, a floor, a bed. For reasons we will look at later, Nature appears to have gone to some pains to coax upright, bipedal human beings into adopting a horizontal position when they finally mate.

The first major bodily modification in this new phase concerns the distribution of blood supply, which begins to be diverted from the organs and muscles to the skin surface. The body accordingly feels hotter to the touch – the pre-sexual radiance that makes lovers appear flushed. Blushing of the face and neck is what appears to the eye, since the participants' clothes are normally still on at this point, but there is more widespread reddening of the skin, especially in the female. It spreads, not unlike a measles rash, from the stomach and upper abdomen, to the upper part of the breasts (which are already starting to swell perceptibly with the onrush of blood), and then to the upper chest, the underside of the breasts, the shoulders and the elbows. Men's skin reddens with the sexual flush in similar areas, plus the forearms and thighs.

The swelling of the breasts in women as intercourse approaches is caused by vaso-congestion – the effect of arteries pumping blood into the area faster than the veins can drain it. Similar swelling and hardening also starts to occur in the man's penis, which may well be fully erect under his clothing by this time, although his erect organ serves more as a sexual signal of intent for the moment than as a sexual tool.

Less noticeable is the swelling of soft parts of the nose and expansion of the nostrils; the sensation of breathing heated air as orgasm later approaches is far from illusory. Additionally, the nose should not be disregarded as a sexual instrument. It has long been an evolutionary puzzle why human beings have developed such extraordinarily outsize noses compared to the neater, snub-shaped breathing apparatus of apes. Perhaps there is a correlation here with the outsize penis noted previously.

And more is now known of the enhanced role of smell in both the pre-copulatory and coital phases of human sexual intercourse. Along with the heavy two-way traffic in tactile, facial and verbal communication during pre-sex and sex, there is also a busy chatter of olfactory signals in the form of pheromones, for which the nose is the receptor. The nose, indeed, is something very close to the radar of sexual engagement.

The earlobes are another seemingly peripheral outcrop with a role in the build-up to orgasmic release. When the earlobes become swollen and engorged, they develop an unexpected hyper-sensitive, erotic capacity. Earlobes were formerly considered an uninteresting relic of a time when we had bigger ears, but are now known as a place so sensitive that large numbers of men and women are capable of achieving orgasm purely through earlobe manipulation – especially by mouth, tongue and teeth. The nipples are another only peripherally sexual area which come into play as erogenous playthings in pre-sex; in both genders, but more so in the female, nipples become swollen and erect, increasing in length in women by as much as a centimetre. Tumescence darkens and emphasises the pigmented skin around female nipples, which turn a deeper red.

What, then, of the principal players in this drama, the penis and the vagina? Early on, in the more social stage of copulation, the dinner and a movie stage, if you will, the human penis undergoes one of the most dramatic changes seen in Nature. From a short flaccid organ in its normal state, it expands within a few seconds to as much as double its quiescent length and girth. The scrotum simultaneously contracts, applying constricting pressure to the testicles, which are drawn into the body.

The erection is the site of the greatest volume of vasocongestion in the human body. Within the quite confined space of the penile tissues, the volume of blood in the erect penis increases approximately eleven-fold. This explains the quite remarkable change in the organ's size and rigidity.

At this stage the erect penis presents quite an aggressive and determined sexual statement. But, as the feminist writer Naomi Wolf has pointed out, the erection is also lost very easily. Slight distractions such as a sudden noise, especially by children who might be about to threaten privacy, can rapidly subdue a male erection. (Women's response to distraction, surprisingly in the light of their greater susceptibility earlier in pre-coitus to 'losing the mood', is less sensitive now, according to Wolf; women will tend to be the more resolute in trying circumstances to see an orgasm through to fruition.)

Erection unrelated to orgasm is common in men; a variety of Internet sources ('Jenn Even', an American e-zine sex advisor on *www.sagazette.com* is just one) suggest that the average Western man has eleven imagination-stimulated erections a day, and it would be reasonable to suppose that something like that applied to early man. But blood engorgement pre-intercourse also starts to affect less obvious, but equally indicative, extremities of erectile or semi-erectile tissue in the body. The facial lips of both men and women also become swollen and red; lipstick, along with rouge, is a socially acceptable mimicking of these pre-sexual changes.

The walls of the vagina also become engorged with blood, often at an early, flirtatious stage of a sexual encounter. This onrush of blood quickly – as with the penis, within seconds sometimes – stimulates lubrication of the vaginal tube. There is an accompanying lengthening and expansion by up to four inches of the inner parts of the vagina as the female enters the phase of high sexual excitement. Her blood, too, is now pouring into the vessels of the pelvic and genital area and expanding the nerve bundles throughout. Muscles around the clitoris, the opening of the vagina and the anus swell.

The clitoris also becomes enlarged, sensitive and protuberant, although its distension is obscured from view by more general swelling of the labial hood. The clitoris will not, when penetrative intercourse finally proceeds, be directly stimulated by the penis. In some women (but not all) it will retract entirely as orgasm approaches. But, so long as the penis has a reasonably broad diameter, it will in theory stretch the outer third of the vagina and tug down on the labial hood with the rhythmic pressure of the male's thrusting movements, thus providing, albeit by a far less than mechanically satisfactory means, the friction for easier orgasm, or for the beginnings of easier orgasm. Spongy sacs inside the external female genitals also swell noticeably in preparation for penetration, the outer labia reaching two to three times their normal size. The inner labia, while turning a bright red, expand to a similar degree, by as much as a hundred per cent, to the point where they sometimes extend beyond the outer labia.

Blood pressure, pulse rate and respiration all increase dramatically as copulation approaches. Glands have continued to secrete in even greater amounts the mucoid fluids flowing from the walls of the vulva and facilitating entry of the penis. (These are the same lubricants necessary in childbirth to ease the passage of the baby through the vaginal canal.) The female's breasts have also by now swollen by anything up to 25 per cent of their normal dimensions, becoming firmer and more rounded as they grow, while the testicles have, in some men, nearly doubled in size to try to keep in proportion with the shaft of the penis.

Even with all the ingredients of this sexual banquet nicely simmering and ready to be served, copulation proper may still be delayed by another tempting course – of oral stimulation. Oral sex, even if it is still technically illegal in some US States and taboo in many other cultures and subcultures, is apparently more widely acceptable and practised than it has ever been in the recorded past. Both homosexual and heterosexual lovers like to use it as a means of expressing deep, intimate feelings.

Oral sex need not be considered a decadent indulgence. It has been used very successfully over the generations as a method of sexual interaction when a male has difficulty attaining an erection, or when intercourse is painful for either partner. Oral sex has also been an emotional and physical lifeline for people with disabilities such as spinal cord injuries. Heterosexual couples, additionally, have found it an unexpected delight in cases where conception must be avoided, or in late pregnancy, or after childbirth when intercourse proper might be dangerous or painful.

Starting with kissing on the mouth, which is both oral and sexual, use of the lips and tongue for sexual purposes soon extends, predominally for the male, to the breasts and nipples of the female. (From there, it is only a navigational matter of inches to the more contentious business of each sex using the mouth to fellate the genitals of the other.)

Cunnilingus, the stimulation of female genitals by lips and tongue (from the Latin *cunnus* [vulva], *lingere* [to lick]), is performed by either a man or another woman. The partner gently separates the outer and inner labia with the fingers or tongue, then licks, sucks and teases the clitoris, sometimes rapidly flicking the sides of the clitoris shaft. Some women enjoy a slow, steady rhythm, moving backward and forward to the vaginal opening, sometimes with deep insertion of the tongue just before orgasm. Manual stimulation may be employed simultaneously. Very light biting, sucking or nibbling actions usually occur. Some women enjoy having their partner blow a little air into the vagina, but too much can be dangerous, causing infection or even embolism.

Even the sexually neglected nose may come into play in oral sex. As a partner's tongue plays with the vagina's inner and outer labia, the nose is perfectly placed to be simultaneously stimulating the clitoris. It may only be modesty that prevents male fashion from highlighting nose size in men to match the blatant display of sexuality seen in women's lipstick. There again, it cannot be entirely his singing and personality that make Barry Manilow the sex symbol that he is; could it be that, *inter alia*, his millions of adoring female fans dream idly of what Manilow might be able to achieve with that outsize proboscis?

Men can have mixed feelings about cunnilingus. With visual stimulation so important for them, most become sexually excited by the intimate view of the genitals and the intriguing sight of the vagina opening like a flower. The taste of the vaginal secretions may also be extremely stimulating. Other men feel slightly obligated to perform cunnilingus, with hygiene, religious, intimacy and other reasons holding them back from actually enjoying it. Research by US sexologists Drs Jennifer and Laura Berman suggests that men who perform cunnilingus are more goal-oriented and do it primarily as 'warming up' pre-intercourse tactic rather than as an end in itself. In Thailand, many men believe they can suffer *choak suay* – bad luck – by engaging in cunnilingus. This might seem a rather novel excuse, but squeamishness does not explain why Thai men also feel it is unlucky to walk under laundry containing women's skirts or underwear. Other cultures have dressed cunnilingus up as a near art-form; on Ponape, in South Pacific Micronesia, some men like to place a fish in a lover's vulva and slowly lick it out.

Cunnilingus is also, it should be remembered, a primary method for reaching orgasm for most lesbians. Women who perform cunnilingus tend to be more effective in giving pleasure to another woman, knowing as they are bound to what feels good. They are inclined to approach the activity less hastily than men do, and to prolong it for its own sake.

Fellatio (from the Latin *fellatus*, past participle of *fellare*, to suck) is the oral stimulation of the male genitals, primarily the penis, by a woman or another man. The penis may be inserted into the mouth to a depth of a few centimetres, or practically to the base of the shaft. The sensitive corona, or tip, of the penis is the most common focus of fellatio, although the shaft, frenulum, perineum, scrotum, and sometimes the anus or immediate surrounds can also be involved. All these areas are sucked, licked or tenderly nibbled. The term 'blow-job' appears singularly inappropriate because no 'blowing' is performed, but it most probably derives from the more understandable term 'below' job.

Until AIDS, at least, fellatio was the most common sexual activity practised by homosexual men. It is certainly the most requested act by prostitutes' 'punters' – as well as being their own preferred method since it is easy, quick and portable, requiring as it does neither premises nor preamble. As in the case of empathetic women performing cunnilingus, men are often better at it than members of the opposite sex. Some men who consider themselves heterosexual are happy to be fellated by another man because they do not feel as if they are performing a homosexual act, as they would with anal intercourse.

According to a psychologist specialising in sexual studies, Dr William R. Stayton of Widener University in Pennsylvania, most homosexual men giving fellatio swallow the semen, which adds to the erotic pleasure for the recipient, whereas most women do not, citing the taste (fishy for some, salty for others, bitter for others still and pleasantly spicy for a handful of real enthusiasts), worry about gagging on such a large object in the mouth and a common belief that semen is fattening – it actually contains only about 5 to 15 calories per ejaculation. At an average time of four minutes from the height of the pre-orgasmic state through to the end of orgasm, we can typically expect to burn 25.6 calories per coition, meaning even with a healthy helping of semen on the side, sex is still a reasonably slimming activity.

Just as the taste of vaginal secretions varies from woman to woman, so does the taste of semen vary from individual to individual. Diet is a factor; asparagus, in particular, gives semen a strong, bitter flavour; dairy products are almost as bad because of their inherent bacterial putrefaction. The gag reflex, which may also be stimulated by ejaculation in the mouth, may be overcome for some fellators by grasping the base of the penis with a hand and thereby feeling in control of the depth of penetration. Some skilled 'deep throat' fellators manage to learn to recondition their gag reflex by a process of slowly taking the penis deeper into the throat.

The attitudes of women to fellatio range from the idea that

it is perverted and unhygienic to the view that it is normal and pleasurable, and even a preferable method of sexual activity because it is free from the risk of pregnancy. Most men love being fellated by a woman, but some worry that if they ejaculate in her mouth, they will choke her.

Is oral sex as applied to the genitals a natural, if optional, part of the pre-copulatory sexual routine, or a contrived, add-on behaviour calculated to provoke orgasm where none has previously occurred, or alternatively to repeat orgasm in a refreshingly different way after copulation? And is it possible, while we consider this, that our distant ancestors ever discovered fellatio and cunnilingus?

To the first question, the answer is 'either'; oral sex is ingenious, but cannot be said to tax the intellect or imagination too much. The sensitivities of the more primitive American States notwithstanding, it may not be unreasonable to describe it as a fairly 'natural' practice. What is interesting about oral sex from the point of view of the history of the orgasm is that it is a further example of the non-reproductive but pleasure-seeking complexities that have evolved with human sex. Masturbation and, more importantly, contraception, both of which we will examine later, are further variations on a behaviour that, it is easy to forget, is essentially reproductive.

Along with contraception in particular, oral sex, despite having only recently become mentionable in polite society, is one of the boldest statements of the primacy of orgasm in human life. Not only is it the most stark example of reproduction-free sexual pleasure, but it could even be said to act as a mild contraceptive if practised before copulation proper, by allowing the bulk of the male's sperm a non-productive nemesis, ensuring that the next body of semen he produces during the same sexual encounter will have a lower sperm content.

Oral sex has, surely, to merit a special position in the panoply of human sexual behaviour. A British writer, Paul Ableman, put up a sturdy case for its essential humanity in 1969 in *The*

Mouth and Oral Sex – the last book in Britain, interestingly, to be the subject of an obscenity trial when it was published.

'One can imply, by performing oral/genital contortion, that nothing about the loved one is offensive,' Ableman wrote. '. . . the proximity of the excretory apparatus to the genitalia can be conceived of as a chivalrous challenge, the acceptance of which expresses some such sentiment as: "See how much I love you – if I am prepared to do this." Putting one's mouth to the genital regions, breaching the hygiene taboos and conventions of decency, is in fact a gesture of intimacy that transcends coitus itself.'

Whatever the position was in past eras, in cultures where the taboos against it have withered or died, oral sex is extremely popular. Modern research cited by William R. Stayton indicates that in Western societies some 80 percent of single men and women between the ages of 25–34, and 90 percent of those unmarried and under 25 years of age, participate in oral sex. Stayton notes that from the 1950s onwards studies strongly suggest that the practice is more prevalent among better-educated and younger people.

This willingness among the better educated to engage in oral sex has come about, one would imagine, as a result of a modern intellectual rejection of what are seen as overly strict hygiene obsessions. Progressive twentieth-century thinkers on sex from D.H. Lawrence to Dr Alex Comfort have despised hygiene at the expense of sexual rapture as suburban and *petit bourgeois*. As for prehistoric humans and their take on oral sex, there is obviously a dearth of evidence in the form of cave paintings or artefacts, but informed guesswork rather suggests that both Mr and Mrs Ug will have discovered it one way or another, in the absence of any hygiene hang-ups over bringing the excretory organs into close proximity to the organs of breathing and eating. It is probably only when religion began poking its snout into human groins that taboos against oral sex took root.

There may have been isolated taboos concerning wastage of

sperm, but this again is improbable since it is not likely that prehistoric people were aware that sexual intercourse and child production were connected. Surviving primitive peoples often still do not acknowledge any relationship between the two.

Above all, prehistoric humans will have had time in abundance to discover oral sex. Prehistoric nights were long and dark, and there is no reason to suppose that couples did not sleep in a huddle together, if only for warmth. It is beyond contention, surely, that at some stage man realised that the mouth on a face bears a distinct similarity to the mouth of a vagina, and then had a hunch that it might be interesting for woman to apply her mouth to his penis, and vice versa.

It has been suggested that proximity to animals, if nothing else, will have prompted a curiosity about oral sex in early man. Paul Ableman postulated that: 'a relatively common way in which children discover the possibility of pleasurable contact between mouth and genitalia, is through accidental contacts with animals. This form of initiation is much more common to girls than boys, the obvious reasons being that a dog or cat may perform cunnilingus spontaneously but not fellatio.'

A less obvious form of orgasmic pleasure via oral sex occasioned by the proximity of animals was the subject of a slightly shocking speech in the British House of Lords one June evening in 2003. Lord Lucas of Crudwell and Dingwall, a conservative peer, proposing an amendment to a new Sexual Offences Bill brought their Lordships' attention to an old agricultural practice he knew of called avisodomy. This is sex in which a chicken's anus is employed as a makeshift receptacle, or the still nastier practice of using a decapitated chicken's throat. His description, as chronicled by *Hansard* is something of a collector's item. Avisodomy, he told the hushed benches, is 'the practice of breaking a hen's neck at the moment before penetration so that you benefit from the spasms that the animal undergoes afterwards.' Lord Lucas professed himself against avisodomy, but was concerned that it might avoid being outlawed because the chicken was clinically dead at the time its conspirator was enjoying his orgasm.

The probability of prehistoric cunnilingus having existed or not might be more problematic. Cunnilingus would not seem to be an obviously instinctive behaviour, and neither is it as immediate an option as fellatio. But knowing as we do that the prehistoric female probably owned a working clitoris, we can assume that she discovered how manipulating it can be highly pleasurable. Transmitting that information to men will not have stretched rudimentary communication skills too far (even if it does for some couples today). And the non-existence of knowledge about either hygiene or conception will, arguably, have made it still more likely that men and women would ultimately have put two and two together.

Whether the pleasurable diversion of oral sex is followed or bypassed, most sexual encounters ultimately result in penetrative intercourse. Yet just as orgasm, the apex of the sexual pyramid, is far briefer and more acute than its preamble, the penetrative phase of sex is, today, generally much shorter than the pre-sex, wooing and foreplay phase. Did prehistoric humans spend time on foreplay? Although we can have no evidence one way or the other, it would not be unreasonable to surmise that the whole copulatory sequence today is much longer than it was for our most distant ancestors.

But what will almost certainly have been the same then as now is that the length in time of the consummatory phase of intercourse will have been dictated, as it is today, by the amount of thrusting movement it takes before the male reaches orgasm. This can vary from a matter of a few seconds to an hour or more, the latter in the case of highly controlled males using deliberate and quite intellectually demanding delaying strategies; it is usually difficult for penetrative intercourse to proceed immediately after male orgasm, although a degree of tumescence may persist, enabling the male to continue, albeit with some effort and discomfort to him.

At the moment of the male orgasm (which we will deal with first for the very good reason that it often occurs first unless measures are taken to avoid it) the vital functions reach some of their highest peaks outside those attained in moments of extreme danger. Heart rate at the sexual climax can accelerate to 180 and blood pressure a bottom (diastolic) figure as high as 250. Extra heart beats and skipped beats are not uncommon, especially in people who are not in good physical condition. Respiratory rate may increase to 40 per minute. There is facial grimacing and breathing becomes a desperate gasping for air, with a rhythmic moaning or groaning as a counterpoint. The testes withdraw upwards to their maximum elevation.

Erotic sensations travel from the various outposts of erogenous skin to brain at high speed – 156 mph to be precise. As muscle tension and blood-flow to the pelvis approach their peak before dispersing in the final orgasmic reflex, there are spasms in the buttocks, tingling in the fingers and toes, and mounting muscular tension (*myotonia*) in the neck, legs and arms. Most (but not every) male orgasm involves ejaculation, which internally occurs in two stages. During the first stage of emission, seminal fluid is expelled from the vas deferens, seminal vesicles, and prostate gland into the base of the urethra near the prostate. The collecting fluid is felt as a consciousness of imminent ejaculation. Some men can have an orgasm at this point through rubbing the prostate gland.

In the ejaculation stage, the seminal fluid is propelled by the muscular contractions of orgasm into the portion of the urethra within the penis and then expelled from the urethral opening. The mechanism of the ejaculation is a series of powerful muscular contractions from the pubic or pubococcygeal (PC) muscle group that supports the pelvic floor, running from the pubic bone in front to the coccyx, or tailbone at the back, and within the testes and scrotum, then continuing through the epididymis, vas deferens, seminal vesicles, prostate gland, urethra, penis, and anal sphincter. These contractions, a maximum of three or four in number, at 0.8-second intervals, have the effect of expelling the seminal fluid along the urethra at a speed that has been measured at 28 m.p.h. The initial contractions can be followed by two to four slower 'aftershock' spasms. The length of a male orgasm is typically from 10 to 13 seconds.

Even though friction from the passage through the penis obviously slows sperm down from its 28 m.p.h. starting speed, the initial volley is occasionally, especially when there is a build-up of semen, still fast and powerful enough on exit to project a body of seminal fluid from roughly a fraction of a centimetre to a few centimetres. The young male capacity for sexual self-mythologising will often expand this masturbatory projectile capacity to implausible ceiling-scraping proportions, however. In Argentina, it is said, boys have for generations gathered in parks to compete to see who can project their semen further while masturbating. The evidence suggests this can hardly have been a great spectator sport.

Pornographic films, in which an impressive (and often faked) 'cum shot' (otherwise known as 'the money shot') is *de rigueur*, increase the pressure on males to exaggerate their own projectile capacity. It might be noted that once inside the vagina, where it is designed to go, semen has a fight ahead akin to swimming in treacle. It slows down to a less than snail-like 0.0011 m.p.h., which means it spends five minutes travelling the six inches or so to the cervix.

The amount of ejaculate expelled, typically one to two teaspoons (2½ to 5cc) per orgasm, differs between men, the volume (as well as the potency) varying with the frequency of orgasm. A man produces 125 million sperm each day, and each orgasm expels between approximately 80 million and 800 million. In the average lifetime, a man produces 14 gallons of ejaculate, enough to fill the fuel tank of the average-sized family car. The semen's chemical composition, spermatozoa aside, is inconsistent even in individuals. About 8 per cent of the solution consists of dry solids, the majority of the sperm is a mix of more than thirty elements and compounds from nitrogen to fructose, cholesterol and vitamins C and B12. Interestingly, the average ejaculatory volume contains about 60 percent of the American recommended daily intake for vitamin C.

Just as the truth about ejaculatory distance is fogged by vanity and nonsense, ejaculatory volume is also a contentious area, undermined by mythology and bragging. The line taken by many sex websites (often those selling quack potions) is that women are impressed by voluminous ejaculations (although less so than by penis size) for psychological reasons to do with their perceptions of virility and the likelihood of their being fertilised. The 2½ to 5cc ejaculatory norm is even said to be on the low side, many women regarding anything less than a tablespoon as an unacceptably small load. The one to two teaspoon figure, it is asserted, is misleading because ejaculatory quantity is related to sexual excitement, and men are not inclined to be very excited in laboratory conditions.

So far as the truth on both distance and volume is concerned, it unsurprisingly lies towards the bottom end of the middle range. As far as one can tell from subjective evidence, most men have smaller ejaculations than they would wish for, or imagine a partner hoping for; additionally, the true distances of schoolboy and dormitory ejaculation stories are rarely a fraction as great as stated, and tend in the vast majority of cases towards the minuscule. Even the demands of 'the money shot' are relatively modest in terms of projectile distance; if semen were to be ejected at a ridiculous speed, the camera would never be able to catch it from the requisite few inches' filming distance. Fluid mechanics dictate anyway that semen, so long as it is healthy, is simply too viscous and the penis too inefficient a firing instrument to send it spurting aloft like an ornamental fountain.

While, classically, the male orgasm has been saddled with a monolithic definition implying that it is interchangeable as both term and concept with ejaculation – the male sexual climax has probably always been experienced differently by individual men and by those men at different times.

Although human males have a trickier task maintaining erection than most mammals in that they depend entirely on hydraulics – most animals, including all primates except man, have a bone called the *baculum* to shore themselves up – men still do not inevitably lose their erection after orgasm or ejaculation. A controversial (and lately fashionable) body of opinion exists that orgasm and ejaculation in men are quite separate functions; that physiologically, ejaculation is simply a reflex that occurs at the base of the spine, an involuntary muscle spasm resulting in the ejection of semen and felt only in the penis, whereas orgasm is somehow much more than that, an unspecified 'whole body experience' produced by clenching muscles throughout body to avoid the penis being sensitised.

Multiple orgasm, in which ejaculation is not necessarily involved, is according to some modern research and ancient texts on sex something of which men are physiologically capable only by the application of learned techniques. Researchers William Hartman and Marilyn Fithian of the Center for Marital and Sexual Studies in Long Beach, California were the first sexologists to present scientific data on the existence of multi-orgasmic men. They monitored the orgasms of 282, of whom 33 proved multi-orgasmic. Their most prolifically orgasmic subject was an athletic young man who consistently managed sixteen orgasms or thereabouts in less than an hour. Sex researchers Beverly Whipple and colleagues, writing in the *Journal of Sex Education and Therapy*, have for their part reported on a man who had six orgasms in thirty-six minutes with no erection loss.

The secret to men achieving multiple orgasms, according to Hartman and Fithian, is nothing more spiritual or arcane than learning to control ejaculation via the PC muscle, also known as the voluntary urinary sphincter muscle, that starts and stops the flow of urine. Once strengthened, it can provide the same sort of control over ejaculation. 'Just prior to the moment of ejaculatory inevitability, you clench the PC tight and hold it until the urge to ejaculate passes – roughly fifteen seconds,' Hartman reports.

Another modern researcher, Barbara Keesling, who has worked as a surrogate sexual partner, has moreover identified three distinct patterns of male multiple orgasm: one, she calls non-ejaculatory orgasm (NEO) in which a man has an orgasm but inhibits ejaculation using the PC muscle, and only allows himself to ejaculate ('release the hounds') after several orgasms. Keesling's second model is multi-ejaculation, in which a man has several partial ejaculations in succession. Her third pattern is for the man to have one intense orgasm and ejaculation, followed by less intense 'aftershocks'. All of these patterns, says Keesling, can occur without loss of erection. Other researchers speak of a phenomenon called 'injaculation', whereby semen is retracted by force of will into the bladder instead of out through the penis. ('Injaculation' is one of the Holy Grails of practitioners of 'Tantric sex', an offshoot of Buddhism which explores sexuality, as we will see later, as a way of transcending the limitations of ordinary life. But it is worth noting that backwards-flowing semen is elsewhere regarded as a male sexual dysfunction.)

Men who can achieve multiple orgasm, Keesling reports, say – *say* being the important word, since one always suspects a measure of one-upmanship in this field, as in ejaculatory volume and trajectory measurements – that they feel energised rather than depleted after orgasm, and that their climaxes are stronger and more intense. She writes: 'I describe orgasms on a continuum from a localised genital sensation that is mildly pleasurable to a full-body orgasm with intense psychological sensations and all the fireworks – the kind of orgasm one of my clients calls "the psychedelic jackpot that lights up the universe".'

The most significant feature of an apparently sophisticated sexual technique practised by men, however, is not that it may or may not be more imagined than real; nor that it may be destined to become a debunked myth to equal that of Freud's distinction between clitoral and vaginal orgasm, which, as we shall discuss in a later chapter, is today a thoroughly discredited theory; nor that Tantric and Taoist methods of orgasm delay are predicated on the arguably vain, macho misconception that hours of thrusting is what women actually want from sex; nor that such techniques are a retrospective attempt to 'feminise' the male orgasm now that men have belatedly realised that women have the richer orgasmic experience of the two genders.

The point is that, useful or not, ejaculation withholding is an acquired skill, whereas the majority of men's sexual pleasure is highly instinctive. Erection is an involuntary, hydraulic phenomenon, which cannot normally be willed. Ejaculation usually results from intercourse with a minimum of effort.

It would seem to be stretching our belief in the intrinsic 'naturalness' and instinctive character of sex to imagine that prehistoric men were as interested as were subsequent cultures in developing the (arguably) female-pleasing ability for extended erection and multiple orgasm. Boys never have to be taught to masturbate, whereas, even today, large numbers of women live and die without knowing such a thing is even possible. It follows that for sex to be a shared pleasure, the will to make it so — plus study and application — are of primary importance.

Whether on the other hand sexual dysfunction afflicted the prehistoric male human is an interesting question. Simple erectile dysfunction is strictly a little outside the scope of this book, since both orgasm and ejaculation can easily occur without erection, and for some men, a flaccid or semi-flaccid penis provides a more satisfactory masturbatory experience than a full erection.

However, averaging out a variety of studies, we can gauge that some 80 per cent of men in the modern world are *at some time* unable to get an erection; and given that most women greatly enjoy the sensation of penetration even if it does not lead *directly* to orgasm, erectile failure must after all rate as both a dysfunction and, equally, as an indication to a woman that something is either medically wrong with a partner or psychologically wrong with his attitude to her. In these respects, it seems plausible that both erectile and orgasmic dysfunction were probably far from unknowns to the real-life Flintstones.

It should also be noted that for men there is scope for debate over what comprises premature ejaculation. It is only within the framework that designates 'satisfactory' heterosexual intercourse exclusively as penetrative sex culminating in simultaneous orgasm – the unobtainable Holy Grail for the huge majority of people – that premature ejaculation as it is classically delineated becomes a true handicap to a loving and mutually orgasmic sex life. Strictly speaking, a man who ejaculates after ten or fifteen minutes – an heroic performance by average standards, but nevertheless insufficient for most women to have even the slim possibility of a 'natural' orgasm – is suffering from premature ejaculation. A lot of feminists would say so, just as a lot of misogynist men would (and do) say that any woman who fails to orgasm through penetrative sex, howsoever perfunctory, is, *ergo*, dysfunctional.

In modern times, male sexual dysfunction is the stuff of girls'-night-out jokes and saucy seaside postcards, but it deserves a more sympathetic approaching. The female orgasm may be a subtle and complex phenomenon, but the sensitivity of the male response can be, and generally is, underestimated. In 1994 the Massachusetts Male Ageing Study, at its time the largest ever epidemiological survey of male sexual functioning, revealed that 52 per cent of American men between the ages of 40 and 70 had minimal, moderate or complete sexual impotence.

In a typical modern Western population at any given moment – and one can speculate as ever about the implications this may hold for earlier eras – epidemiological surveys of male sexual functioning suggest that 13–17 per cent of men of an age to be sexually active suffer from decreased or non-existent libido; 7–18 per cent are unable to get an erection; 28–31 per cent of those who can suffer from premature ejaculation as they themselves see it; 7–9 per cent of those with fine, long-lasting erections find themselves nevertheless prey to anorgasmia, the inability to orgasm at all; and 15 per cent are anxious either from their own or their partners' perspective about their sexual performance. The incidence of erectile dysfunction in men increases with age; at 40, about 5 per cent suffer the condition; at 65 and older, the incidence is 15–25 per cent. But although sexual vigour in men declines with age, a man who is healthy, physically and emotionally, should be able to sustain erection, and enjoy sex regardless of age; impotence is not an inevitable part of ageing.

The male sexual performance is clearly a rather more delicate flower than is generally acknowledged. Ageing aside, any of the above symptoms can be brought on by problems in any of the following areas: ill health, psychological wellbeing, medical treatments, smoking, family, societal and religious beliefs, and neurological, vascular or endocrine systems. Ejaculatory disorders come in three varieties: premature ejaculation, retarded ejaculation and retrograde ejaculation – the propulsion of semen through the urethra back into the bladder rather than out through the tip of the penis, aka 'injaculation'.

Sexual dysfunction in men can additionally be quite paradoxical because the mind exercises such a huge influence on sexual function. It is commonplace, for example, for a man to be unable to get an erection because a woman is unattractive to him. But with such a partner, he may easily be able to achieve and maintain a fine erection – only to have his underlying lack of true sexual attraction betrayed by anorgasmia. And just as easily as he is able to put in what appears (until he fails to ejaculate) to be a championship performance when he does not actually find a woman very attractive, he may equally suffer a disastrous premature ejaculation precisely because he finds his partner extremely beautiful. Su-nii-ching Fang Nei Chi, the seventh-century author of a book called *Secrets of the Bedchamber*, recognised the potential for men to use to their advantage this ability to maintain an erection with an unattractive partner. He advised: 'Every man who has obtained a beautiful crucible will naturally love her with all his heart. But every time he copulates with her he should force himself to think of her as ugly and hateful.'

Masturbation also confuses the picture of male sexual dysfunction a little. Many men can easily orgasm through masturbation but have difficulty in heterosexual intercourse. This may be because they use far heavier pressure in masturbating than is normal in vaginal intercourse. Until they learn to orgasm with lighter pressure, they may well have problems in ejaculating during coitus. Equally, masturbation may be an easier way than intercourse for an anorgasmic man to ejaculate because he suffers an emotional dysfunction over intimacy.

Then there is the fraught question of whether a lot of sexual dysfunction *seen* as the woman's predicament is actually a male failure. What has been called 'the dissatisfaction theory' holds that a great deal of female sexual dysfunction – 'frigidity' as it was charmingly called until recent times – is not caused by psychological factors, hormone deficiency, diminished pelvic blood flow or any one of the usual suspects; it results from nothing more than inadequate genital stimulation, by men.

A host of factors, from religious observance to shyness to simple lack of communication, can result in men not knowing how to stimulate a woman so that she becomes aroused; this leads to unsatisfactory sex and in turn to lack of sexual interest, depression, and aversion to sex. The self-evident fact that young, healthy, apparently balanced women experience sexual dysfunction is probably the clincher for this view.

As for whether male sexual dysfunction was 'treated' in any way by prehistoric man, we are particularly clueless. We know, as we will see in a later chapter, that the ancient civilisations were very aware of the subject and had any number of supposed folk remedies for it. But as far as prehistoric man is concerned, we have no idea if dysfunction was considered a problem, because we do not know if functioning *correctly* was thought particularly desirable.

3

Herstory

'I'll have what she's having'
Director Rob Reiner's mother's line in *When Harry Met Sally*

In theory, if the tumescent male, caveman or not, can control himself enough to thrust for long enough inside the female, typically for between ten and twenty minutes, the female will automatically reach an explosive orgasm, a consummatory release as intense and tranquillising as the male's. Her muscular actions will have become largely involuntary from the moment of penetration, her vaginal contractions harmonising with the thrusts, her eyes losing focus in ecstasy.

In practice, things are a little more complicated. But as the mythical Tiresius discovered at the cost of his sight, even if reaching orgasm for a woman is a longer and more painstaking process, a sensation that does not simply happen spontaneously but has to be consciously coaxed and cultivated, the neurological reward is much more intense, long-lasting, satisfying – and almost immediately repeatable.

How long does it or should it take? In laboratory experiments – not the most seductive atmosphere for all but the more imaginative among us – a minority of women need to stimulate themselves for an hour or more before reaching orgasm, although, with experience (of the clinical atmosphere of the lab, that is), this time generally reduces. The average time for most women to reach orgasm in the laboratory is twenty minutes. The shortest time recorded in the research laboratory for a woman to reach orgasm is fifteen seconds, but such alacrity is extremely rare.

Laboratory observation of orgasm, as with clinical studies of any physiological process from sleep to behavioural experiments, is obviously only suitable for a small proportion of people. It may well be that subjects prepared to take part in sex experiments are of a mildly exhibitionist temperament, but this should not affect the physiological processes under examination. And physiologically the female orgasm as observed scientifically is not dissimilar to the male, even to the extent of encompassing an analogue to erection (the swelling and stiffening with arousal of the area surrounding the urethra) and to ejaculation, with the secretion at the moment of climax of a small amount of pale milky-coloured fluid, consisting of the plasma-like product of the Skene's glands surrounding the urethra and the paraurethral gland, mixed in with traces of lubrication, male ejaculate and urine. A major component of female ejaculate is lubrication that pools in the back of the vagina and is expelled by the contractions at orgasm.

Since it is not widely known that female ejaculate exists, even though women are as capable as men of wet dreams, it is generally mistaken for urine alone, and for this reason, in the tense minutes preceding orgasm, women tend instinctively to brace and tighten the vaginal walls and bladder. After she hits the plateau, however, a woman involuntarily releases the tension and experiences swift muscular relaxation; the small amount of ejaculate she was holding in, plus some male ejaculate if any is present, now exits, albeit at a more stately velocity than the male ejaculation.

Some women report that this ejaculatory outward passage of liquid through the urethra during orgasm, combined with the flow of accumulated lubrication from the vagina, contributes to a more powerful and intense sensation, perhaps of the same order of concentrated physical climax men feel from liquid rushing through the penis. It is one of the most intriguing features of orgasm that we cannot know this for sure because nobody in history has experienced both forms of orgasm. But women's descriptions of orgasm rather suggest to envious men either that females do get a rather better deal from their more complex anatomy – or that they merely feel less reticent about describing their feelings. 'An altered state of consciousness,' 'euphoria', 'a spiritual experience,' and, 'an oceanic roaring,' are among the more mellifluous of women's representations of their orgasms. The 'oceanic' metaphor interestingly echoes Freud's belief that such a feeling lies at the root of all religion.

Contrary to the modern trend in thinking that the male and female orgasm are relatively similar in the way they feel, descriptions by women sent questionnaires by cultural historian Dr Shere Hite for her 1976 book *The Hite Report* vigorously support the view that the female orgasm is more powerful and interesting than the male. Hite's book was proclaimed as revolutionary, although strictly speaking what it did more than to pioneer was to popularise material on female sexuality that had been presented more dryly in the 1950s by Professor Alfred C. Kinsey, a zoologist at Indiana University, who before undertaking his sociological studies of human sexual behaviour – work that was the Western world's declaration of sexual independence from hundreds of years of mischief and mythology – was a world authority on the gall wasp.

Both researchers demonstrated that the manipulation of the clitoris was the best or, usually, only way women had orgasms. Hite concluded: 'Not to have orgasm from intercourse is the experience of the majority of women', while Kinsey held that, 'The techniques of masturbation and of petting are more specifically calculated to effect orgasm than the techniques of coitus itself'. But while Kinsey's solidly academic doorstop works were surprise post-war bestsellers, Hite's slicker style and somewhat arousing anecdotal accounts made her a media favourite.

'The orgasm itself reminds me of a dam breaking. I can feel

contractions inside me and a very liquid sensation. The best part is the continuing waves of build-up and release during multiple orgasms,' reported one of Hite's respondents. Another wrote: 'My vaginal and clitoral area gets absolutely hot and I seem to switch into a pelvic rhythm over which I have no conscious control; every contact with my clitoris at this point is a miniature orgasm which becomes more frequent until it is one huge muscle spasm!' Another still: 'First, tension builds in my body and head, my heart beats, then I strain against my love and then there is a second or two of absolute stillness, non-breathing, during which I know orgasm will come in the next second or two. Then waves, and I rock against my partner and cannot hold him tight enough. It's all over my body, but especially in my abdomen and gut. Afterwards, I feel suffused with warmth and love and absolute happiness.'

Needless to say, ejaculation is not the major component of female orgasm. For women, sexual Nirvana is primarily a variable peak muscular experience accompanied by involuntary, rhythmic contractions of the vaginal walls, the uterus, rectal sphincter and urethral sphincter, all allied to the partial dissipation of the muscular tension of sexual arousal and the parallel release of vaso-congestion – but, critically, not as complete a release as men undergo.

It remains a moot point whether the vagina itself is sensitive to sexual friction. Kinsey was adamant on this: 'The vagina walls are quite insensitive in the great majority of females . . . There is no evidence that the vagina is ever the sole source of arousal, or even the primary source of erotic arousal in any female,' was his verdict. Germaine Greer, however, in *The Female Eunuch*, published in 1970, wrote: 'It is nonsense to say that a woman feels nothing when a man is moving his penis inside her vagina. The orgasm is qualitatively different when the vagina can undulate around the penis instead of a vacancy.' If only as a modern woman, not merely as a pioneering theorist on female sexuality, Greer might be expected to know better than the decidedly odd figure of Kinsey, whose position is, as will be examined later, today being critically reappraised in a rather negative light.

The muscular vagina certainly has a fundamental role in sexual response, whether or not it is sensitive. As orgasm approaches, there is swelling of the vaginal barrel's contracting outer third, and during orgasm itself there is a two- to four-second muscle-spasm in this region, followed by rhythmic convulsions at intervals of 0.8 of a second. There are from five to eight major rhythmic contractions in each orgasmic experience, followed by nine to fifteen minor ones. The total duration of these muscle contractions has been measured as lasting between 13 and 51 seconds, although women reporting their subjective perceptions state that orgasm lasted between 7 and 107 seconds. During these contractions, the heart races and the arms and legs spasm more or less out of control. There is slight expansion of the inner two-thirds of the vagina, a contraction of the uterus and, frequently, strong muscular contractions in many parts of the body. Respiratory and heart rates approximately double.

The woman's face may remain composed quite rigidly until the beginnings of her orgasm, at which point her composure breaks and the features become mobile and distorted. Her breathing in and out loudens and accelerates and the nostrils flare, while extra saliva flows in the mouth, which can make her tongue react by lolling and contorting. The pupils of the eyes dilate and temporary photophobia along with the delirious pleasure of the moment may cause her to shut them. As the climax of orgasm continues, disjointed words, highly charged with emotion, may come from her mouth, coalescing into a single long cry expressive of sublime pleasure.

One orgasm is seldom sufficient. Women are naturally multi-orgasmic. 'A woman will usually be satisfied with three to five orgasms,' stated William Masters and Virginia Johnson in their 1966 book *Human Sexual Response*. 'That is,' they added, 'if a woman is immediately stimulated following orgasm, she is likely to experience several orgasms in rapid succession. This is not an exceptional occurrence, but one of which most women are capable.'

An interesting variation on the multi-orgasmic woman theme emerged in 2003 when Professor Sandra Leiblum and psychologist Sharon Nathan of the Centre for Sexual and Relationship Health at the MMDNJ-Robert Wood Johnson Medical School in New Jersey, reported a previously unknown affliction they called Persistent Sexual Arousal Syndrome, PSAS. This is *constant* desire for orgasm that has to be dealt with one way or another for the women to function normally. PSAS, Leiblum said, has been identified in some 50 women worldwide. One woman suffering PSAS reports that she needs as many as 800 orgasms a day.

Her gynaecologist, the afflicted woman wrote in the London *Sunday Telegraph* told her, she was, 'every man's dream'. 'I wanted to punch him. I said, "How would you feel if you had a permanent hard-on all day long?" That shut him up.'

The physiological similarities between male and female orgasm notwithstanding, there is also an enormous difference. It is not merely that males have fewer orgasmic contractions – three or four at best – or that, unlike her mate, the female's genitals retain their engorged blood on orgasm so she can climax again and again if she wishes. Nor is it just that the contractions of female orgasms last longer, while men's most intense pleasure lies in the first few spasms. The biggest difference is, rather, that orgasmic feelings in men are localised in the immediate genital area of penis and testicles, while for women orgasmic sensations are felt throughout the pelvic area.

Anatomically, the female orgasm is a more widespread phenomenon than the male. And that is why sexually aware women have such an enormous appetite for orgasm and the capacity to climax repeatedly without tiring. As Dr Mary Jane Sherfey, a radical feminist psychiatrist, asserted in a 1973 book, *The Nature and Evolution of Female Sexuality*, because of the way women are constructed, it is wrong to expect them to be satisfied by one orgasm: the more orgasms a woman has, the better they become, and the more she wants. The greater scale and complexity of the female orgasm, it is thought, is also the reason why it is so important in the evolutionary scheme of things.

Sherfey argued that the female capacity for sexual pleasure is more fundamental than even the extraordinary existence of the clitoris – a bundle of nerves designed solely for sex – would suggest. The woman's hedonistic potential is, she explained, a function of the structure of a woman's lower pelvis, the whole of which she described as an extensive 'erotic network', in which the clitoris, labia and perineum, the outer vagina, anal region and G (Gräfenberg) spot (of which more shortly), all serviced by four or five dense masses of veins and nerves, congregate into one integrated, hyper-responsive sexual organ with the potential to out-perform the simpler male sexual apparatus.

The combined array of female genitals has similarly been described by other researchers as forming an 'orgasmic crescent' of erectile tissue. Dr Sherfey also asserted that the network of blood vessels which creates the female's extreme potential for sexual responsiveness becomes more complex after each birth and with age. In other words that, again contrary to what might seem necessary to a strict Darwinian, women are biologically designed to get keener on sex the further they advance into their childbearing years – and beyond.

The size and range of organs involved in the woman's 'erotic network' makes it unsurprising that at various times theorists have posited that there is more than one kind of female orgasm. Freud's 'vaginal orgasm' is discredited. Sigmund Freud theorised, or more likely fantasised, that females were biologically inferior to males and live their lives tormented by feelings of inferiority and 'penis envy'. His theory was all the more outrageous because it was accepted as the gold standard for most of the twentieth-century. Freud asserted that clitoral orgasm was a sign of sexual immaturity, suitable only for young girls. A mature woman, he pronounced, should be able to have a quite superior vaginal orgasm. By making the distinction, of course, he gave legs to the male-dominated ('androcentric' or 'phallocratic' in the argot of gender studies) notion that a woman must respond to the kind of sexual activity that happens to be the most swiftly gratifying for a man. If she failed to achieve orgasm from penetration alone, she was worse than anatomically defective, which would suggest a level of blamelessness; she was sexually *infantile*.

The Harvard palaeontologist and biologist Stephen Jay Gould, in a celebrated essay, 'Male Nipples and Clitoral Ripples', pointed out the pain and anxiety that Freud caused to women, which remained fashionable – a cult, almost – as late as the 1960s. Gould wrote: 'As women have known since the dawn of our time, the primary site for stimulation to orgasm centres upon the clitoris. The revolution unleashed by the Kinsey report of 1953 has, by now, made this information available to men who, for whatever reason, had not figured it out for themselves by the more obvious routes of experience and sensitivity.' Gould suspected further that part of the reason for Freud's assertion about infantile and mature orgasms may have flowed from male vanity – from an angry rejection of the idea that a woman might get sexual pleasure without the need of a man.

But the ascendancy in the late twentieth century of the clitoris as the real site of female sexual pleasure has not resulted in the female orgasm becoming as one-dimensional as the male's in women's view. If women's experience of orgasm could be neatly encapsulated, it would be that all orgasms are ultimately *centred* on the clitoris, but that they can start in a host of places within the 'orgasmic crescent'. The G-spot, the bean-shaped mass of sensitive tissue found in some women between the back of the pubic bone and the top of the cervix, and which in many is the focal point of sexual arousal, is just one of the these orgasmic centres. (The link between the G-spot and orgasm is far from clear; but it certainly seems to work extremely well for millions of women.)

A 1972 article in the *Journal of Sex Research* identified three fairly distinct forms of human female orgasm – vulval, uterine and 'blended'. The vulval orgasm is what we know better as the clitoral orgasm, characterised chiefly by involuntary contractions of the PC muscle; the uterine orgasm is the result of stimulation deeper inside the vagina; the blended orgasm is the deeply satisfying combination the two.

This natural facility of women for a kind of multimedia sexual pleasure, which can only be assumed to have been the case for all humankind's existence, has led over the millennia to understandable confusion on behalf of 'consumers' of sex, both female and male. While our prehistoric ancestors must surely have chanced upon the many and varied forms of female sexual gratification, they are unlikely to have tried to rationalise or tabulate them.

Successive cultures, however, did just that, and never more so than in our own scientific age, in which researchers, aware that there is more than just the clitoris involved (and that anyway the clitoris does not respond especially well to stimulation immediately after orgasm), have adopted an alphabet soup approach to mapping the alternative female pleasure points necessary for the three to five climaxes the average woman needs to be sexually satiated; first came recognition of the G-spot, then the U-spot – the sensitive opening to the urethra; and then the X-spot on the cervix ('Better than the G-spot and easier to find,' according to the slogan of its 'discoverer', Chicago sexologist Debbie Tideman). And Barbara Keesling has found an interesting nearby area in the upper rear of the vagina known as the cul-de-sac or fornix. 'Incredible' orgasmic sensations, she reported in *Psychology Today* magazine in 1999, can be achieved by a phenomenon called 'tenting', in which, when a woman becomes highly aroused, muscles and ligaments surrounding the uterus lift it up to allow penetration into the small space *behind* the cervix. Keesling has also found that stimulation of the PC muscle surrounding the opening of the vagina is successful in enhancing orgasm.

Three major points about the basic and unchanging sexual Nature of the female human animal can be drawn from the discoveries of successive generations of sex researchers in different times and cultures.

The first is that the clitoris is not simply a female penis. As Germaine Greer has written: 'If we localise female response in the clitoris we impose upon women the same limitation of sex which has stunted the male's response.' Not only are there plenty of other orgasmically sensitive areas, both below the waist and elsewhere, but sexual enjoyment for women is even more of a generalised, 'whole person', psychological experience, that, most crucially, continues after the last muscular spasm of orgasm. As Dr Greer puts it again: 'The male sexual ideal of virility without languor or amorousness is profoundly desolating; when the release is expressed in mechanical term it is sought mechanically. Sex becomes masturbation in the vagina.'

The second is that the complex and time-consuming nature of women's orgasm (as opposed to the user friendliness of the male) is not a design flaw, or some psychological quirk of fickleness or awkwardness. There is an underlying logic behind it all. Mary Jane Sherfey's view is that the female need for long, drawn-out foreplay and psychological scene-setting is biological, that nature, not culture, has determined that women must be gently, sensitively and lovingly led by the hand towards sexual intercourse. Since women do not have an oestrous cycle, this psychiatrist contends, they need caressing and encouragement to stimulate the blood flow to the pelvis they need for their reproductive system to gear up to best advantage. (There are other, parallel rationales to be discussed later for the strength, length and power of the female orgasm, as distinct from its typical starting difficulties.)

The third conclusion to be drawn from the mysteries of the female orgasm is balder and more disturbing to generations of men. It is, simply, that penile penetration is rarely involved other than in a peripheral role with the attainment of orgasm for women. They may get plenty of psychological fulfilment *conducive* to orgasm from penetrative sex, but, according to every serious study and the vast majority of anecdotal evidence, it is downright unusual for a woman to reach orgasm solely through the friction of conventional sexual intercourse – even if she gets substantial pleasure from the feeling of penetration and simply having the man's erect penis inside her.

It is undeniable, meanwhile, even when ambiguities over anatomy and questions of copulatory etiquette and gender politics are factored out of the sexual equation, that not all women have orgasms. Statistics understandably vary, the matter of the female orgasm being more subjective than the binary certainties of the male, but the most comprehensive and methodically sound investigation into Americans' sexual practices ever conducted, the 1994 survey *Sex In America: A Definitive Survey* by members of the University of Chicago's National Opinion Research Center, confirmed that it is women who experience far greater problems of sexual satisfaction and interest than men. This study and others indicate that 40 or more per cent of women suffer from some type of sexual dysfunction. Most of these women are between the ages of 25–50.

Most women, it appears, can attain orgasm with clitoral stimulation, but only about 50 per cent of women who can orgasm ever *claim* they can reach climax during coitus. Around 10 per cent of women never achieve orgasm, whatever the situation or degree of stimulation. Women can be orgasmic throughout their lives, and 85 per cent retain sexual desire after the menopause, for once a woman learns how to reach orgasm, she rarely loses that capacity. However, sexual activity tends to decrease after the age of 60 because of lack of partners and untreated bodily changes such as atrophy of the vaginal mucosa.

A pooled global survey in 2001 of 27,500 men and women aged 40–80 in thirty countries, *Global Study of Sexual Attitudes and Behaviors*, carried out again by the University of Chicago and funded by Pfizer, suggests that a third of all women at any one time (not just of the dysfunctional 43 per cent) lack any interest in sex, a third are unable to orgasm, a third have only occasional orgasms, 21 per cent do not find sex pleasurable, 20 per cent have trouble lubricating, and 14 per cent experience pain with intercourse.

Orgasmic disorder in women, it is generally accepted today, may be lifelong or acquired, general or situational. Because of the widely varying definitions of female orgasm – an orgasm can mean different things to different women – the diagnosis of sexual dysfunction in women is frequently problematic. The most clinical definition is that, with increased blood flow, the vagina and uterus contract and make orgasm impossible. But much of the female sexual anatomy even today remains unknown, particularly the nerves and blood vessels affecting sexual function. An even greater variety of causes can affect and nullify orgasm in women than in men.

Among the factors which can make orgasm difficult or impossible for women are medical diseases, minor ailments, depression, medications including antidepressants, stress, psychosocial difficulties such as financial, family or job problems, family illness or death, physical, sexual abuse or rape (currently or in the past), smoking, cycling (unlike horse riding, which can stimulate orgasm, bicycle seats can cause perineal pressure and reduced blood flow), anger, ignorance of genital anatomy and clitoral function and of arousal patterns and techniques, anxiety, association of sex with sinfulness and of sexual pleasure with generalised sense of guilt, more specific guilt (such as felt by a widow with a new partner or a woman engaged in an extra-marital affair), fear of intimacy, concern about reputation, fear of unwanted pregnancy, hormonal changes, mood disorder, fear of 'letting go' and losing control, fatigue, time pressure, religious taboos, social restrictions, sexual identity conflicts, sexual inexperience, different sexual preferences from a partner and other conflicts, and poor sexual communication.

The Canadian QueenDom.com website revealed in a 1999 poll of an unusually large sample of 15,000 sexually active adults in the US, Canada and the UK, that the simple matter of self-consciousness inhibited orgasm in a large proportion of women and an equivalent or larger number of men, too. While 46 per cent of anorgasmic women blamed the problem on lack of confidence in their appearance, 70 per cent of men having trouble with orgasm admitted that the difficulty was that they get hung up about their looks when they have sex.

Grunting, groaning and facial grimacing during the latter stages of sex (the critical endgame known colloquially by some, because of the facial expressions involved, as 'the vinegar strokes') can make us feel embarrassed and fail to orgasm – 61 per cent of women and 72 of men feel they 'lose the plot' when they start becoming too demonstrative, the figures showed. Equally, if a partner seems to be thinking about work or football scores during sex, the chance of orgasm for many vanishes. Anger with a partner over some unresolved emotional issue was shown to be a reliable showstopper, as was the wrong kind of physical assertiveness. Here, however, the statistics are a little double-edged: 60 per cent of women and 52 percent of men explained that when orgasm eludes them, it is because their partner is too rough; conversely 50 per cent of women and 63 per cent of men said that it happens because their partner is not rough enough!

New sexual dysfunctions are regularly identified. According to recent work by Jim Pfaus, who studies the neurobiology of sexual behaviour at Concordia University in Montreal, for instance, some women confuse what is called sympathetic arousal, as evidenced by increased heart rate, clammy hands, nerves and so on, with fear. As Pfaus explains: 'That makes them want to get out of the situation. Psychotherapy is a common treatment for the condition, although if anxiety is a factor, patients may also be prescribed Valium. But then Valium can actually delay orgasm.'

Cultural differences, especially male machismo, also come

into play in anorgasmia. In a 1985 survey, 60 per cent of working-class women and 50 per cent of professional women in Puerto Rico admitted to faking orgasm to avoid a vanity-fuelled interrogation from their male partner. In South Africa, the Sexual Dysfunction Clinic at Johannesburg Hospital has treated black men worried about their inability to sustain an erection for an unrealistic length of time, or to have sex up to four to five times a night. In Brazil, where sexual expectations are generally perceived as high, poor female field workers treated for anorgasmia in 1990 were found to have neither expectations nor a scintilla of sexual knowledge. Over a third of the women were unaware that the sexual act was normal in marriage, although they knew that prostitutes and other 'bad men and women' engaged in *sacanagem* ('the world of erotic experience'). They were under the impression that all sex was immoral and indecent and that their husbands were insane for desiring sex.

The caveat mentioned earlier, that orgasm can mean different things to different women, deserves more than cursory attention. There is, by some accounts, a significant phenomenon among women of mis-perceived orgasmic dysfunction. This may well spring from the reluctance among many women, open as they are to frank discussions with their peers about sensitive topics, to talk as openly about their experiences and expectations of orgasm.

A urologist, J.G. Bohlen, working in the early 1980s, made the remarkable finding that there was minimal correlation between the perception of orgasm by women and physiological signs of it as measured in the laboratory. Some women he monitored said they had experienced orgasm when no muscle contractions had occurred. Other sex researchers have also reported that, in tests, some women can have what they are satisfied is an orgasm while lying perfectly still and without contractions.

Conversely, Hartman and Fithian monitored a group of 20 female therapy clients who claimed they were not orgasmic.

Three-quarters, however, were found to be undergoing the classic physiological responses associated with orgasm. Once the women had these changes highlighted for them, all but one were able to identify it for themselves as an orgasm the next time they were monitored. Significantly, many of the subjects had read up widely on orgasm, but decided what they had did not seem to feel what it was supposed to be like. It is as if the modern mythology and cult of orgasm has placed the sensation on such a pedestal – created such an aspirational 'superbrand' of it – that women perfectly capable of orgasm refuse to believe they are having a legitimate one and must instead be experiencing an inferior imitation brand. Either that, or they simply discover that, for their taste, orgasm simply is not all it is cracked up to be.

The judgement that a woman is anorgasmic – and the above strongly suggests that, at some level, it *is* a judgement – is also subject to the cultural wind blowing at any particular time in history. One early male sex researcher, E. Elkan, argued in 1948, in an attempt to place female orgasm in an evolutionary context, that 'fixing' mechanisms such as hooks and barbs have evolved in lower species such as snakes to allow the male time to inseminate the female. In species where males do not have such capture mechanisms, there are behavioural immobilising mechanisms such as skeletal contractions to ensure insemination. Elkan went on to argue that since orgasm is not one of these mechanisms and therefore does not occur in animals, women should regard orgasm as a gift and not part of their due. An anorgasmic woman, therefore, should be no more worried about it than if she were unable play the piano.

In a very different age, the 1980s, the libertarian Professor of Psychiatry Thomas Szasz wrote in a book, *Sex: Facts, Frauds and Follies*, that women have learned that being sexually self-affirmative is 'unfeminine', and hence, are unable to discharge their sexual tension through coital orgasm. This socialisation argument, that women simply *learn* to be less orgasmic than men, has wide currency today. 'Such women are now called anorgasmic,' contended Szasz, 'but men who cannot weep are not called alachrymal. The former condition is thought to be a sexual dysfunction, but the latter is not considered to be a lachrymal dysfunction.'

A number of interesting explanations for the mystery – and frustration – of female anorgasmia have been garnered from experts by QueenDom.com. Peg Burr, a Californian sex therapist avers there: 'My guess is that anorgasmia relates to a lack of efficacy and control in one's life. Orgasm requires becoming vulnerable and open. This openness is based on an intact sense of self which does not feel threatened (engulfed, or overpowered) by sexual union. Persons who are rigid and/or controlling have great difficulty allowing themselves to be vulnerable and completely orgasmically responsive with another person . . . Women have less personal power in (and over) their own lives, due to social roles which teach them to be passive and non-assertive. They therefore may (unconsciously) exert control where they can, over their own bodies, and unfortunately, limit their own sexual pleasure.'

A practical and pragmatic analysis of anorgasmia – and a possible solution to it – comes from Dr Judith Schwambach, the Indiana-based syndicated sex advice columnist. 'By far the most common culprit I have observed in my practice is a weakened female PC [pubococcygeal or pubic] muscle. Without a strong PC, most women require direct clitoral stimulation to experience orgasm. A very weak PC may be unable to provide sufficient vaginal tightness for the man to easily achieve orgasm.' The advice, says Dr Judith, is 'Kegel exercises', available from a variety of medical sources, to tone up your PC.

Bryan M. Knight, meanwhile, a Canadian hypnotherapist and proprietor of the Web domain *http://hypnosis.org* puts his explanations for anorgasmia more crisply still, in four bullet points: If there is no biological cause, then possible reasons are, 'The woman was sexually abused as a child. She has a need to feel in control. She's having sex with an inconsiderate or unknowledgeable person. She'd be responsive to a woman.'

Other suggestions from QueenDom experts to combat the problem include self-hypnosis, making sure you are not already having orgasms but simply not recognising them as such, managing stress, avoiding alcohol and drugs before sex, not worrying about losing composure or dignity, never faking orgasm – and 'being a little greedy: when you know what you like, ask for it. Your pleasure is your partner's delight.'

But is it still possible that the human female, ultimately, is just less well designed for sexual and orgasmic pleasure? Is it something in the plumbing? Or is the most important thing we have to understand about sexual delight that men and women desire it equally, are equally capable of it – but are designed to approach it via separate routes?

The sociobiological case for the latter is put eloquently by an Illinois clinical psychologist, John B. Houck. 'The best strategy for men to increase their gene pool,' explains Houck, 'is to father as many babies as possible with as many women as possible, trusting that some will survive to adulthood and produce more offspring. This leads a man to be prepared to have as many orgasms as possible. In contrast, the best genetic strategy for a woman is to form a relationship with a man and get him to protect and provide for her and her children, since without modern fertility drugs, she can usually have only one child a year, and needs to extend every effort for those children to grow to adulthood and reproduce. This leads a woman to be focussed on her relationship with a man, to be turned on when she feels safe, protected and provided for, and not turned on when she doesn't.'

Houck concludes: 'From whatever perspective we come from, it is clearly more important for most women to have more time to feel safe, protected, loved, cared for, and special, in order for them to reach orgasm. Foreplay for many women starts a day or two before the sex act, with the man showing them special attention and love, which begins to put them in the mood for love. Men usually do not need such a long time to get ready.'

4

Afterglow

'Orgiastic potency is defined as the capacity for complete discharge of all damned-up sexual excitation through involuntary pleasurable contractions of the body'

Wilhelm Reich, *Function of the Orgasm, 1942*

Many ancient cultures believed their orgasms were mystical experiences, and there can be little doubt that such a perception had its roots in the accumulated folk wisdom of the ancient people's own distant ancestors. It should be of no wonder, really, that the rapturous sensation of the immediate aftermath of orgasm was revered as something on a parallel with a religious experience from the moment human beings began to develop spirituality – the belief, often prompted by times of crisis, that there is meaning, purpose, inspiration and answers about the infinite to be had in life.

The most common word that even atheists exclaim when they have an orgasm is 'God!' It is that easy, *in extremis*, with the oxytocin and other pleasure-inducing chemicals flowing, to confuse an exceptionally pleasant bodily sensation with an awed, revelatory, mystical metaphysical feeling of harmony with the universe.

Even today, there is a wide and sometimes slightly woolly literature arguing the case for orgasms as a mystical experience. Long before the current 'Tantric' cult, a lot of psychedelic bric-à-brac from the sixties had pioneered a school of thought that the ancient Hindus, Mayans, Aztecs, Egyptians and so on had better and more meaningful orgasms than modern, consumerist man. Here, for instance, is Elizabeth Gips, a leading voice of the sixties 'counterculture' in her memoir called *Scrapbook of a Haight Ashbury Pilgrim: Spirit, Sacraments and Sex in 1967/1968*. She recounts thus a particularly splendid orgasm she experienced on New Year's Eve 1966: 'Male and female are one body that is no body in the time before time when God/me gave birth, created itself. An orgasm beyond orgasm that shakes loose streams of energy which become space, stars, planets, trees, bugs and people. RAPTURE. Am God, energy or whatever, me/you/they. Everything. Created creator.'

The importance of a neo-mystical feeling of post-orgasmic rapture is discussed a little less excitably by clinical psychologist John B. Houck. 'The spiritual dimension of sexuality is very important,' he says, 'since sexuality can lead people into powerful spiritual experiences, including the ecstatic experience of unity with the divine, with each other, and with all of creation.'

To those brought up in the modern Christian world in which sexual pleasure was, as a matter of policy, to be imbued with guilt as a way of asserting man's superiority over animals, thinking of sexual and religious rapture in the same breath is, of course, anathema – which is obviously what attracted the hippy movement to it. But in pre-Christian days, most notably in the Old Testament millennia, there was nothing un-Godly about enjoying sex; in fact, to do so was rather religious. An historian, R.C. Zaehner, in a 1957 book *Mysticism Sacred and Profane*, wrote: 'There is no point at all in blinking at the fact that the raptures of the theistic mystic are closely akin to the transports of sexual union, the soul playing the part of the female and God appearing as the male. The close parallel between the sexual act and the mystical union with God may seem blasphemous today. Yet the blasphemy is not in the comparison, but in the degrading of the one act of which man is capable that makes him like God both in the intensity of his union with his partner and in the fact that in this union he is co-creator with God.'

Or as George Ryley Scott, an historian, sociologist and anthropologist, put it in 1966 in his book *Phallic Worship*: '. . . the more abstract, intangible and symbolic becomes the cult, the more likely is sexual indulgence to prove the only possible outlet for what would otherwise result in a sense of frustration . . . Once it was thoroughly realised where lay the responsibility for the pleasurable Nature of the sex act, it was perfectly natural that the organs concerned in this sensation should be treated with the greatest respect and adoration . . . sexual indulgence had a magical effect. It was always, and is, to some extent, even to this day, imbued with mystery.'

This worship of sex among the very earliest civilisations and neo-civilisations is seen in phallic symbols such as the men with enormous, erect penises carved into chalk downs in Southern England. A visceral, Neanderthal reverence for the bodily joys to be had from the penis (and the vagina, to a lesser extent) survives, it might be argued, in the toilet-wall daubings of less articulate members of modern societies, where both men and women of limited potentiality draw dripping, orgasmic penises (vaginas for the more artistic) as a statement of – what? Exultation? Rapture? It barely matters, but there is a remarkable constancy in the phenomenon from culture to culture – a feeling that these are authentic, albeit inelegant, echoes of voices from the distant past.

We now know that the pleasant mental and physical experience of orgasm is primarily a chemical one, caused by the release of a cocktail of neurotransmitters throughout the nervous system, namely the two catecholamines, noradrenaline and dopamine, plus the indoleamine, serotonin. Serotonin and dopamine in particular are said by scientists to release the pleasure-stimulating compounds called endorphins in the brain. These essentially pain-relieving molecules have a similar structure to the morphines.

That numbing, mildly narcotic, tranquillising sensation alone, the sense of comfort and temporary disconnection from reality, may well explain the eternal appeal of orgasm to humans. The famous French description of orgasm as '*le petit mort*' – the little death – is a little hyperbolic, but is no more than an elaboration of the same theme, acknowledging as it does the edge of self-destructiveness that characterises any voluntarily taken narcotic. Algernon Charles Swinburne, for example, a poet who, rarely for a Victorian, celebrated carnal passion, could scarcely separate the idea of sexual satisfaction from pain and death. Sexual yearning cannot even be satisfied on earth, according to Swinburne's poetry. And a slight (yet slightly gratifying) feeling of loss, of emptiness and sadness – or of a minor death to the dramatically inclined – is for many people as much an integral part of the sexual climax as the preceding euphoric sensation.

Some sexual theorists have argued that the perceptual link between orgasm and death is traceable to the elevated state in which orgasm in a rare few cases is accompanied by a loss of consciousness. Another explanation for the *petit mort* idea is inherent in a rare but persistent folk belief, of uncertain origin but found in cultures from Europe to the Far East, that a person is born with a certain number of orgasms in him or her, and that when the last is used, the person dies.

The death and orgasm link pervades popular culture. It is found throughout Shakespeare, as we will see later. A Swedish-born German poet, Ernst Moritz Arndt, in an eighteenth-century letter to one of his sexual conquests, writes: 'A divine fire consumes my soul, I melt, I die of rapture, I am quenched like lightning in the radiance of the day.' It appears too in common ditties, such as this anonymous male orgasm poem of the same period: *Stand, Stately Tavie*:

Stand, stately Tavie, out of the codpiece rise,
And dig a grave between thy mistress' thighs;
Swift stand, then stab 'till she replies,

Afterglow

Then gently weep, and after weeping die.
Stand, Tavie, and gain thy credit lost;
Or by this hand I'll never draw thee, but
against a post.

The root of the death-orgasm idea may be nothing more complicated, after all, than the fact of the male erection literally 'dying' after ejaculation. This action may also account for historical taboos against masturbation. Ideas of sperm dying a pointless death are one thing, but could the ancient suspicion of masturbation be nothing more than a disapproval of suicide? If a little bit of us dies for each orgasm, is not a 'fruitless' ejaculation then a minor suicide attempt?

Men and women have not always resorted to florid images of life and death when describing orgasm. W.H. Auden memorably, and more effectively than the death-by-orgasm brigade, described the sexual craving as 'an intolerable neural itch'. But the satisfaction of having scratched that itch certainly makes humans behave very differently from the way they do in the run-up to intercourse – and very differently from any other occasion. First they fall into a limp but exquisite state of relaxation, the pain-relieving endorphins providing relief and release, along with gratifying feelings of diminished control and, in some, an altered state of consciousness.

The man's erection will, in all but a handful of cases, have subsided within seconds of ejaculation, leaving him temporarily spent and incapacitated for further sex, and sometimes in a little pain from the end of his penis and his testicles. If anyone is liable to want to recommence sex quickly, it is the woman: Sherfey wrote in a 1966 essay, 'A Theory of Female Sexuality', 'Theoretically, a woman could go on having orgasms indefinitely, if physical exhaustion did not intervene.'

Physically, there is an unravelling of nearly all the signs of the excitable pre-copulatory fever pitch and a return to the normal, quiescent physiological state, with an overlay of sleepiness. The sexual flush abates, vanishing in reverse order to its sequence of appearance. There will often be an outbreak of copious sweating in both sexes, regardless of how much physical effort went into the preceding sexual activities. This sweating is related more to the intensity of the orgasm than the real amount of aerobic exercise involved. A film of perspiration forms on the back, chest, shoulders and thighs, the palms of the hands and soles of the feet, the face, forehead and upper lip. Sweat may also run from the armpits.

Psychologically, overwhelming emotional feelings, with outbursts of laughing (or weeping), are normal post-orgasm, and both sexes will typically enter a giggly, agreeably childlike state that makes them more receptive to the oxytocin effect, with its attendant feelings of attachment, tenderness, affection and calm – all highly beneficial to bonding a relationship.

But that parallel feeling of undefined sadness after orgasm, the post-coital *tristesse* encompassing mild disappointment mixed with anticlimax, is never far from the surface, as we know from common experience as well as much literature and popular song. Some of this post-orgasmic feeling of exhaustion is physiological in nature, the physical concomitant of feeling 'spent'. But there is an important cultural or social feeling that frequently impinges here – one of regret.

Post-orgasmic regret, the kind of mournful sensation that seems to be alluded to in a thousand miserable Country and Western songs, may be occasioned by a simple factor such as the copulation having been illicit or adulterous, especially in social systems where this can lead to grave, even fatal, legal consequences. But there are more subtle and complex factors at play, too.

Women may find themselves wondering quite why they had sex with the man lying beside them, and even be seized by the idea that they have 'wasted' themselves on an unworthy mate. The man who seemed attractive hours ago has lost his lustre with the dissipation of sexual fervour and the resultant orgasm.

The woman's occasionally inescapable feeling of post-orgasmic regret seems from common experience to be at its most intense when she has had sex with someone she is not committed to, either because the relationship is new and unsure – or because it is old and worn out. The regret is quite separate from the familiar disappointment that ensues (all too often) from her not having reached orgasm at all in a sexual session. Neither is it necessarily associated with guilt stemming from religious or social injunction or indoctrination against sex, especially illicit sex.

Anecdotal accounts again suggest that such a regretful sentiment is more likely to come instead from a sensation of misdirection – that the woman has followed her perceived biological destiny to be as sexually predatory as the male, but she is not quite sure why she has done so. Being as driven as any man towards orgasm-seeking may be the true state of the human female, or it may be a modern feminist construct. But a 1962 sex manual, *Woman and Love*, by Dr Eustace Chesser, could have touched closer to the truth of the matter than we like to think in a more liberated age. Chesser argued that women are biologically programmed to view intercourse not as the pursuit of a single, isolated, hedonistic moment, but as a series of such moments to be celebrated, embracing 'courtship, wooing, sex act, conception, gestation, childbirth, lactation and maternity'.

In a tougher, more feminist-inclined world than Chesser's, post-orgasmic regret has become something of a political football in the long-running grudge match between angry men and angry women. The point of contention is whether post-orgasmic regret on a woman's behalf is sufficient grounds for a complaint of rape. This is something of a one-sided argument, because most women, naturally, would contend that 'rape is rape' and can never be mistaken for consensual sex. A small but noisy 'men's movement', however, believes a minority of women interpret their post-coital regret as an indication that they were raped, when, perhaps, they were only a little ambivalent about a sexual encounter after it was over.

Furthermore, now that the 'post-coital regret' basis for a rape complaint has become a popularised theory, women are starting to allege that male detectives investigating rape claims are openly suspicious that they really had sex consensually, but because they were, for some reason, disappointed by experience, they have decided to press charges.

In a speech to a gender issues forum in Seattle in 1993, a men's activist, Rod Van Mechelen, put the controversial case thus: 'In the US and Canada, misandrists – those who hate men – are working hard to make it easy to convict men of rape if they have sex with women who are under the influence of alcohol, women who experience post-coital regret, or with women who were reluctant to have sex.

'Years ago, when I was in college,' Van Mechelen continued, 'my buddies used to speak of a secret fear. The fear that they would go to a bar one Friday night, get drunk, and then wake up the next morning in the arms of some female equivalent of a "dirty old man". I had friends this happened to, and they felt pretty bad about it. Like these women had taken advantage of them. Were they raped? No. College boys can't be raped because only women will be protected by law.'

A variant on post-orgasmic regret is reported in gay male culture, too. Dan Savage, writer of a sex-advice column, 'Savage Love', which is syndicated is the US alternative press, has described a liaison gone badly wrong due to regret. 'He was a beautiful AmerAsian boy,' Savage recounts in an interview, 'and once I put him in drag he looked like a girl. It was one of those experiences where you go, "OK, this isn't my boner, but this is never gonna happen to me again." We didn't do anything that was unsafe and I didn't do anything in which he didn't take the lead because I didn't want to spook him. But he had the post-orgasmic regret that people can get when they're just starting out. The moment he came, he wanted to leave. He left in such a hurry that he left behind his panties and his watch. I called him and he said, "Don't call me." I said, "I'm just calling to tell you . . ." And he hung up on me before I could say, ". . . you left your watch in my apartment." It was pretty weird'.

Heterosexual men, while a little less prone to post-coital regret – or even a great deal of thoughtfulness – commonly suffer a complementary sensation nonetheless, that once the oxytocin-induced post-orgasmic 'coupling moment' has passed, they urgently need to be elsewhere – a feeling that has been charmingly described as 'pork and walk syndrome'. Those who feel compelled physically to vanish post-orgasm are thought by some psychologists to be anxious that they have not suddenly been trapped in an unwanted relationship, that they have not, to use a phrase that crops us frequently throughout the history of sex, 'swallowed bait'.

Women sometimes experience the post-orgasmic 'swallowed bait' slump, but rather than experiencing a need to distance themselves from the lover, become overwhelmed instead with a feeling that they have been seduced or used. Their anxiety may manifest itself as a sudden, urgent need for verbal reassurance that they are loved. Some researchers observing small children have suggested that these post-coital emotional patterns start early. Boys, it is said, tend to avoid eye contact with maternal figures as a way of stating their need for separation, while girls seek eye contact for reassurance – a yearning arguably reflected later in life by the desire for close contact with lovers.

If one sensation, however, were to characterise male post-orgasmic *tristesse* it would be a non-specific perplexity, a kind of generalised numbness, a spell during which the man cannot quite recall just why he was so anxious hours or minutes before to have sex with the woman. The novelist Jonathan Franzen captures this moment exquisitely in *The Corrections*: Alfred's wife, the neglectful and cold Enid, is crying after having brief and highly unsatisfactory sex with him. The moment is all the more traumatic because Enid is pregnant and neither she nor Alfred believes it is right for them to have sex. 'Why did wives choose night to cry in?' Franzen writes. 'Crying at night was all very well if you didn't have to catch a train to work in four hours and if you hadn't, moments ago, committed a defilement in pursuit of a satisfaction whose importance now entirely escaped you.'

A lot of this feeling has to do with men's ability to exaggerate in their minds the attractiveness of a female partner – until they have actually had sex with her, whereupon, post-orgasmically, the 'scales fall from their eyes' in an equally exaggerated, hormone-fuelled manner. Studies carried out in singles bars, and quoted by the University of Texas psychology professor David M. Buss in his 1994 book, *The Evolution of Desire*, suggest that men find the same women progressively more attractive and desirable over the course of an evening. This increase in perceived attractiveness appears to be independent of how much or little alcohol men have consumed – the proverbial 'beer goggles' which can, in crude common parlance, turn a dog into a fox.

It may instead be, as Buss puts it: 'attributable to a psychological mechanism sensitive to decreasing opportunities for casual sex over the course of the evening. As the evening progresses and a man has not yet been successful in picking up a woman, he views the remaining women in the bar as increasingly attractive, a shift that will presumably increase his attempts to solicit sex from the remaining women in the bar.

'Another perceptual shift may take place after men have an orgasm with a casual sex partner with whom they wish no further involvement. Some men report viewing a sex partner as highly attractive before his orgasm, but then a mere ten seconds later, after orgasm, viewing her as less attractive, or even homely,' writes Buss. 'It is not unreasonable to believe that mechanisms attuned to reaping the benefits of casual sex without paying the costs have evolved,' he concludes.

It is important to mention at this point that the healthy, fulfilled physical sensation men and women can safely be assumed to have experienced in the afterglow of orgasm since the beginning of time has a basis in modern medical research.

Orgasm, it seems, along with being rather pleasurable, is good for you, subverting for once the maxim that everything enjoyable is either illegal, immoral or fattening.

An orgasm is said to be as mentally and physically beneficial as a five-mile jog, thanks to the raising of heart rate, which is equivalent to an energetic aerobic workout. Its less obvious benefits have been shown by various research efforts (of differing degrees of scientific rigour, one suspects) to be manifold. Orgasms apparently improve our breathing and circulation, cardiovascular conditioning, strength, flexibility and muscle tone. They are also said to result in bright eyes, facial glow and shiny hair, and to relieve the symptoms of menstrual problems, osteoporosis and arthritis. Orgasms may even help us lose weight. The adrenaline released by orgasm allegedly helps break down glucose and prevents it being stored as fat.

Two studies – one of couples in Edinburgh, the other of middle-aged men in Wales – have suggested that regular sex, either within or outside stable relationships, may be a significant factor in preventing premature death from a variety of causes, and can additionally make individuals and couples look as much as ten years younger than their real age. There seems to be no specific anti-ageing mechanism involved, but rather an especially benign combination of emotional and physical wellbeing.

Additionally, Australian research published in 2003 suggested that frequent male orgasm helps prevent prostate cancer. A team led by Professor Graham Giles, Director of the Cancer Council Victoria's Cancer Epidemiology Centre in Melbourne, concluded from a study of over 2,000 men that the more ejaculations men aged between 20–50 enjoy, the less likely they are to develop prostate cancer. Men who, in their twenties, ejaculated more than five times a week were a third less likely to suffer aggressive prostate cancers when they were older.

The Melbourne study attracted particular attention because it specified that masturbation was more likely to act as a cancer-preventative than intercourse, because of the way sexual infections tend to increase the chance of prostate cancer. Professor Giles speculated that masturbatory ejaculation prevents a key carcinogen in prostate fluid, 3-methylcholanthrene (also found in cigarette smoke), from building up. 'It's a prostatic stagnation hypothesis,' he told *New Scientist* magazine. 'The more you flush the ducts out, the less there is to hang around and damage the cells that line them.' Associate Professor Anthony Smith, of the Australian Research Centre in Sex, Health and Society at La Trobe University in Melbourne, commented to the magazine: 'If these findings hold up, then it's perfectly reasonable that men should be encouraged to masturbate.'

Then there are the immune system benefits from orgasm that are believed to contribute to disease resistance. The endorphin release following orgasm has been shown to boost immunoglobulin A (IgA), an antigen found in saliva and the mucus in the nasal passages, and which binds to bacteria, prompting the immune system to attack them. The emotional and physical bond wrought by orgasm through the agency of oxytocin forges, additionally, a strong basis for maintaining good mental health.

The flow of endorphins resulting from orgasm, too, may also have a positive mental health payoff, relieving anxiety and depression and increasing vitality. Some research has suggested that women's mere contact with semen through the vaginal wall can help them experience a kind of post-coital bliss.

A Canadian professor of psychiatry, Philip Ney, documented in 1986 in an obscure journal called *Medical Hypotheses* a depressed woman whose symptoms subsided after she had sex. Then, in 2002, a study of 293 female students aged 18–35 by a psychologist at the State University of New York, Dr Gordon Gallup, found that those who always had sex without a condom were quantifiably happier according to a standard psychological questionnaire, the Beck Depression Inventory, than those who had protected sex or abstained.

Those who used condoms sometimes, Gallup found, came

in second on happiness scores, while the least happy women were those whose partners always or usually used condoms. Women who were not having sex scored between the 'always' and the 'usually' groups. Such obvious distorting factors as the effects of oral contraception were apparently ruled out by the experimental design, leading to the psychologists' conclusion that mood-altering hormones in the semen, especially testosterone, oestrogen and prostaglandins, were entering the woman's body topically through internal tissues and going on to act as an antidepressant. Gallup reported at the time of announcing his findings that a similar test on a group of 700 women had shown the same trend.

Gallup expanded on his findings after reporting them in a Cambridge University journal, *Archives of Sexual Behaviour*. Among his rather startling associated discoveries were that women whose partners do not use condoms become more depressed when their relationship ends, that agitation and irritability increase with condom use, as do suicide attempts – and that women not exposed to sperm take longer to move into a new sexual relationship, while those whose partners 'go bare' become involved more quickly. He has also become convinced that contact with semen becomes a chemical dependency for some women.

Gallup's findings, it has to be said, are seen as distinctly fringe by the medical profession. Some medics have seen them as too simplistic, others as an open-Sesame to an era of unsafe sex. It is also interesting to observe that Professor Ney in British Columbia, who claimed the original discovery, is closely involved with the religious anti-abortion movement in Canada and the US. This is not to suggest for a moment that he is necessarily carrying a Catholic anti-contraception torch, but it seems worth taking note of all the same.

As for physical *dis*benefits of orgasm, beyond simple and temporary soreness in the genitals of both sexes due to the mechanics of intercourse, they are few. The only widely documented ill effect of orgasm is the rare condition of 'orgasmic cephalalgia', an intense, explosive headache that can begin at the moment of climax in both sexes. The headache can last for several hours, and occurs in masturbation as well as coitus. Its cause is not known, but it is assumed to be related to increased blood flow through blood vessels during sex.

5

The Way We Were

'Sex is great fun . . . it is the dance of the genitals; it is an ecstatic swinging in the arms of the beloved. It ought not to be too intense; it must not be degraded by possessiveness or defiled by jealousy'

Anthropologist Verrier Elwin
on sex among the aboriginal Muria people of Northern India.

Sexual intercourse culminating in ejaculation is unlikely to have struck many early prehistoric humans as the cause of pregnancies that only became visible several months later. Even if we conjecture that some thoughtful individuals may have wondered whether the placing of the penis in the vagina was entirely unconnected with the emergence of babies from the very same place, we can be quite confident that the all-important link between sex and babies did not exist to any significant extent in the early human being's mind – a state of affairs that would not recur, strange as it may seem, until the age of the Pill caused intercourse once more to be perceived by both sexes as a consequence-free, pleasurable, pastime.

The situation is unlikely to have changed much when the first belief systems formed in human societies. George Ryley Scott, in his *Phallic Worship: A History of Sex and Sexual Rites*, states: 'Without exception, the worship of sex by all primitive races originated in the pleasure associated with coitus, and not in any clearly conceived notion that intercourse would produce children. The sex act gave pleasure to those engaging in it, and by analogy it would give pleasure to the gods. Man could think of no part of himself for which he had greater regard than his sexual member, and no part of woman for which he had greater reverence than her pudendum.'

As it would again in the still inconceivably distant 1960s, an obscuring of the connection between orgasmic bliss and the burdens of childbirth and parenthood tilted the balance of power in human societies rather in favour of women. Women supplied orgasms to men and could withdraw them more or less at will; and what was more, they could magic up babies, by all appearances of their own will and making. Men did not even monopolise hunting; women shared that too.

Given that the history of civilization has partially been the history of the sublimation of women's sexual desires, there is an assumption among academics, and not only feminist ones, that aeons ago 'cave' women could also take the initiative and be as demanding, predatory and promiscuous as men would later become. At the end of the second millennium AD, sisters began 'doing it for themselves', in the sense of taking the sexual initiative – and congratulated themselves on thinking of such a revolutionary notion. But it is always worth remembering that things had been like that once before in the sparse, isolated outposts of human settlement that comprised the prehistoric world. When there was no more than one person per thousand square kilometres, women seemed to be the sole creators of new human life. And to this day, fascinatingly, in preliterate societies around the world, where the sex/babies connection is unknown or misunderstood, women, sole suppliers of the delicious snack for the sensations known as orgasm, retain appreciably more power in their societies than in more knowledgeable cultures.

Yet the more civilised we have become, the less instinctive, perversely, sex seems to have become. The sex act, although inborn in animals and intuitive in more primitive human groups, is far from instinctive in civilised man, and, like hunting, seems to become progressively less instinctive as civilisation grows more complex.

Civilisation entails successive overlays of morality, complex economics, religious constraints and politics, and each veneer of culture seems to make the simple enjoyment of sex a more complicated business. In the first civilised era, the Neolithic age, men – thanks to their observations of the farm animals they were beginning to spend their life tending – appear to have realised at last that as fellow humans they played an equal, if mysterious, part in the production of children. We know that women in the first organised agricultural human settlements were inclined to be in charge of crops, while men looked after the herding. We assume consequently that after many generations, males noticed how their ewes could graze happily for ever without getting pregnant, but how with the introduction of just one ram to the flock, all the females would become pregnant. We surmise accordingly that it dawned on men how creatures with penises like their own possessed some awesome power. But the more thought that went into the subject subsequently, the more confused we seem to have become.

Following the somewhat startling revelation that they were essential to the baby-making process, men seem to have seized every chance to reclaim power and make themselves the dominant gender. Subsequently, they more than compensated for their longstanding inferior status to women by imbuing their penis with the status of a life-giving totem that demanded worship – a sexual consensus that in some ways still pertains in most societies. Men's idolisation of the penis is also evident in the widespread male homosexuality that would later pervade the classical world. We have no particular evidence of homosexuality among prehistoric peoples. But given that it exists in many animals, and was a positive cult in most ancient cultures, notably, as we shall see, the Greeks, it would be perverse to imagine that preliterate humans were not also interested in homosexual activity.

The shock discovery by men that they, too, made babies led directly to the cult of virginity in women, which would be of such core importance to the Abrahamic faiths, and especially underwrite Christianity after the story of Mary's supposed *post partum* virginity was written retrospectively into Jesus Christ's CV by Mark and Luke in their Gospels. The more pagan early religions, nevertheless, seem to have imbued sex, rather than the lack of it, with a special reverence – even without knowing what it was for.

In preliterate, pre-agricultural tribes today, where understanding of the male role in conception is incomplete, female virginity is of little or no account, as we will see later. But in societies where the male's contribution to fertilisation became known, men swiftly began to develop a keen interest in the concept of a child being 'theirs'. Evidence of an emerging nuclear family structure with clearly defined property rights has been discovered at Neolithic Çatal Hüyük in Turkey, the world's earliest known city. Houses in this settlement of some 10,000 people were grouped tightly together. Neighbours were close at hand to be coveted and gossiped about. Sex in such a social place became other people's business and, accordingly, something for which one had to seek privacy because who was doing it with whom had community significance. The architecture, archaeologists have found, reflected the new, settled, semi-public pattern of people's sex lives; bedrooms with a clear sexual purpose are seen for the first time. They are adorned with erotic works of art – an engraving of a woman with her legs apart and a penis entering her here; another of a naked woman carried between two leopards there.

Women did not benefit from the Neolithic revolution. It meant the emergence of farming and property possession, but also possession of women and their precious fertility. It was now that the veil was invented and men began to control female sexuality. If a field or an ass could be 'mine', it followed that so could a child and its mother. Paternity as a concept dovetailed neatly into the emerging idea of property rights. What, after all, would either woman or child be without the power of the man's semen? One would be barren, the other non-existent. Sex was now becoming more than mere pleasure, it was business too. The transition from nomadic hunter gatherer to commercial farmer saw women's role dwindling to one of breeding machine. Men at the same time, significantly, were spending more time sitting around and having great thoughts while the crops grew, the animals grazed and the women and children did the donkey work. Intellectual life, philosophy and religious thought undoubtedly have their roots in this period 8,000 years ago.

Even the method of having sex seems to have changed in this era. From figurine evidence, Dr Timothy Taylor, a Bradford University archaeologist who has studied Çatal Hüyük closely, guesses that standardisation of houses and the new, urban, semi-industrial lifestyle signified the standardisation of sex too. The position now favoured, he believes, was with the woman on all fours – a depersonalised mode copied from the farm animals.

Amid such a social structure, it followed that a premium, both social and financial, would became attached to a woman who could be guaranteed on betrothal to be a virgin; it was not so much fastidiousness about other men having touched her as a more-or-less cast-iron warranty that any child of the marriage would be the husband's and nobody else's, and therefore the correct person to inherit the husband's land and animals when he died. This discovery of paternity was a breakthrough for the development of stable families, civil laws, cities and societies. But what disappeared as part of the newly serious equation was the idea of sex as an enjoyable activity, particularly for women. The whole body of laws and customs across several cultures requiring suppression of sexual desire, strict containment of orgasmic pleasure and repression of woman, can be traced back to the Neolithic discovery that men were needed to make women pregnant.

Property considerations have also shaded notions of love in various later cultures. Among the Bedouin, for instance, romantic love is said to be considered a shameful flaw, because people are supposed to love their family and dynasty rather than a mere spouse. The Bedouin are accordingly horrified by public displays of affection between man and wife.

For pre-Neolithic man, we can surmise from such scraps of evidence as exist that sex was purely for pleasure and that orgasm as an end in itself was, if not a matter of consuming, life-or-death importance, still a very special thing. The physical satisfaction of intercourse and the sublime release of orgasm seem to have been sufficient to induce reverence for sex and the bodily parts involved in the act. This veneration was manifested in phallic cults that were the precursors of religion, but also in the worship of the vagina, which may co-incidentally have functioned as the birth canal, but was also the specific site of a very special and unique form of physical pleasure for both sexes.

One of the earliest pieces of evidence that sex for pleasure may have been a preoccupation for Stone Age people exists in a group of pinkish-red ochre lumps with a cross-hatched pattern scratched on their surface. The ochre was found in Blombos Cave, near Stilbaai on the Cape coast in South Africa, and is believed by some anthropologists to be 70,000-year-old lipstick. The world's oldest known evidence of symbolism in human culture, if lipstick is what it is, their theory has it that the pigment was used by females to make their mouth look more like vaginas – and by so doing, signal to males whether sex was on offer or not.

Some 50,000 years later, however, comes a far less ambiguous example of evidential bric-à-brac supportive of early human cults of sexuality – a rash of more than 200 clay figurines of female bodies dating from the Ice Age of 30,000 to 20,000 years ago and found in the past hundred years from Russia to France. These so-called Venus figurines are faceless, but have large breasts, a rounded belly and a prominent vagina detailed down to labia and even, on one, a clitoris – an especially fascinating find to keen students of the orgasm.

A clue of sorts to something – it is unclear what – is that there are no equivalent male figures; apart from the sketchy graffiti of cave paintings (a large number of which were of vaginas), the only male depiction found to date from the same post-Neanderthal, pre-Neolithic period was, entertainingly for today's women, a puppet. So Venus figurines may have been primitive fertility goddesses, or teaching aids for women to instruct girls in the art of becoming a woman, or art for its own sake, created by men in the same tradition as heterosexuals, at least, have today, that the female form is more attractive than the male nude.

The figurines could, additionally, be nothing more than early pornography, three-dimensional centrefolds for men away from their mates to masturbate to; why, after all, should Palaeolithic men on hunting trips have been very different from modern men on business trips masturbating in hotel rooms? So, were the Venus figurines revered goddesses or cheap pin-ups? Guessing their purpose has exercised generations of academics. The fact that statuettes are often found broken in waste dumps, just as porno magazines tend to be found in hotel room rubbish bins, rather suggests the latter. But we are still in the realm of relatively uninformed conjecture.

A far clearer insight into whatever part orgasmic pleasure played in the lives of prehistoric peoples can be gleaned by basing our imaginative reconstruction of ancient sex lives on the habits of primitive peoples that have survived more or less intact from the Stone Age. The theory of the Blombos Cave lipstick, for example, was based by its anthropologist authors on similar body-painting sexual rituals they identified in some surviving traditional societies.

Likely as it is to provide the soundest evidence available to us of sex as it was in the real-life Garden of Eden, the study of ancient, preliterate tribes has not always endeared modern researchers to their contemporary society. Alfred Russel Wallace, the Victorian follower of Darwin studied the peoples of the Amazon, who were universally derided at the time as 'savages'. Wallace thoroughly discomfited Victorian society, already traumatised by the assertion of Darwinites that humans were descended from monkeys, by further noting that there was less distance than you might think between Amazon Basin 'savages' and us.

The obsession with orgasm and its accoutrements that is evident from a brief trawl around contemporary societies, both Western and Eastern, would not be unfamiliar to any visitor from a preliterate society – and, by extrapolation, would be equally recognisable to a prehistoric time traveller. He might also recognise the confusion, misinformation and mystification that continue to surround sexual issues in so many societies, our own included. If in subsequent chapters a sense of congratulating our own culture for making sexual knowledge so universal becomes evident, it is only partially deserved; the extent of enlightenment over sexual issues in modern societies is all too easy to exaggerate.

A Pacific island tribe, for instance, the Trobrianders, investigated by the anthropologist Bronislaw Malinowski for his 1929 book *The Sexual Life of Savages of North-West Melanesia*, have no word for father. This follows from their having no concept of paternity – even though they fully understand the sexual reproduction of animals. They laughed at Western missionaries who urged them to introduce Christian marriage as a way of legitimising the notion of fatherhood. To Trobrianders of the time, it seemed (unless, and it is far from impossible, they were pulling Malinowski's leg) sex existed solely as a source of pleasure.

To this day, many Australian aborigines, too, tend to dissociate sexual intercourse from childbirth. When the facts of life as we know them were explained to a woman of one aboriginal tribe, anthropological folklore has it, she was derisive about such nonsense, responding with scorn, 'Him nothing.' As late as the 1960s, the Tully River people of North Queensland believed that a woman became pregnant because she had been sitting over a fire on which she had roasted a particular species of black bream given to her by the prospective father. Another Australian tribe avowed that women conceived by eating human flesh. Not far away in Papua New Guinea, members of the Hua people are still said to contend that a man can become pregnant, but will die in childbirth. (This has recently been deconstructed as a probable description of a malnutrition disease, kwashiorkor, one symptom of which is a grossly distended stomach.) And the Bellonese of the Solomon Islands believed, until the influence of missionaries persuaded them otherwise, that the sole function of sex was pleasure, and that babies were implanted in women by ancestral gods.

Even without any clear notion of how human reproduction works, though, the idea still seems to have taken root in a wide variety of primitive societies many thousands of miles apart that the sperm produced by orgasm is a very special and powerful substance. If the Sambia of New Guinea, studied in the 1970s by the anthropologist Gilbert Herdt, are anything to go by, our ancestors may have been quite obsessed with semen. These transient jungle farmers believe sperm is the most important element in the production of children – and that it is also precious and in very short supply. They believe that a boy is born with an internal organ that will eventually produce both semen and growth, but it must be supplied with semen from older men before it can do so. To make the most of this dwindling resource, they harness an unusual mix of homosexual and heterosexual practice designed to produce brave, strong men who are capable of having babies with women.

The key act by which the male Sambian's supply of strong sperm is ensured, is regular male-on-male fellatio. A set of rules determines who the semen donor will be (the fellator's sister's husband is desirable; the father is not acceptable). Boys from about the age of ten try to accept semen on a daily basis by performing fellatio on a proper donor. After six to eight years as an acceptor, a boy becomes a donor.

'Among the New Guinea Sambia,' Herdt observed, 'an aberrant bachelor is one who does not offer his penis to be sucked by pre-pubescent boys.' An older teenager, he explained, arranges a meeting with a younger boy at a quiet jungle rendezvous, puts his penis in the mouth of the younger male and is brought to orgasm. The crucial part of this custom is that the younger boys swallow the semen produced. The older, fellated boy demonstrates his superior status by standing during this procedure, while the fellator represents his inferiority by kneeling in front of him. Curiously, however, the fellat*ed* professes to derive little sexual pleasure from this otherwise fairly standard homoerotic duet. For the convention is that the younger fellat*or* is the prime beneficiary of the ritual, acquiring as he ingests his peer's precious emission valuable, high-potency fuel for his future seminal needs. The fellat*ed*, furthermore, donates his sperm at a certain risk to himself, for even strong, mature males in Sambian philosophy can find their semen depleted if they are over-generous with it.

The Sambian sperm rite seems like an institutionalisation of homosexuality, and adolescent and adult males in this society turn out commonly to be stimulated erotically by images of and fantasies involving younger males' mouths. Not surprisingly, young Sambian fellators do not report receiving sexual pleasure from performing their duty; yet as they approach puberty, some of them begin to find it stimulating and get erections while doing it. (Sadly, there is little they can do about these erections since masturbation is taboo because of the waste of semen; the Sambians consider spillage caused by involuntary orgasms in the form of wet dreams to be the result of bad spirits coming and seducing them while asleep.)

However, pornographic websites originating both in the West and Asia discuss what they call an oral or mouth orgasm as something women can enjoy with considerable intensity while fellating. Some women on the more literate of these sites describe it as similar to 'an electric current' beginning in the mouth and taking over the rest of the body. A dissenting sexological voice on one fairly standard, non-pornographic sex advice site which advocates the mouth orgasm (*www.after-glow.com*) argues that it is the result of women unconsciously compressing their thighs or rubbing their legs together while fellating, thus stimulating the clitoris and inducing a conventional climax. A similar mechanism could well be operating for young male Sambian fellators.

Despite such overtly homosexual initiation traditions, almost all Sambian men marry heterosexually – with the slight twist that their early sex life involves exclusively oral sex, the husband now being fellated by his wife, who swallows his semen in the belief that this will enhance her sexual maturity and, in particular, her milk flow when she is lactating after childbirth.

No wonder, then, given the survival of such rituals amongst a preliterate population, that the renowned Duke University anthropologist Raoul Weston La Barre, who studied mainly Native American and South American peoples, concluded that to our ancestors, sperm, bone marrow and brain matter were one unified life-substance – a conviction that he held went back to the Neanderthals a quarter of a million years ago.

Yet what we now know to be misconceptions about the finer details of human reproduction continue to mislead even peoples who acknowledge the concept of fatherhood. Elders of the (until recent years) headhunting Sema people in Nagaland, in north-eastern India, maintain that pregnancy is the cumulative effect of having sex many times. The Sambians, again, similarly hold that babies are formed in the womb by regular doses of semen from a male.

And more than twenty other tribal societies from Paraguay to Tanzania studied by a Pennsylvania State University anthropologist, Stephen Beckerman, believe a child can, and ideally ought to, have more than one father. Until the middle of the last century, Canela women, in Amazonian Brazil, had sex consecutively with up to forty men in festive rituals designed both to ensure conception and to blur paternity.

Yet any uncomfortable image of a misogynistic 'gang-bang'

mentality in this culture can safely be dismissed; men were very much under the thumb of women in Canela society. When they wanted to have a child, they would choose their favourite five or so lovers to help their husband with the job of fertilisation. Every bit of semen was believed to contribute to the baby, and mothers-to-be would select for a variety of desirable traits among her contributing lovers, from sexual skills to good looks to a good singing voice. The multi-fathering tradition among the Canela was only finally extirpated by missionaries, who convinced them that it was morally wrong by such propagandistic methods as translating the Bible into the Canelan language – and by traders in such items as pots and pans, who taught them the entirely alien concept of personal property.

Married women of the Barí tribe in Columbia and Venezuela who are bearing a child still proudly announce a list of 'fathers'. Elderly Barí women will reportedly chuckle and nudge one other as they recall their lifetime's roll-call of lovers. Among the Curripaco people in the same area, conception is regarded as a process that requires a great deal of co-operative work by many men. But they still put on what Beckerman's collaborator, Dr Paul Valentine of the University of East London, South Africa, describes as 'a smug look' when they describe this process – an indication that multi-father 'conception' represents the ultimate coincidence of work and pleasure.

Multi-fatherhood is a puzzling belief; but curiously, Charles Darwin himself would not have disagreed with it. It was still not understood as late as the nineteenth century that fertilisation was brought about by a single sperm acting in one instant. Though the question remains as to whether the supposedly unsophisticated people who express such ideas to anthropologists are in fact having a joke at the researchers' expense, testing the extent of Western credulity, being diplomatic about sensitive or embarrassing matters – or being discreet about the delights of having many orgasms a day, something that, for all they know, 'refined' Westerners do not, or cannot, enjoy.

It can hardly be insignificant that it was not until decades after the bulk of pioneering anthropological work was done in Papua and New Guinea that Herdt felt free to investigate the Sambians' homosexual practices. Even in academic research in the mid- to late-twentieth century, it was not considered decent in Western society to probe such a matter. The Sheffield University zoologist Tim Birkhead, author of a 2000 book on the central evolutionary topic of sperm competition, makes the same point in his preface – that the scientific study of reproductive matters is mostly recent because, for a long time, it was not considered respectable.

Another reason for isolated peoples to cherish the idea of multi-fatherhood – beyond the fact that it allows everyone to enjoy a great deal of sex – is that the tradition has sensible and stabilising economic and social benefits. Barí children with more than one father are more likely to survive into adulthood, with gifts of food from several sources in times of scarcity. Barí males also tend to die young from malaria and tuberculosis. 'You know that if you die, there's some other man who has a residual obligation to care for at least one of your children,' explains Beckerman. 'So looking the other way or even giving your blessing when your wife takes a lover is the only insurance you can buy.'

We are still some distance away, too, from being free from our own curious tribal customs. When the Tully River aborigines speak of women becoming pregnant by the action of barbecuing fish, they are not stating anything more outlandish than a Western Catholic would when expressing the belief that Christ was born to a virgin. As one modern anthropologist has commented: 'If we believe in the Virgin Birth, we are devout; if others do, they are idiots.'

Educated atheists in the modern world can also find themselves uncertain about whether certain sexual behaviour is taboo or not. According to a Glasgow midwife interviewed by the author some middle-class Western mothers, for example, follow the practice (well known in the South Pacific, as we will see) of altruistically masturbating or fellating very young baby boys in the hope of developing and enhancing their infants' sexual feelings – or even of sending them to sleep. But they carry out this ancient folk practice somewhat guiltily – aware *inter alia* that for a father to do the same to an infant girl would be sufficient to guarantee him a lengthy jail sentence.

Alongside confirmation that we are still a long way from being comfortable with our sexuality, there is parallel evidence, too, of creditable sexual refinement among some primitive peoples, suggesting that any idea we may cherish of a brutal, male-centred 'caveman' form of sex among our distant ancestors is misleading. Less developed societies with a tradition of encouraging males to delay their orgasm while affectionately stimulating the female are as well documented as those cultures that have less sexually democratic traditions.

Some indications of the existence of good prehistoric sex may be gleaned, for instance, from anthropological studies of the !Kung bushmen of the Kalahari and of the Muria, an aboriginal people on the plains of central India. These peoples, although separated by huge distances, are remarkably similar in their sexual beliefs and practices – and therefore can plausibly be relied upon to retain echoes in their sexual culture of humans as they were at the dawn of mankind.

The exclamation mark in the !Kung name represents a popping sound in their language made by forming a vacuum between the tongue and the top of the mouth and then snapping the tongue down – although it might equally be a commentary on the fact that the male !Kung, unusually, manage to be semi-erect at all times. These priapic Bushmen believe that babies are produced by the combination of menstrual blood and semen, which poses the question of how many little !Kung are ever conceived since when women are menstruating they are highly unlikely to conceive. But the tribe is immensely positive about sex, equating it with food as both a medium of survival and pleasure. !Kung women demand their orgasms; if a man has 'finished his work', they say, he must continue until the woman is satisfied, too. They hold that if a girl grows up without regular sex, she will lose her mind and end up eating grass and dying. The !Kung do not perform oral sex, but both men and women masturbate energetically.

In Murian society, as studied in the mid-twentieth century by the Oxford-educated anthropologist Verrier Elwin, sex is a duty performed by men for women, whether married or not. A devout Christian who went to India as a missionary, Elwin converted to Hinduism, married a Muria woman, Kosi, and lived among the people for much of his life. He became convinced that the tribal traditions were superior to those of the 'civilised' societies, as there was no sexual inequality and responsibilities were equally distributed.

'The Muria have a simple, innocent and natural attitude to sex,' Elwin wrote. '[They] believe that sexual congress is a good thing; it does you good; it is healthy and beautiful; when performed by the right people [a male and female who are not taboo to each other], at the right time (outside the menstrual period and avoiding forbidden days), and in the right place . . . it is the happiest and best thing in life. This belief in sex as something good and normal gives the Muria a light touch. The saying that the penis and vagina are *hassi ki nat*, in a 'joking relationship' with each other, admirably puts the situation. Sex is great fun . . . it is the dance of the genitals; it is an ecstatic swinging in the arms of the beloved. It ought not to be too intense; it must not be degraded by possessiveness or defiled by jealousy.'

A Muria proverb collected by Elwin was, 'Woman is earth; man cannot plough her' – meaning, according to Elwin, that just as a single plough cannot break up the earth, no man can really satisfy a woman. Sexual pleasure, Elwin concluded, is regarded by the Muria as a woman's right, her compensation for the pains of menstruation and childbearing. And the Muria woman uses her sexuality as a means of dominating and subjugating males. On the other hand, it might be noted by sceptics that, citing tribal tradition, Elwin took another woman, Leela, the daughter of a village head as an occasional alternative to Kosi.

Four thousand miles away in the Amazon basin, sex is the major leisure pursuit for the jungle-farming Mehinacu people, who were studied in the 1970s by the anthropology professor Thomas Gregor, of Vanderbilt University in Nashville. Sexual maxims collected from Mehinacu tribesmen by Gregor include: 'Good fish get dull, but sex is always fun', and, 'Sex is the pepper that gives life its verve'. Sex for the Mehinacu ends the instant the male ejaculates, which does not sound promising. The Mehinacu are well aware that the clitoris hardens during sex and that it is the seat of pleasure. But they have no expression or word for female orgasm, which suggests that they do not have them. (In the West, at least we have developed a term – i.e. female orgasm – even if it is not of the most elegant.) They also have no concept of romantic love. After hearing a Portuguese song on the radio, Gregor wrote, one of the Mehinacu asked him: 'What is all this, "I love you, I love you"? I don't understand it. I don't like it. Why does the white man make himself a fool?'

In the Melanesian islands of the south-west Pacific, female orgasm is first achieved by mutual masturbation, with penetrative sex only starting just before a simultaneous climax. Another Pacific people, on Mangaia, southernmost of the Cook Islands, meanwhile, enjoys a profoundly erotic culture. Young boys on the island are instructed at the age of thirteen or fourteen in the erotic arts by older women. A typically 'good' girl has had three to four lovers between the ages of thirteen and twenty; and all women are said to orgasm, usually several times, during intercourse.

Mangaia's sexual culture was known to nineteenth-century anthropologists, but was obviously an awkward subject for them to explore in their own prevailing anti-sexual culture. At the time, the Mangaians were better known for being rather dour and truculent; they gave Captain Cook a particularly fierce reception when he 'discovered' the island in 1777. So it was left to Donald S. Marshall, an American anthropologist, to make the island famous for its relaxed attitude to sex. His landmark essay, 'Sexual Behavior on Mangaia,' appeared in *Human Sexual Behavior. Variations in the Ethnographic Spectrum*, in 1971.

Young male Mangaians, Marshall noted, learn several techniques of intercourse, plus cunnilingus, kissing and sucking of breasts, and are taught always to bring their partner to orgasm several times before allowing themselves to ejaculate – and only then in time with one of the partner's climaxes. The boys' instruction ends with a practical intercourse session with an older woman. Mangaian boys are expected to pay attention to their sex lessons, too; men who prove sexually inattentive are prone to have their partners swiftly leave them for more skilful males, and the women will often go out of their way to ruin the failed lover's reputation *en passant*. Mangaian women have several partners before they marry.

Girls on Mangaia are also coached by older women; but it is believed that while orgasm 'must be learned' by a woman, this can only ultimately be done with the help of a skilled man. Perfect sex on the island consists of no more than five minutes of foreplay, followed by fifteen to twenty minutes of energetic thrusting, with active female participation and encompassing, for her, two or three orgasms. The female's final orgasm should coincide precisely with the man's. The typical eighteen-year-old Mangaian couple make love three times a night, every night, until their thirties, when the weekly average drops to a mere fourteen.

Western sexual researchers have since become a near-permanent feature of Mangaia, a fair definition of a South Pacific paradise, with 1,000 inhabitants and a handful of discerning tourists. The island's sexual traditions (now largely Judaeo-Christianised) have been of particular interest because they exist outside any religious construct; they seem to reflect a joy in bodily pleasure for its own sake rather than the indulgence of a religious duty.

Mangaia's sex life has been also been cited as an ideal for a more liberated attitude to children's sexuality in Western societies. Dr Alayne Yates, Professor Director of the University of Hawaii Division of Child and Adolescent Psychiatry, wrote in a 1982 book, *Sex Without Shame: Encouraging the Child's Healthy Sexual Development*, that on Mangaia: 'infants are special people, rocked and indulged by all family members. Bare genitals are playfully or casually stimulated and lingual manipulation of the tiny penis is common . . . Privacy is unknown, as each hut contains five to sixteen family members of all ages. Adolescent daughters often receive lovers at night and parents "bump together" so that young children may be awakened by the slapping sound of moist genitals. Although adults rarely talk to children about sex, erotic wit and innuendoes are common.

'At the age of three or four,' Dr Yates continues, 'children band together and explore the mysteries of the dense tropical bush . . . Sex play flourishes in the undergrowth and coital activity may begin at any time . . . The young boy is taught at puberty by older males . . . [he] is coached in techniques such as the kissing and sucking of breasts. He is told about lubrication and trained in methods of bringing his partner to climax several times prior to his own ejaculation.'

For anyone with a mind, however, to believe that caveman sex, after all, was probably peremptory and brutal, there is a thread of evidence in the apparently liberated sex life of Mangaia. Marshall detected more than a trace of a darker side, too.

Although it seems that the women of Mangaia are their men's equals in their desire for casual sex, Mangaian men, according to Marshall, still succeed in being far more promiscuous than the women. 'The average girl has had at least three or four lovers between the ages of thirteen and twenty whereas the average boy has had over ten,' wrote Marshall, adding that boys often travel to neighbouring islands to expand their experience. So while the Mangaia culture may be more promiscuous than

Western cultures, men still allow themselves within its generous confines to be more libidinous than women. Marshall additionally discovered that Mangaian men believe they naturally desire sex more frequently than their womenfolk. As a result, he concluded, 'some husbands beat their wife into submission'.

Since Marshall, although Mangaia is still routinely used as a standard argument in favour of extreme sexual liberality, there has been a parallel process of scales falling from researchers' eyes. The island's violent side now attracts more academic attention than its sexual free-for-all. Sociologist Murray Strauss of the University of New Hampshire headed up a more recent report, published as part of a project on intra-family violence for the US National Institute of Mental Health, with this statement from a Mangaia woman: 'How do I know that he loves me if he doesn't beat me?'

6

The Evolutionary Paradox of Orgasm

> '*Sex and the City* star Kim Catrall and husband Mark Levinson have split up shortly after forty-six-year-old Kim said he was the first lover to give her earth-shattering orgasms. The thrice-married star now claims he has become obsessed with sex. They are co-authors of a book called *Satisfaction: The Art of the Female Orgasm.*'
>
> December 2002 news item quoting
> US gossip columnist Cindy Adams

The chapter that follows is in many respects the core of this book. It seeks to examine how the sexual desires and orgasmic sensations of the two genders have diverged; how this might account for some of the ways our societies have become ordered; whether women, freed from cultural restraints and taboos, from physical and social conditioning, are naturally as promiscuous and gratification-centred as men; whether the very different Nature of the female orgasm from the male has developed as a Darwinian adaptation, with specific reproductive benefits, or whether it is a pointless if pleasurable biological quirk.

'The desire for intercourse is the genius of the genus,' wrote Schopenhauer. But what a complex genius it turns out to be. By virtue of a series of devilishly clever evolutionary tricks, or perhaps due to sheer happenstance shaped by cultural factors, or by the deliberate design of a devious God, women and men have quite different sexual desires, different sexual experiences and different sexual aims. They probably always have had different expectations from sex, since the dawn of humankind. And as men and women are aware, they do not actually need one another to enjoy orgasm.

Yet from prehistory up to the present, most members of these two very different tribes have continued to seek out one another's company and spend their lives broadly together, centring a large part of their shared emotional existence on an activity, sexual intercourse, of which they have a very different experience. The genius of the genus is that the huge majority of the world's population is not homosexual – that, for one reason or another, the complex attraction of otherness has always managed to outweigh the easy pleasures of sexual like-mindedness, and the furtherance of the species has thus been assured.

It is a close-run thing whether the most striking disparity between the male and female yearning for orgasm is emotional or physical. On the emotional front, it is axiomatic, if not everyone's experience, that women fall in love first and later discover lust, while men fall in lust and only subsequently learn to love. Put another way, there is a broad, cross-cultural, popular perception, accurate or not, that women set out with a generalised longing for romance, affection and security that only finds proper fulfilment with the relief of a localised neural desire in the pelvic region; whereas men set out with a localised neural desire in the pelvic region that only finds proper fulfilment in romance, affection and security.

The most basic physical disparities between the male and the female orgasm are the most conspicuous. Take the obvious point, as delineated by Kinsey in his 1948 debut on the sexology scene, *Sexual Behaviour in the Human Male*, that, 'Men have orgasms essentially by friction with the vagina, not the clitoral area, which is external and not able to cause friction the way penetration does.'

Or take another major conclusion of the same work, that, 'For perhaps three-quarters of all males, orgasm is reached within two minutes after the initiation of the sexual relation.' (A time some men will regard as quite impressive!) Or the demonstrable fact that the average female has to be twenty-nine before she can match the orgasm rate of a fifteen-year-old boy – a remarkable inequality given that, according to Stephen Jay Gould in *Male Nipples and Clitoral Ripples*, male and female humans are anatomically very nearly the same creatures. Typical male orgasm also lasts no more than a couple of seconds, while in women, climaxes of up to a minute are known.

'Males and females are not separate entities, shaped independently by natural selection,' Gould argued. 'Both sexes are variants upon a single ground plan . . . The external differences between male and female develop gradually from an early embryo so generalised that its sex cannot be easily determined. The clitoris and penis are one and the same organ, identical in early form, but later enlarged in male foetuses through the action of testosterone. Similarly, the labia majora of women and the scrotal sacs of men are the same structure, indistinguishable in young embryos, but later enlarged, folded over, and fused along the midline in male foetuses.'

For women, despite such similarities at the manufacturing stage, sexual intercourse is a hopelessly inefficient way of producing orgasm. According to Desmond Morris in his 1967 book *The Naked Ape*, at the age of fifteen, peak child-bearing time for our distant ancestors, only 23 per cent of twentieth-century females have experienced orgasm in any form, masturbatory or through sex. By the age of twenty – on the elderly side for our forebears – Morris claims only 53 per cent of women have known orgasm, while the figure only reaches 90 per cent at the female sexual peak of thirty-five.

More modern research does not show any significant improvement in those dismal 1960s statistics. The Queen Dom.com poll of 15,000 respondents supported the existence of a fundamental difference between men's and women's experiences of orgasm, especially during intercourse. Women were still much less likely to orgasm. Only 42 per cent reported that they 'get there' most of the time, while 26 per cent of females said they experience orgasm 'rarely or never' during intercourse. The 'during sex' figure for men was 89 per cent, with only 2.5 per cent of men reporting problems reaching climax. 'To put this into perspective,' the website commented, 'women are ten times more likely to have problems with orgasm during sexual intercourse.'

The masturbation figures showed an improvement on Morris's data from forty years earlier, probably reflecting a relaxation on the taboo of practising and/or admitting to masturbation. In 1999, 77 per cent of women said they could masturbate to orgasm always or most of the time, 10.5 per cent, rarely, almost never or never. The corresponding masturbation 'success' figures for men were 92 per cent and 2 per cent. Looking less positively on the same figures, it was still the case that even when masturbating, orgasm is five times more likely to be to some extent problematic for women than for men.

Men, as the QueenDom figures plainly reflect, are virtually assured orgasmic climaxes – and additionally can orgasm from the age of ten or eleven, and often much younger, unless they suffer from some physiological or psychological disability. More often than not, however, the male mechanism is far too swift and efficient to give a female partner even a slender chance of a 'classic', penetration-induced orgasm. Orgasm for women, as a result, is far more often produced by a masturbatory mechanism of some kind than by straight reproductive intercourse. But as if to compensate for this physical mismatch, Nature has intriguingly made the female orgasm produced by masturbation far and away the more intense.

It is in the arena of the emotions, however, that the gap between male and female orgasmic expectations and feelings widens still further. Whether this was the case for our preliterate ancestors is a matter for conjecture, but the evidence again from surviving primitive tribes gently nudges us towards the conclusion that there has always been such a gulf to a greater or lesser extent.

Everyday experience and anecdotal evidence strongly suggest that human males *tend* to a significant extent to have a high interest in orgasm, placing less importance on relationships, coupling, security and monogamy, while females *tend*, again, to prioritise relationships, coupling, security and monogamy over mechanical orgasmic satisfaction *per se*.

Research data supports this proposition. In the QueenDom poll, respondents were asked how important orgasm was to them. The site editor comments: 'We asked this question from two different perspectives: how important is it that YOU reach orgasm and how important is it that your PARTNER reaches orgasm? Only 10 per cent of women say that it is extremely important that THEY reach orgasm, but 41 per cent of women say it is extremely important that their partner reaches orgasm. In other words, it is four times more important that her partner achieves orgasm.' (This is not to decry the importance men place on their partners' orgasms; 48 per cent of male respondents reported that it is extremely important that their partner reaches orgasm, which suggests they place *more* stress than women on mutuality of pleasure. At the same time, though, 27 percent of men said in the poll that orgasm was extremely important for them. So what the figures reflect is probably a greater *all-round* interest in orgasm as an essential of sex for men than for women.

Women's more compromising (not to mention accommodating) Nature is accepted even by radicals like Mary Jane Sherfey, who acknowledged in her 1966 essay 'A Theory of Female Sexuality' that: 'A woman may be emotionally satisfied to the full in the absence of any orgasmic expression.' (She did add in *The Nature and Evolution of Female Sexuality*, that: 'The woman usually wills herself to be satisfied because she is simply unaware of the extent of her orgasmic capacity.')

Feminist theorists find women's tendency not always to prioritise orgasm for its own sake a little uncomfortable. Their models of sexually voracious, promiscuous females in prehistory may, for all we know, be perfectly accurate. Yet survey evidence in the modern world stubbornly refuses to support such a view. Shere Hite in *The Hite Report* found that whether or not they regularly achieve orgasm in intercourse, the majority of women who responded to her questionnaire cited affection, intimacy and love, rather than orgasm, as their principal reason for liking sex. Neither did women report orgasm to be the most important bodily gratification resulting from sex; the preferred physical sensation by a significant majority was the moment of penetration.

For many women, then, orgasm is an overrated sensation. A British journalist, Lucretia Stewart, has written: 'Personally, I have always thought it was possible to exaggerate the importance of orgasm, believing that in many cases it's better to travel hopefully than to arrive. As any fool knows, the best stage of a love affair is the early stages, when you are in a state of constant, frenzied, unsatisfied desire. Once a man knows how to satisfy you there is no mystery, and within lie the seeds of boredom.'

Previously, Madeline Gray, in a book called *The Normal Woman* – which set out to assure women at the height of the sixties Sigmund Freud cult that they were normal even if they had not experienced 'mature' vaginal orgasm – had been considered rather controversial by writing: 'So female orgasm is simply a nervous climax to sex relations . . . It may be thought of as a sort of pleasure prize that comes with a box of cereal. It is all to the good if the prize is there, but the cereal is valuable and nourishing if it is not.'

It would seem that for women across cultures, during an individual act of sex, the journey – from wooing to scene setting to foreplay – is all important, while the consummatory end is very much a secondary goal. But, in any overview of a woman's sex life, her ultimate ends – enduring love, security, intellectual parity, intimacy, trust, a good environment for the children and so on – are more important than the sum of individual journeys, of isolated moments of gratification. This apparent reality has contributed to the common – and possibly correct – view of women as 'traders', who exchange sexual favours for security rather than sex for sex.

The opposite applies on both counts for males. Men are able to offset almost all the ends women value most for relatively few, and brief, orgasmic 'highs', often with partners who are unsuitable over a wide spectrum of parameters – the most extreme case being prostitutes.

For a zoologist such as Tim Birkhead, this disparity between men and women has a biological rationale; males, for Birkhead, are interested principally in how many eggs they can fertilise, women in who fertilises them. The different emphases of the genders are no better illustrated than in the pleasures each sex chooses deliberately to forego when joining monastic orders. Monks give up the male's powerful craving for the frustration-releasing physical 'sigh' of orgasm and subsume it (or so we are led to believe) in prayer; there is no approved way of dissipating the desire for female companionship. Nuns, on the other hand, forego relationships and 'marry' Jesus instead.

Even the deliberate adoption for cultural or political reasons of a more male-pattern sexual lifestyle seems not to diminish in women some inexorable, atavistic drive towards romance and stability. Tarvis and Sadd, summarising an enormous, 100,000-response 1977 survey by the American *Redbook* magazine, made a revolutionary finding for that time: that women who were more adventurous and experimental sexually had fewer orgasms than women who, if they had sex at all before marriage (which many did not), had it with the man they loved and subsequently married. Among the *Redbook* respondents who had had a series of one-night stands, for example, 77 percent said they never reached orgasm, compared to 23 percent among the women who had sex frequently with stable partners. Tarvis and Sadd quoted Doris Lessing's *The Golden Notebook*: 'And what about us? Free, we say, yet the truth is they get erections when they're with a woman they don't give a damn about, but we don't have an orgasm unless we love him. What's free about that?'

Donald Symons, Professor of Anthropology at the University of California, Santa Barbara, and author of what is widely acknowledged to be the most authoritative book on sex, *The Evolution of Human Sexuality*, prefers a pithier maxim from W.H. Auden to sum up the disparity between men and women's view of sex: 'Men are playboys, women realists'.

Another anthropologist, Lionel Tiger, delves into the lyrics of a classic pop anthem for his illustration of male sexual thinking and how similar it is to our imaginings of a caveman's sexual repertoire. 'In the story told in "Paradise by the Dashboard Light", the raucously brilliant song by Jim Steinman that was recorded by Meatloaf, a man is seducing a woman. The process is described (by the baseball hero and announcer Phil Rizzuto) in adolescent male terms – getting to first base, second base, rounding third, heading for home. Suddenly, the music stops. The woman calls an abrupt halt. She heatedly queries: 'Do you love me? Will you love me forever? Do you need me? Will you never leave me? Will you make me so happy for the rest of my life? Will you take me away and will you make me your wife? . . . I gotta know right now.'

'"Let me sleep on it," exclaims the evasive, equivocal male. She refuses sexual access until he promises to be with her "till the end of time". In the bitter final verse, the new husband desperately shouts, "Now I'm praying for the end of time." The promise of the relationship and of the duct of pleasure at its center is not matched by the real quotidian life within which they exist. Here, as elsewhere, the female has the more serious and complex task of assessing this pleasure and its role in her life. She must form and re-form the larger picture and adjust it for the passage of time and the varying circumstances in which she and her partner might spend their time.'

Not all men can accept that they are, in effect, ejaculation addicts, semen-spraying drunks, who inexorably seek the palliative null point of post-ejaculatory relief at whatever cost in broken promises, social embarrassment or any of the other penalties resulting from careless or cavalier sexual conduct. There are plenty of pop songs from a male point of view that attest to how much better 'making love' is to 'just fucking.'

But there seems no getting away from the conclusion that it is primarily for women that shaking off inhibitions and responding sexually to the point of orgasm depends on being in love and feeling comfortable with their lover. The female trait throughout history of body decoration and adornment – with great attention given to make-up, clothing, jewellery, dieting and so on – would suggest that women are more interested than men in attracting a superfluity of sexual partners. But the female's extensive concentration on her appearance is a decoy; it has to do with factors other than the signalling of sexual availability, continuing as it does long beyond the successful discovery of a mate.

There is little by way of parity either between the volume of sexual activity each gender finds necessary or desirable. Women have what, to men, seems a remarkable ability to shut down their sexual urge for long interludes in order to preserve monogamy and exclusivity in their relationships. With men in prison, at sea or at war, women seem to function perfectly well without sex for months or even years at a stretch. This ability to shelve sexual needs happens to fit in with human hunting/working/warring patterns perfectly, with men in a variety of ancient societies frequently living away from their mate for long periods.

Additionally, nature has ingeniously devised and we have simply discovered a monogamy-preserving safety fuse in masturbation. It works; but as an echo of the generally more frantic sex drive of the male, masturbation is typically of a rather different tempo for each gender. Anecdotal accounts suggest that it is unremarkable for a typical man on a hunting/business trip to masturbate once or more daily, women, every couple of days or far less. There is, again, an interesting disparity between masturbatory routines. Women can normally masturbate from stone cold to orgasm in about four minutes. Men, however, as women do in 'real' sex, often need to get into the mood either by careful mental scene-setting, or the visual stimulus of pornography, after which orgasm is easily and swiftly attained.

Either way, though, the *self*-masturbatory shortcut is just satisfying enough, if crucially short of skin-to-skin and mind-to-mind contact, to make it, for most people, a wholly fulfilling intercourse substitute. This is paradoxical in itself, because the mechanical orgasm acquired in women by masturbation is widely acknowledged to be more intense than anything achieved through intercourse alone. While it is said jokily that masturbation is better than sex because you can sleep with whoever you want, both sexes in reality seem programmed to enjoy orgasms more in a duet than as soloists, and this makes sense. For if the uncomplicated, emotion-free joys of auto-masturbation were ever to become a full analogue for intercourse, it could threaten the continuation of the species.

For Lionel Tiger, nonetheless, masturbation provides the most arresting and bizarre example of incongruity between male and female orgasm. This is, for Tiger, the existence of dildos and dildo-shaped vibrators. Men have never needed such an aid to masturbation since they have hands as built-in 'dildos'. But even though fingers can help, women have historically appreciated a little artifice. What fascinates Tiger, however, is that while dildos are broadly penis-shaped, they bear only the most tangential, fleeting relationship to an authentic penis, what with all the natural bumps and ridges smoothed out and a vibrating mechanism, of all things, inserted. The finest human-surrogate aid for a woman to achieve orgasm, then, is a machine that bears the scantest resemblance to any human bodily part. The classic male masturbatory *aide memoire*, the inflatable plastic woman with a fake electronic vagina and stick-on pubic hair, is, by contrast, a positively sentimental object celebrating the female form.

These may seem marginal points, yet they are not so. For it is almost as if our bodies were wilfully mis-designed for the propagation of the species. For other scientists too, the core sexual discrepancy between men and women is a basic matter of design. Stephen Jay Gould argues that the clitoral orgasm is a paradox for Darwinian biology. 'Evolution arises from a struggle among organisms for different reproductive success,' he writes. 'Sexual pleasure, in short, must evolve as a stimulus for reproduction. This works for men since the peak of sexual excitement occurs during ejaculation – a primary and direct adjunct of intercourse. For men, maximal pleasure is linked with the greatest possibility of fathering offspring. In this perspective, the sexual pleasure of women should also be centred upon the act that causes impregnation – on intercourse itself. But how can our world be functional and Darwinian if the site of our orgasm is divorced from the place of intercourse? How can sexual pleasure be so separated from its functional significance in the Darwinian game of life?'

In the few decades that such matters have been a suitable subject for serious discourse, three distinct, cogent and fascinating theories have been put forward to explain the central problem Gould articulates.

The first, classical theory is the one advanced by Desmond Morris in 1967 and supported by a wide range of researchers, from psychologists to sociologists to psychiatrists to even Masters and Johnson. Their hypothesis is that we display more intense sexual activity than any other major species of primate, and we are also unique in the length of our courtship and pair bonds. This is the result of male and female reproductive attributes having co-evolved, each counterbalancing adaptations in the other. The female orgasm is, then, more or less uniquely human and is 'designed', so to speak, to be complementary to the male orgasm by being much harder to bring about. The female orgasm, according to this view, has evolved as an adaptation to enhance the monogamous pair bond and make family life more rewarding. This is because only a long-term, stable male partner will know through familiarity how to make a particular woman climax properly.

The second theory, conceived and embraced by feminists and, to some extent, by the political left wing, also holds that the female orgasm is an evolutionary adaptation, but that it is supposed to be triggered by nothing more elaborate than straight intercourse; if it is not, furthermore, there is either something abnormal about the woman or inadequate about the man. The female's ability to multi-orgasm without the subdued 'refractory' period the male goes through after ejaculation is additionally, to this school of thought, evidence of an almost insatiable sexual desire, an 'aggressive eroticism' as Mary Jane Sherfey puts it. For these theorists, monogamy is unnatural and an instrument of political repression. Post-agricultural civilisation, in this view, is another term for the growth of patriarchy and consists of the ruthless subjugation of female sexuality.

The third view of the female orgasm, proposed by Donald Symons and heartily backed by Steven Jay Gould, is that a whole nexus of anatomical, social, cultural and emotional factors make female orgasm the subtle phenomenon it is. Theirs is something of a middle way, proposing as it does that female orgasm is the happy coincidence of an existing, but minor, bodily quirk resulting from the similarity of the sexes in the womb – an echo, in other words, of the male orgasm – and a cultural artifice no more adaptive than a learned ability such as reading and writing. The curious ability for human females to orgasm, to Symons and Gould, is magnified by our unusual preference for face-to-face copulation. The cultural, learned aspect of female orgasm is similarly amplified by our propensity for love at an emotional and intellectual level.

Anthropologist Helen Fisher points out that, despite its name, the missionary position is not a Western imposition but the preferred copulatory posture in most cultures. (It is seen in pre-Columbian American pottery, for example, says Desmond Morris in *Manwatching*.) So, argues Fisher, it seems that the peculiar arrangement of forward-tilting vagina and face-to-face sex may have evolved as it has for the very reason that it encourages social copulation, where partners can see each other and communicate with intimacy and understanding.

Desmond Morris's key point in describing the first theory, which is overwhelmingly the most widely accepted, is that because the female orgasm is unnecessary for procreation, it has in the past been regarded as a pleasure-seeking indulgence. However, in reality it is 'a unique human evolutionary development of the utmost importance', which helps to ensure and maintain pair-bond maintenance.

The evolutionary reason for this having come about, he argues, is the size of our brains and the inordinate time they take to develop from babyhood to maturity. Because our young take so many years to become independent, then, a child's chance of survival is best helped by a large social group sustaining him or her. We need consequently to maintain stable families, and the best way of doing that is by providing human couples with the pleasurable incentive of female orgasm – so long as this is a fickle and scarce commodity attained only after practice with a partner who knows a woman intimately at a bodily and an intellectual level.

One of the prerequisites of this 'higher' form of intercourse humans possess is our form of face-to-face, personalised sex. This practice also happens to facilitate (a bit) the stimulation of the clitoris during penetrative sex, and has accordingly changed our entire sexual morphology at some point in our transition from four-footed scurrying to knuckle-dragging to bipedalism. As opposed to other primates, humans have evolved displaying virtually all their erogenous zones and sexual signals on the front of their bodies. The vagina seems at some time to have taken up its human forward and downward-tilting attitude to facilitate further the progress of face-to-face intercourse The process of reconfiguring humans for their preferred sexual position also required some interesting decorative flourishes. By the time the frontal position was adopted, for example, early humans had shed their fur.

Why this happened half a million years ago is an interesting question, especially since nakedness plays such an important and unique role in human sexuality; we are the only creature that has a distinctive, bare-skin 'sexual' mode, or has at least, in most cultures, sexualised lack of clothes to suggest availability. It has long been thought that our hairlessness evolved as an aid to keeping cool during the day after we came down from the trees and began hunting on the hot African plains; but this would mean Nature allowed humans to trade the marginal benefit of hunting in comfort for the very considerable disadvantage of freezing at night, which would make little sense.

The latest thinking is that the disappearance of our hairy covering was a clever evolutionary strategy to deal with the time-consuming, but necessary, business of controlling ticks and fleas. Getting a little sweaty under one's fur while chasing dinner is not species-threatening; but being driven frantic with itching 24/7 as well as contracting dangerous, parasite-borne diseases is enough to destroy a population. Other primates, it is easily observed, have almost a full-time job grooming themselves and one another; but, generally, they only have to stretch up to a nearby branch for lunch. The intellectually ambitious early man's bigger brain required high-protein sustenance, and his schedule was simply too busy with chasing it to be spending time picking out ticks. The cold was still problematic some nights, but not the stuff of extinction since we had discovered fire and were using shelters.

This, of course, leaves the conundrum of why we still have head, beard and pubic hair. Head and beard hair almost certainly survived as sexual attractants. Pubic growth, it could be argued, has a cushioning function, and could equally have remained as a navigational aid in pitch-dark shelters. Another current theory is that pubic hair helps store and disperse sexually enticing odours.

According to the Morris *et al* theory, the female orgasm has an ingenious by-product, too. Female monkeys, once inseminated can wander off on all fours without losing any seminal fluid to gravity. But if a woman gets up after sex, gravity will ensure that some or most of the sperm she has received escapes. The satiating nature of orgasm, however, has the side effect of keeping the female horizontal and exhausted for a while, hence motivating her to retain semen in the vaginal canal for longer and giving a better chance for fertilisation. Finally, as Morris does not mention but other theorists in the same vein do, the satisfaction of orgasm stimulates a woman to seek more coitus, and thus further raise the chances of a successful conception.

The neat Morris theory offers a plausible explanation, additionally, for a number of social benefits in humans, all of which tend to reinforce the idea that we are designed to live in monogamous couples and stable families. The most important of these social advantages seems to provide a highly satisfactory explanation for the most perplexing of the disparities between the male and the female orgasm – the level of mechanical contrivance necessary for the great majority of male/female couples to achieve anything approaching mutual orgasm.

Is then orgasm a *natural* process which can only work by the non-instinctive, learned application of fingers, toes, noses, mouths, whatever, to the clitoris, and by males making a determined effort to think of the previous week's football results in an attempt to avoid ejaculating too early? It all sounds highly *un*natural, which doubtless has led to the taboos that surrounded such forms of manually and mentally assisted sex in so many cultures for so long. Yet the Morris view asks us, and cogently, too, to accept that it *is* part of a grander design; that nature *means* shared orgasm to be difficult.

His argument is that in prehistoric societies, where women were in a minority of two to three – a reality confirmed by skeletal remains – and also lived on average eight years fewer than men, they had the luxury of choosing which men to have sex with. It is stretching a point to suggest that women were consciously in the position of choosing a father for their children, because that link, as we have previously noted, was unlikely to have been made.

But it is not unreasonable to assume that sex was seen as a special, valued activity, and if a women was going to do it, she would surely be more likely to do so with a male who showed some sign of wanting to make the experience a pleasant one for her too? A male who had taken some time to learn, by whatever means – from an older, more experienced male, from the guys in the hunting gang – how to raise sex from a pleasant diversion to a proper hobby. A male who, to put it baldly, had learned a thing or two about foreplay and had the brainpower to work out, or learn, what a clitoris was for. It is likely that such attentive males have for thousands of years attracted the pick of the female population, and, in addition, had the more psychologically satisfying sexual experience *themselves* of feeling they are satisfying a woman.

The mild muscular 'clamping' feeling to a man's penis of a woman experiencing orgasm while he is inside her is pleasant yet pretty marginal, much as is the slight sensation some women get from a man ejaculating inside them. But the feeling of the penis being gripped is also clearly one not welcomed by a minority of men; notions of the vagina having teeth or being lined with thorns have existed in many cultures as the justification for a fear of castration. However, the sense of accomplishment that a man feels in the rare situation of supplying a woman with an orgasm by penetrative sex is tangible.

Failure, especially when repeated, to induce climax in a woman by any method is, conversely, very damaging to the typical man's prestige. He is painfully aware of having failed a woman in an event whose success he believes depends on him.

There is another perspective on men's fear of the vagina and of orgasmic, sexually active women in general. There are obviously many psychological and political explanations for this, but a simpler reason cannot be overlooked. It is this: that women who want sex, rather than being prepared merely to accept or endure it, are in the male's mind more likely to expect and demand a high standard of performance. An inexperienced man fears that this can mean only one thing – unless he is able to maintain an erection for some implausible, Olympian duration, he will be subject to scorn and derision. The more sophisticated man will know that what will almost always be more welcome to women is a level of mental engagement combined with a deft action of finger or tongue in the right place; his failure then can only be down to laziness, which is optional, not unavoidable.

The sexually adept male may still however be mistaken; Donald Symons suggests that the male's concern with the female orgasm can be 'based on the misconception that it plays the same role in her sexual experience as it does in his own'. But concern it is all the same. (Symons also makes the interesting point that prostitutes who feign enjoyment of sex are the more highly prized and expensive for so doing – evidence that the desire for some semblance of an equitable sexual exchange rather than a mere one-way traffic in pleasure runs fairly deep in men.)

Another facet of the evolutionary importance of pleasure-giving in sex by men for women is pointed out by Lionel Tiger. Even women on 100 per cent effective contraception, he points out, are still notably choosy about who they mate with; the atavistic, selective tendency asserts itself in most circumstances. And the durability of the ancient erotic and emotional imperatives also shines through in the orgasm-only mating patterns of *homosexuals*. Lesbians, Tiger observes, mirror their reproductive, evolutionary role even in patently non-reproductive relationships. They are, he says, far more discriminating than men, typically having four to twelve partners in a lifetime. 'Even when there is no ultimate cause of reproduction,' he concludes, 'male and female homosexuals still act as if there is.'

The mechanically tricky, emotionally enigmatic nature of the female orgasm can thus be seen in a remarkable new light: as a selective mechanism for women to choose mates not as an animal would, by body size, ferocity and aggressiveness, but by qualities such as intellect, sensitivity, kindness, reputation and popularity – plus a little dexterity with finger or tongue, for added spice. In animals, cruder parameters are the most important predictors of a male mate's providing ability and his genetic quality; but in humans, social qualities are an infinitely better augur of genetic success in the future.

No wonder, then, that Darwin said: 'The power to charm the female has sometimes been more important than the power to conquer other males in battle.' Or that the clitoris has been referred to by anthropologists as the 'mate meter'. Or that our friends the Mehinacu in the Amazon basin have a saying that the clitoris is, 'the nose that sniffs out sexual partners'.

We are perilously close, in this tidy model of evolution, to seeing the clitoris and its picky preferences as something closely akin to one of the primary motor forces in the promotion of human intellectual progress: men who are intelligent and sensitive get more chances to inseminate women; they pass on their own genetic advantage to their children; the traits of intelligence and sensitivity are selected for; humanity moves up the intellectual ladder stage by stage; the survival of the smartest thus ensures the continuance of the species.

And once you start applying this model to other aspects of human sexual behaviour as it has both evolved so far and developed culturally, everything makes sense. The woman's near-unique, permanent sexual availability, a true evolutionary adaptation if ever there was one, is in perfect harmony with a monogamous imperative. So would the much-discouraged, and apparently self-defeating, practice of faking orgasm, which most women and not a few men have very likely done since ancient times. Additionally, the persistent idea that it takes more than a single sex act to produce a baby, which, as we have seen, is strongly believed in many preliterate societies, is a conviction quite consistent with the maintenance of monogamy and the painstaking development of a loving, skilful sex life. (Tribes we studied in the previous chapter who allow a series of men to 'help' inseminate a woman could be regarded as monogamous too, so long as a broad definition was allowed, in which *all* the men of a village constitute the tribe's combined manhood, rather than a group of competing males.)

A seemingly peculiar research finding like the mood-elevating effect semen appears to have on women, which makes no obvious evolutionary sense, can also be seen in a certain light to possess a logic of its own. It is rewarding to the male if a woman feels happy after orgasm, as well as advantageous – especially, one might say, if he is not planning to stay for breakfast. Men frequently see their partner's orgasm as a reflection of their virility. It is equally rewarding for women to receive a mild dose of a natural antidepressant at the very *après*-afterglow moment their mind might be inclined to stray to the possibility of having just become pregnant.

Even committed sceptics on evolution will point to sexual characteristics in humans that they attribute to culture, but that might equally be the product of Darwinism at work. The American orthodox rabbi and prismatic thinker Shmuley Boteach asserts that God has ordained it that womanising men tend to be low-standard lovers because they never have to be imaginative or creative, but instead can rely on just running through the same old sex routine again and again. But whether it be it God or evolution's doing, Morris's or Boteach's theorising, it remains a constant that being good at sex is a surefire way to keep the pair-bond alive.

Morris's theory of the female orgasm is satisfying and inspired. If the likes of both sexes were congruent, we humans would experience a very dull kind of coupling. But in its curiously unbalanced form, each sex has a sufficient interest in and identification with the other's perspective while simultaneously being interestingly different. This makes the coupling fulfilling and, above all, long-lasting, which is the principal reason for it.

And the Morris theory has proved intellectually leakproof enough for some of the West's most brilliant minds. The philosopher Jacob Bronowski, for one, found it a compelling idea. 'The preoccupation with the choice of a mate by both male and female, I regard as a continuing echo of a major selective force by which we have evolved,' Bronowski wrote. 'All that tenderness, the postponement of marriage, the preparation and preliminaries that are found in all cultures, are an expression of the weight that we give to the hidden qualities in a mate. Universals that stretch across all cultures and divides are rare and tell-tale. Ours is a cultural species, and I believe that our unique attention to sexual choice has helped to mould it.

'It is irresistible to speculate on this idea that women's sexual pleasure has been an evolutionary boon. Perhaps its function is to give females a physical incentive to seek out mates who are attentive, and a disincentive to stay with a selfish partner. Distinguishing between partners in this way might prove an advantage in the effort to find a protective, nurturing male who might help to rear the young. Such a mate also benefits in the biological sweepstakes. In short, making women feel good may help men to win the Darwinian contest of supremacy.'

The Morris argument nonetheless came under sustained attack – as did the second, feminist theory of female orgasm – by the postmodern voice of Donald Symons. He thought the Morris case to be a little over-stated and designed, consciously or unconsciously, to provide an evolutionary justification for the tradition of marriage. Symons derided as wishful thinking the notion that the clitoris is automatically stimulated *to orgasm* (as opposed to simply feeling quite nice) during intercourse. Furthermore, if orgasm were an evolutionary adaptation, there would, surely, be very few non-orgasmic females alive – nor many men suffering from premature ejaculation – because orgasmic females and super-controlled males would long since have been favoured by natural selection.

The second theory, of the sexually insatiable female held back from the chance to orgasm constantly only by patriarchal society, was dismantled with rather more relish by Symons. Morris's 'pair bond' and Sherfey's 'insatiable female' theories are contiguous, differing only in their conceptions of the ideal – for Sherfey a paradise of endless sexual indulgence, for Morris a nirvana of sexually intense monogamy.

Both theories, for Symons, exist only in the human imagination. Not only is there, for him, no evidence at all of insatiable, multi-orgasmic women in pre-agricultural peoples, but they are still a rarity today. Even on sexually liberated Mangaia, men are still acknowledged to be more keen on sex than women. Furthermore, Symons argues, a tendency towards sexual insatiability would have interfered with early woman's genuinely adaptive activities of food-gathering and preparation, and childcare. As the celebrated anthropologist Margaret Mead also noted, Symons says, in cultures like that of Samoa, where foreplay is encouraged, all women orgasm; but in cultures where foreplay is forbidden, or you have to be clothed to have sex, or the lights have to be out, or all odours obscured by deodorants, the potentiality for orgasm may be universally untapped. 'It is important to realise,' Mead concluded, 'that such an unrealised potentiality is not necessarily felt as frustration.' Or, as Symons noted acidly: 'The sexually insatiable woman is to be found primarily, if not exclusively, in the ideology of feminism, the hopes of boys and the fears of men.'

Symons mentions, but does not support, two views that the

female orgasm is actually *dys*functional. The first holds that orgasm may lessen the chance of conception. Extreme excitement in the final phase of sex constricts the outer third of the vagina by vaso-congestion; this has the effect of bottling up and retaining semen. But the vaso-congestion relaxes on orgasm; Masters and Johnson accordingly recommended women hoping to conceive *not* to orgasm.

This is not considered to be correct today. Psychologist David Buss of the University of Texas states in *The Evolution of Desire*: 'Women on average eject roughly 35 per cent of the sperm within thirty minutes of the time of insemination. If the woman has an orgasm, however, she retains 70 percent of the sperm and ejects only 30 per cent. Lack of an orgasm leads to the ejection of more sperm. This evidence is consistent with the theory that women's orgasm functions to suck up sperm from the vagina into the cervical canal and uterus, increasing the probability of conception.' It might also be noted that if orgasm could reduce the possibility of a successful fertilisation, it would surely follow that it would be easier for women to become pregnant through unsatisfactory intercourse – or even rape – than by means of loving sex. (To be fair, though, there is no data on whether bad sex is more or less productive than good sex.)

The second argument for the proposition that orgasm might be reproductively dysfunctional is this: if the sensation is so pleasurable that it becomes a desire independent of thoughts of child conception or welfare, it could undermine women's reproductive efficiency and the creation of the best circumstances for raising her children. The same might equally – in fact, more accurately – be said of the male orgasm; that pure physical pleasure is not the most appropriate sensation to accompany something as serious and burdensome as the conception of a new life.

If, as such a construction suggests, this meant men had no control at all over their urge for orgasm, it would be a flaw in human design. We would then have to consider whether the best way to make sure men understand the gravity of what they are doing when they have sex would be to make sexual intercourse painful and unpleasant, and only permissible after fulfilling much irksome bureaucracy and form-filling. But, of course, one of the key features that marks us out from animals is that both males and females know what sex leads to, and that we have, additionally, the sophisticated psychological mechanisms of romance, love and affection come into play to ensure that, to a large extent, babies are wanted by both parents.

The view of female orgasm that Symons favours is that the phenomenon is not an adaptation at all but a relic, a kind of echo of the male orgasm, just as the male nipple is an echo of the female. While the same body cannot be both male and female, the brain can be wired for both potentialities. Kinsey noted in 1953 that there was a marked similarity between the male and female orgasm, but the most telling evidence was Masters and Johnson's discovery that orgasmic contractions in both sexes come at the precise same 0.8 seconds apart.

Symons speculates further that the female orgasm can only have been retained through evolution (having existed in the first place as a functionless throwback like the male nipple) because it was of such importance to men. Female multi-orgasm, he claims, may be an incidental effect of women's inability to ejaculate. It may even be a direct imitation of the pre-adolescent male's orgasm. Boys who masturbate pre-puberty are able to achieve repeated ejaculation-free orgasm (and this is recorded as early as five months) in a short period of time. Their capacity to repeat this 'dry' orgasm without loss of erection diminishes in adolescence and decreases again in mature manhood. Stephen Jay Gould draws a conclusion from this that many women will find awkward: 'The reason for a clitoral site of orgasm is simple – and exactly comparable with the non-puzzle of male nipples. The clitoris is the homologue of the penis – it is the same organ, endowed with the same anatomical organisation and capacity of response.'

The Symons/Gould camp, although currently in the ascendant, is far from immune itself to post-post-modernist criticism. Helen Fisher differs from them on the question of female multiple orgasm, and why men do not have the same sublime capacity. Comparing the female orgasm to the male nipple is, for her, invalid. The male nipple, Fisher points out, is inert, whereas the clitoris is a highly sophisticated little instrument that produces a massive physical sensation and emotional experience – an altered state of consciousness, no less. It also has a signalling capacity. Men like women to climax because it reassures them that they have satisfied their partner, and that she is less likely to seek sex elsewhere, and the genetic material he has implanted therefore has more chance of surviving than his neighbour's. 'Female orgasm boosts the male ego,' Fisher reasons. 'Why else would women fake orgasms?'

Even woman's notorious failure of orgasm, Fisher says, may have been selected for millennia ago. Women tend to climax when they are relaxed, with attentive and committed partners. That is why, to the moralists' delight, surveys such *Redbook*'s tend to confirm that women have better sex with husbands than with secret lovers. Street prostitutes, she adds, are said to climax less than call girls with clients who are prepared at least to mimic 'real' lovers – confirmation for Fisher, as for Morris, that woman's fickle orgasmic response is a mechanism evolved for her to sort worthy partners from unworthy.

A male anatomical reality, meanwhile, again suggests persuasively that the female orgasm *is* a pointless but pleasant relic rather than the product of ruthless evolutionary selection. It is the bothersome question of how on earth men with small penises have managed to survive the rigours of evolution? Desmond Morris's assertion that human beings have evolved to be the sexiest primate is supported by the human penis being so large. But it is not always that large, and if women's universal assertion that size *does* matter – and very much so – is to be believed, small penises should have been bred out of the population hundreds of generations ago by women voting with their vaginas.

Penis size, however, comes in two dimensions, length and width, and it is width that is the critical factor, Helen Fisher explains. Even if a wide penis does not often produce orgasm on its own, it certainly seems to provide a more satisfying sensation to more women, and this should, Fisher says, have ensured that males with thick penises have more lovers and more children and that penises accordingly became thicker over the generations. Without fossilised soft-tissue evidence one way or another, it is impossible to know if penis size has increased down the generations. The likely reality is that the most minuscule or string-like penises have died out, but that, through the millennia, men have generally been able to make up for any deficit in their penis size with such attractive attributes as charm, kindness, bravery, power and brains.

One very important problem that tends to be by-passed by all these theories of the orgasm, however, is whether women really are as inclined towards monogamy as tends to be assumed. The received wisdom among evolutionary psychologists and their like is that men, with their implausibly large testicles pumping out sperm by the billion, are inveterate philanderers, while women, with a finite supply of eggs that require a significant investment, are naturally reserved and choosy about whom they sleep with, the question of sexual pleasure being very much secondary in their lives. Men, as a result, are biologically inclined to be promiscuous, while women concentrate on sifting through potential mates in search of ideal father material.

David Buss demonstrated this divergence between the sexes experimentally in a 1993 study of college undergraduates. The women in his research typically stated that they would ideally like around five sexual partners in a lifetime. Men's idealised figure was in a range from 18 up to 1,000. Women invariably had high standards even for one-night stands. For short-term partners, however, Buss concluded, men 'are willing to go down to the tenth percentile, as long as she can mumble'.

Even what Symons cuttingly calls the 'Sherfian paradise' – of endless wild sex for women, the older they get – does not necessarily imply female promiscuity. It is telling in the extreme that only one of the four female lead characters in *Sex and the City*, the aspirational manifesto for so many twenty-first-century Western urban women, is truly promiscuous. Sherfey would take heart that the wanton Samantha is the oldest of the four women, but it can be assumed that, informally or otherwise, the show's creators did their market research, and that they must have been persuaded that true promiscuity, as opposed to serial monogamy, does not resonate very well with even that show's target female viewers.

Sexual pleasure and proper, non-masturbatory male orgasms have always been a scarce resource controlled by women who, in prehistoric societies, were both in a minority and, like now, sexually available for only two-thirds or less of their life. But it seems probable, too, that women have always had affairs; for every philandering male, after all, there has to be a philandering female.

A clue as to the real state of women's sexual faithfulness through the ages may be gleaned from modern data on the sex drive of females. In the 1950s, psychologists Clellan Ford and Frank Beach showed that women around the world begin sexual advances, and subsequent studies have shown that a perceived equality of sex drive is more prevalent than not in the large majority of societies. Helen Fisher regards it is as curious that Westerners still cling at all to the image of man as seducer, woman as submissive. She argues that this is a relic of our agricultural past, when women were pawns in property exchanges at marriage and their value depended on their 'purity'. This meant girls' sex drive was denied, effectively, for financial reasons. Today, economically more independent, women are often sexual pursuers – and not merely in *Sex and the City*-style fiction.

Prehistory (plus the usual healthy dose of informed guesswork) must hold the most important clues as to whether women are naturally as prepared as men to encompass promiscuity in pursuit of sexual pleasure. So what may we reliably garner from what we know of our ancestors? Their sexual morphology seems to have been similar to ours, so they almost certainly practised face-to-face copulation; from that, we can be fairly confident that couples were recognised. But were prehistoric couples faithful?

The consensus among anthropologists is that prehistoric females did mate with more than one male in one cycle. But males at some level understood that their sperm was competing with that of other men, because notions of jealousy grew up despite (or because of) the affection-producing hormone oxytocin. This would suggest that women took advantage of their seller's market in sex, but that they also liked to reserve a particular lover as 'theirs' – quite possibly because of his special skill in giving them orgasms.

From the seemingly instinctive territory-marking habits that survive among some women today – scratching a man's back with their fingernails as they reach orgasmic ecstasy is one such – we may intuit that women, too, were capable of sexual jealousy. There is also cultural evidence of formal pair bonding, which leads to the contention that relationships developed in prehistory as a norm in spite of a promiscuous desire in both sexes. Unfaithfulness came to be seen as non-ideal, but frequently necessary to ensure the survival of the tribe or the species.

According to Helen Fisher, there is plenty of genetic advantage to women as well as men in having offspring with a variety of mates. Women could assure themselves of extra resources for them and their children, a measure of insurance, and, although they could not have the foggiest idea they were so doing, better genes and more varied DNA for their children's biological futures. 'Hence those who sneaked into the bushes with secret lovers lived on,' Fisher writes in her *Anatomy of*

Love, 'unconsciously passing on through the centuries whatever it is in the female spirit that motivates modern woman to philander.' She does not mention the more obvious construction on this – that it simply felt nice to women to have orgasms with more skilful or varied lovers. But she does consider it perfectly possible that female prostitutes accept money and gifts for sex not necessarily for economic motives, but because they enjoy sexual variety.

Fisher tells how the high sex drive of the human female has led University of California Professor of Anthropology Sarah Blaffer Hrdy to a novel hypothesis about prehistoric female adultery. Hrdy maintains that female apes such as the bonobo have (without quite realising, one imagines) a lot of non-reproductive sex with successive partners, and that this has to do with the females pursuing a brilliantly Darwinian end – to confuse paternity so that every male in the community will act generously towards her and paternally towards her children, thus helping ensure their wellbeing and success.

In humans, Hrdy argues, the same must have happened, the females pursuing sex with a string of males to keep friends. When the first stirrings of civilisation came, with the move four million years ago from tree-dwelling to a bipedal life on the African grasslands, pair bonding evolved, and young females turned from open promiscuity to secret copulation. But there was bound to have been a substantial pleasure reward involved here as well as a genetic one. And, it may be extrapolated from Hrdy, it is only because female unfaithfulness implies not all men give good orgasms that it is such a strong taboo in primitive cultures.

Helen Fisher adds that it is probable that the veil evolved in Muslim societies and the chaperone in places like Andalusia in direct response to this seductive, sexually acquisitive, orgasm-seeking trait in women. The Talmudic requirement for a man to satisfy his wife sexually, which we will discuss in the next chapter, was also very likely occasioned by the ancient experience of women having a strong sex drive. The expressly matriarchal society adopted by the Ancient Jews may have been a product of or a precursor for this acceptance of powerful female sexuality. The female clitoridectomy too may have been designed by jealous men in the African societies where it exists to curb the high female libido. It similarly stands as an interesting commentary on what must have been a sophisticated early understanding of how female sexual feeling works; to believe the clitoris needs removing to stabilise women's rampant sexual desire requires a pre-existing folk knowledge of female orgasm.

Current research, counter-intuitive though it may seem to many, is that the accepted, age-old economic contract between the sexes of female fidelity and guaranteed paternity in exchange for meat for the family does not quite hold up – that 'slutty' female behaviour is good for the species because it improves the gene pool by giving women a variety of men with whom to mate.

Even evolutionary psychologists, who have generally been rigorous in upholding the validity of the 'fidelity for food' bargain, have moved towards accepting that a bit of female promiscuity can give a woman a measure of back-up insurance if the father of her children is killed. The idea that women in such a situation sleep around for the sheer orgasmic pleasure of it has yet to gain universal currency, however.

The institutionally promiscuous concept of woman, prompted by evolutionary logic discreetly to seek out sexual adventure, seems to be supported by research by the Pennsylvania anthropologist Stephen Beckerman, in communities such as the sexy Canela people in the Amazon, and also by Kristen Hawkes, Professor of Anthropology at the University of Utah, who has spent years studying the licentious Aché, a Paraguayan people, and the North Tanzanian Hadza tribe, who also enjoy a richly varied love life.

The till-death-do-us-part, missionary-position couple of Desmond Morris's model is just a tiny part of human history, in Beckerman's and Hawkes' view. 'The patterns of human sexuality are so much more variable,' Hawkes told Sally Lehrman, of the online magazine AlterNet.org, in 2002. The average Hadza hunter, he found, can only bring in a big game carcass once a month, and he is obliged to share his kill with everyone in the community, his wife and children receiving no special bonus. Strong emotional bonds with extra mates help a Hadza woman remain safer in dangerous times. But, again, the idea of these promiscuous women seeking extramarital sex for pleasure alone is not even on the agenda.

'Pair bonding,' Helen Fisher declares in *Anatomy of Love* after some scholarly humming and hahing, 'is the trademark of the human animal.' Even in polygamous 'families' and free-sex communes, she says, men and women have favourite spouses and tend to form *de facto* couples. And in arranged marriages, couples will frequently form romantic bonds retroactively

Fisher cites data that only 16 per cent of the 853 cultures on record require monogamy, whereas 84 per cent of all societies permit polygyny, the practice of men having more than one wife. Polygyny (as opposed to polygamy, which can cut both ways) would not seem to be particularly conducive to female promiscuity. Yet, Fisher notes, in societies where polygyny is allowed, only 5–10 per cent of men actually have more than one wife at the same time. One wife is, in reality, the global norm, as is, for that matter, one husband.

Fisher's conclusion is that it is unclear which sex is more interested in the pleasure of sexual variety, but that overall we can be said to follow a mixed reproductive strategy – monogamy *and* adultery. Kristen Hawkes believes, however, that the promiscuous traditions of our female prehistoric ancestors have come through the selective process to remain quite rampant in modern society. High infidelity, remarriage and divorce rates for Hawkes may have less to do with modernity than with our collective sexual past. 'It makes the variation we're seeing in modern society so much more understandable,' she says.

'If the anthropologists are right, monogamy may well be counter-evolutionary or an adaptation to modern life. Or perhaps the nuclear family has always been more of an ideal than a reality,' Lehrman concluded in her AlterNet article.

7

Orgasm BC

> 'Let his left hand be under my head and his right hand embrace me,'
>
> from the *'Song of Solomon'*

Very early civilisations may have been the Petri dish in which the culture that became large-scale organised religion began to grow. But while the leading brand faiths, with the responsible goal of long-term species survival on their mind rather than the frippery of momentary pleasure, placed a premium on reproductive sex, their predecessors seem to have been less fussy over whether their orgasms were attained by heterosexual or homosexual intercourse, just so long as they were attained.

Institutionalised male-on-male anal sex is thought to have been quite common in preliterate civilisations across the world; typically, in societies as diverse as the Chuckchee of Siberia, the Aleuts and Konyages of Alaska, the Creek and Omaha of the US, and the Bangala of the Congo, the practice was legitimised by a form of religious marriage between a man and a transvestite. Anal sex was as revered as vaginal, and was associated with the worship of androgynous, hybrid male and female gods. Even when formal temples began to appear in Middle Eastern cultures there are said to have been priests who used anal intercourse as a way of being a go-between between cult adherents and their gods.

It was in the more advanced parts of the world some 5,500 years ago, namely the Middle Eastern lands, that pagan cults grew into complex, codified religions, especially among the desert wanderers who were the intellectual *avant-garde* of their day. As they developed, these embryonic new religions had a habit of incorporating a heavy and egregious sexual content – no surprise again when orgasms were still the most rapturous physical and psychological experience most people enjoyed, other than those privileged to see burning bushes, receive tablets of God's word on mountain tops and so on. The desert religions that turned into market leaders and survived to the present day, namely Judaism and Christianity plus their modernist rival Islam, have all retained a disproportionate preoccupation with sex and with the tortured relationship for thinking people of all eras between the essential act of creating life and the hedonistic luxury of orgasm.

While the great religions were still under construction, the pre-eminent culture existed in Ancient Egypt. There, acts of human-like sex were credited with having originated everything, even the universe. The Egyptian creation story was told in different versions in different cities, which was hardly unreasonable since the whole thing was patently a myth and flexible in interpretation. But the common thread throughout these stories is (surprisingly in view of later civilisations' taboos on the matter) masturbation.

One of the Pharaohs' most onerous ceremonial duties in Egypt was to bring fertility to the Nile by masturbating annually into its waters. The tradition supposedly went back to a primary event in the various versions of the creation myth of the time, although if you think about it, the creation myth must have been invented as a post-rationalisation of the Nile masturbating ceremony. The supreme being, Atum, myth had it, arose out of the primeval darkness. He masturbated to form Shu, god of air, and Tefênet, goddess of moisture, while he himself became the sun god, Ra, the Supreme Lord of Egypt. Tefênet's vagina created the morning dew and their incestuous love created the Earth. Or as the Pyramid Texts, Utterance 527, graphically put it: 'Atum is he who once came into being, who masturbated in Ôn. He took his phallus in his grasp that he might create orgasm by means of it, and so were born the twins Shu and Tefênet.'

The Pharaohs' river ceremony was not so much the sanctification of an act of selfish pleasure as a new recognition of the primacy of fertility. In a landscape where the River Nile so visibly and tangibly made barren land fertile and life possible, it was inevitable that fertility would be as important as it was, overriding entirely any thought of pleasure in sex, especially for women. Magic was also widely employed to aid fertility, the lack of which was a person's greatest worry

What is particularly interesting in the Egyptian creation myth is that Atum's construction of the world was clearly a function of his own physical gratification, rather than a stated desire for parenthood; his offspring were, if Utterance 527 is to be believed, an accidental side-effect of his pleasuring himself.

But while following the example of their sexy gods' potent sexuality was a religious duty for Pharaohs, civil servants and ordinary citizens too are now believed to have aspired to lives as sensual as their king's. Material from excavated middle- and working-class houses (much of it hidden away during the Victorian Egyptology boom) shows that regular Ancient Egyptians covered their walls in explicit, exotic paintings and spent lavishly on their appearance, clothes, make-up, jewellery and perfume – and even dildos, a large collection of which is neatly filed away in wooden drawers deep in the British Museum.

A combination of the weather, the self-confidence of their culture and the green fertility of the Nile delta helped make Egypt an exceptionally sensual society. Under the unrelenting sun, women wore little more than a transparent linen shift, female slaves not even that – just a few beads. The men wore a pleated miniskirt, with an easily discarded woollen cloak for evenings. How could sex not be on everyone's mind in such an erotic setting?

It would be wrong, nevertheless, to represent Ancient Egypt as some free-love sexual Utopia; it was in no sense a liberal society. Girls were regularly deflowered in arranged marriages at the age of six. Men opted for anal intercourse or vaginal penetration from behind to avoid having to lay eyes on their wife. But a belief pertained that sex was a part of the human condition and, as such, inherently guilt-free. Cleopatra is said to have fellated a thousand men, including a hundred Roman noblemen in one night; the Greeks referred to her as *Merichane* – 'gaper', 'the ten-thousand-mouthed woman' and *Cheilon* – the 'thick-lipped'.

Love poetry was rife in everyday Ancient Egypt. Divorce, affairs, sexual indiscretion, adultery and womanising were not particularly discouraged or sanctioned against. The master of a household was permitted to have children by the servants. Virginity was not venerated even as an ideal. Contraception was practised, notwithstanding the cultural importance of fertility; the Kuhun papyrus, discovered in 1860, cites a variety of contraceptive methods, including the use of a tampon of crocodile dung smeared with honey and salt. Homosexuality was acceptable; the gods Set and Horus are described in sodomist congress, and the British Museum also holds a painting of two male court hairdressers having sex. Even bestiality was not taboo; the local sun god, Mendes, was often represented as a goat, and it was said by Herodotus that the city's worshippers of both sexes practised carnal intercourse with goats.

Paradoxically, anal sex is thought to have been relatively routine among the Ancient Hebrews, who began the taboo against it as part of their drive to establish a more ethical spirituality than the Egyptians who had enslaved them. The Biblical myth of Sodom plainly illustrates the desire by the earliest Jews to distance themselves from the practice of orgasm-without-responsibility. It is interesting, too, that as part of the de-sexing of humanity in the interests of a higher, more intellectual calling, the Eden creation story was made superficially so un-sexual; the first full-frontal, nude drama appears at first reading to deal with a host of issues, but sex barely seems to figure among them. (It does, in fact, but only to the most sophisticated of readers, and then again, only in the original, not in translation.)

The original writers of the holy texts were actually a good deal more in favour of sexual pleasure than their later translators, even if it is a little fanciful to suggest, as has been done, that the rocking rhythm of the Hasidic student studying the Torah is itself a form of spiritual copulation. Yet the Bible in large part treats sexuality as a gift of God and a central part of being human, and often describes sex without mentioning if the participants are married. There is plenty of what could be described as erotic writing, such as the account in the *Book of Ruth* of Ruth seducing Boaz on the threshing-room floor. The 'Song of Solomon', to those with half an ear for it, is one long erotic poem. The ancient text is laden with words like 'pomegranate', 'vineyard' and 'garden' that are said by scholars such as David M. Carr, Professor of Old Testament at Union Theological Seminary, a non-denominational graduate school of theology in New York, to be deliberately and overtly sexual images. Raisin cakes are claimed to represent aphrodisiacs. One line, 'I will go my way to the mountain of myrrh and to the hill of frankincense', is alleged to describe a man wanting to bury his face in his wife's bosom. Other phrases are less ambiguous: 'Your two breasts are like twin fawns', 'Your lips, my bride, drip honey'. And the Song certainly advises a man as clearly as possible to stimulate manually his lover's clitoris: 'Let his left hand be under my head and his right hand embrace me'.

Naturally, the tendentious anti-pleasure post-rationalisers have been working on de-sexing such material for nigh on 3,000 years. The 'Song of Solomon' is often dismissed as an aberration out of keeping with the rest of the Old Testament.

Many Jews explain it away as an allegory for God's love for Israel, while some Christians excuse it as a representation of Christ's relationship to his followers.

Generally speaking, when Bible translators have happened upon sexual references, they have been assiduous in seeking out neutralising euphemisms like men with a mission to protect unborn generations of virginal Sunday School teachers. Thus is 'penis' changed in every instance to 'thigh'. 'Put, I pray thee, thy hand under my thigh,' Abraham asks his servant in *Genesis*, 'and I will make thee swear by the Lord, the God of heaven, and the God of earth.' (This is a reference to the custom of 'testifying', by which anyone taking a vow places their hand on their testicles.

What the early Jews officially thought of 'testifying' in, you might say, a more proactive sense, i.e. masturbation, has been obscured by the curious business regarding Onan, the son of Judah, who, a simple reading of *Genesis* suggests, was put to death for the practice. This interpretation led to 'Onanism' being proscribed for thousands of years, although in reality the proscription was not taken very seriously.

There was a sense that in desert communities in which the men needed to produce as many sons as possible, and had as many wives as they liked, both masturbation and withdrawal before ejaculation were a waste of the most precious human resource. But masturbation bothered very few moralists before the eighteenth-century, and at least one writer, the anonymous author of a tract entitled *Hippolytus Redivivus* in 1644, claimed it was a sound remedy against the dangerous allurements of women. Why masturbation later became so very taboo, with the fear of it growing quite hysterical in such prudish times as the Victorian era in England, is something of a mystery. But what we can now say with a degree of certainty is that the Onan described in *Genesis*, who prompted Dorothy Parker to name her canary after him because he kept spilling his seed, was no masturbator. Rather, he was a responsible proponent of *coitus interruptus.*

What transpires from reading the sparse Biblical passage concerned is that Onan was required by the Levirate law of Judaism to sleep with his dead brother Er's wife Tamar, to attempt to preserve the family line by providing her with a male heir. Onan, nevertheless, did not want to get Tamar pregnant. As the passage in *Genesis* states: 'Onan, however, knew that the descendants would not be counted as his, so whenever he had relations with his brother's wife, he wasted his seed on the ground to avoid contributing offspring for his brother.' Whether withdrawing his penis from her vagina ('flower') at the critical moment made him history's first recorded gentleman or a prototypical cad is a matter of interpretation. But for practising a basic contraceptive method and thereby failing to provide his dead brother with a son in accordance with Jewish law, he certainly managed to anger God who, we are told, killed him.

Assuming, as one must to maintain a reasonably sceptical outlook, that the part about God killing Onan was made up by scribes, or that he just happened to die anyway, remaining Biblical parables that are overtly anti-masturbation are surprisingly few. This may well be because any sensible person realises that 'wasting seed' does not always involve either masturbation or contraception; any religion, after all, that advocated not having sex with your wife if she was merely having trouble conceiving would not find many adherents; neither, for that matter, would a faith that believed a man with a low sperm count should not even attempt to have sex because the few sperm he had would most likely go to waste. In *Romans*, believers are told not to use their 'members' to do sinful acts because the appendages in question belong to God. *Leviticus* makes passing mention of spilled semen staining clothing and bedsheets. Moses warns obliquely at one stage against 'sowing seeds on rocks and stones on which they will never take root'. More arcane Hebrew writings liken spilt semen to the dead, from which it follows that touching it is as touching the dead. And that is about it.

There is, additionally, not even a hint in the scriptures that women should not fondle themselves, since they obviously have no 'member' to violate, no semen to spill, nor widowed in-law to decline to impregnate. Why should men have been targeted, if not by the Bible itself, then by its interpreters? Based on contemporary beliefs about the Nature of sexual reproduction, there is at least one humane justification for males refraining from masturbation. To the Hebrews, and to the Greeks after them, sperm were regarded as entities that shot fully formed, but sub-miniature, people into a woman, whose role in reproduction was merely that of incubator. Wasting sperm, by this interpretation, was directly analogous to killing millions of people.

In scriptural matters, knowledgeable interpretation is all. And viewed though knowing eyes, the Jewish Torah and Talmud emerge as little short of practical marriage manuals. The early Jews believed one should enjoy the pleasures of life, sex included, with some rabbis holding that at the last day people would have to account to God for every pleasure they had failed to enjoy.

'In Ancient Jewish thought sexual congress is a metaphor for God's creation of, and interaction with, His world,' Rabbi Shmuley Boteach writes in his book *Kosher Sex*. 'Sex is said to bring about the celestial unity of masculine and feminine energies . . . Since our world was created as an arena to demonstrate the unity of God, no other act demonstrates this better than the physical union of male and female, strangers who become lovers, and lovers who are also friends.

'Long ago, well before Christianity enacted legislation forbidding its clerics from marrying or having sex, the ancient Rabbis were giving explicit sexual advice to married men and women as to how they could enjoy pleasurable, yet holy, intimate relations. The Rabbis made female orgasm an obligation incumbent on every Jewish husband. No man was merely allowed to use a woman merely for his own gratification.' The Bible, he points out, conceives of sex within marriage as the woman's right and the man's duty, while the Talmud later – a mere 2,000 years ago – declared that a woman's sexual passion is far greater than that of man. Later still, Nachmanides, a thirteenth-century Jewish scholar, explained in his commentary on the Bible that when God said Eve would long for Adam after eating from the tree of knowledge, her craving took the form of an exceedingly great sexual desire for him.

Jewish law and custom also concur with the likes of Desmond Morris over the special status of face-to-face intercourse as a passion enhancer: 'In no other sexual position do we see a meeting of mouths accompanied by a full integration of all the limbs,' says Boteach. 'Not only are husband and wife locked together in the genital region, but they coalesce in their totality so that even in appearance they become as one. The missionary position allows us to experience something which is quintessentially human.' According to the rabbi's research, Jewish custom that denounces cunnilingus as 'lewd' and fellatio a sinful waste of seed is 'a travesty of the truth' and runs counter to true rabbinic learning.

Although he professes himself anti-feminist, Rabbi Boteach finds for his theory that Judaism is an inherently sexy religion an unexpected ally in the writer Naomi Wolf. Wolf agrees that cultivating female satisfaction within marriage in the observant Jewish community was always considered – and still is – a primary family value. And, like the Taoists later in China, the Jewish mystics the Kabbalists believed that sexual satisfaction given by men to women pleases the Maker, and creates balance, order and harmony on a cosmic scale

'Even within the Judaeo-Christian tradition there are systems of belief in which female desire is valued more highly than we can imagine,' Wolf notes in her book *Promiscuities: A Secret History of Female Desires*. 'The *Zohar*, the Jewish European mystical tradition, charges that, "When the wife is purified [that is, after her ritual bath when menstruation is finished], the man is in duty bound to rejoice her, in the joyful fulfilment of a religious obligation . . . It is his duty, once back home [from a journey], to give his wife pleasure . . . A man should please his wife because "this pleasure is a religious one, giving joy also to the Divine Presence, and it is an instrument for peace in the world".' Islam, similarly, Wolf emphasises, has a tradition of treasuring female sexual desire.

Geraldine Brooks, an American expert on Islam, points out in a study of women in the Islamic world, *Nine Parts of Desire*, that: 'the lessening of women's sexual pleasure directly contradicts the teachings of Mohamed'. In isolated parts of the world, Islam, it seems, has been hijacked latterly by prudes and misogynists, just as its equally sexy rival religion Judaism has. But according to Brooks, in a truer reading of Islamic scripture and tradition: 'Almighty God created sexual desires in ten parts; then he gave nine parts to women and one to men.' And just as the Talmud, read intelligently, comes out strongly in favour of equality of sexual desire, a large body of commentary on the Koran holds identically that Mohammed and his disciples were in favour of female sexuality and women's sexual pleasures within marriage. Mohammed actually encouraged husbands to be attentive in bed: 'When any one of you has sex with his wife, then he should not go to her like birds; instead you should be slow and delaying,' he said. Elsewhere, he refers to sex without lengthy foreplay as cruelty, and chastises a husband for being 'too busy' to make love to his wife.

Just as many Ancient Judaic sexual taboos (and tips, too) were echoed in the later sect of Christianity, the sexual pleasure principles established in Islam by the Prophet and his followers have continued for thousands of years to inform a raft of sexual traditions across a slew of Middle Eastern cultures broadly in the Islamic crescent of influence, but including the Hindu and other worlds, too. The origin of belly dancing was to help bring impotent males to orgasm. Avicenna, the Arabic philosopher and physician of the eleventh century, more properly known as Abu Ali al-Husain ibn Abdallah ibn Sina, of Uzbekistan, Central Asia, maintained that women had sperm, and that this provoked 'a specific sexual itch . . . in the male's spermatic vessels and in the mouth of the womb . . . which is relieved only by the chafing of intercourse or its equivalent'. It was impossible, Avicenna maintained, to suppress female desire and if it was not sated in some way, then women would resort to 'rubbing, with other women, in order to achieve amongst themselves the fullness of their pleasures'.

Female circumcision, or Female Genital Mutilation as it is known today, is the (supposedly) Islamic sexual tradition Westerners struggle with the most. Originating in the Nile Valley it has for some 5,000 years been regarded by many in that area as distinguishing 'decent' women from prostitutes and slaves, the main measure of a family's honour being the sexual purity of its women. Strabo, a first-century BC Greek geographer, first recorded the custom while travelling up the Nile. Instances have been recorded of it in Malaysia and South America, but it has remained almost exclusively the custom of North and Central Africa. In Sudan to this day a girl cannot marry if she is not circumcised – even though circumcision has been officially illegal since 1956.

FGM involves two procedures: clitoridectomy, the excision of all or part of the clitoris, and infibulation (aka 'Pharaonic circumcision'), the sewing together of the labia majora with thread, catgut or thorns. Moreover, women will have the trauma of their original circumcision repeated when they begin to have sex and during childbirth. Husbands are usually required to 'cut open' women before intercourse can take place. Women are usually 're-sewn' after they give birth. Unsurprisingly, FGM is extremely dangerous, with a variety of long- and short-term side effects, some of which are often fatal.

Sorting out what this brutal practice actually is from what it is not is obviously a prerequisite in a study of the orgasm, principally because the practice is such corroborative evidence of the female orgasm's existence and importance in antiquity.

The first thing that needs to be said about FGM is that there is nothing Islamic about it. It is practised by African followers of all religions including Catholics, Protestants and Copts

(Egyptian Christians), and is tolerated by many male religious leaders. Not that local varieties of Islam are blameless. Many Muslim sheiks and leaders preach that female circumcision is a religious requirement. But it is fair to say that Islam has borne the brunt of Western criticism, partly because most of the lobbying against FGM has come from US feminists who are arguably less fastidious about taking on Islam than are African or European women. (They have duly been accused, especially by Africans, of cultural imperialism for campaigning against FGM.) Naomi Wolf, for instance, has been prepared to name and shame: 'Today, of course,' she comments, 'the Koran and its commentaries, the *Shari'a*, are used to condone the most egregious abuses of women, from murder of teenage girls who lose their virginity before marriage and married women accused of adultery, to clitoridectomy and infibulation. The widespread practice of clitoridectomy in some Muslim countries is performed on the basis of the belief that "without it a woman wouldn't be able to control herself, she would end up a prostitute".'

The public relations problem for Islam with FGM is that although the faith in no way condones the practice, a large majority of its adherents *profess* to be devout followers of Islam. It is also unfortunate for mainstream Muslims that FGM's inherent misogyny is not entirely alien to the more fundamental forms of Islam.

In reality, local economics and beliefs about women's health, combined with lack of education for women, are as much to blame for FGM as hardline religious rules, hatred of women or fear of their ravenous appetite for orgasm. It is practically impossible for fathers in the isolated areas where circumcision is rife to marry off their daughters uncircumcised, and a large amount of a family's income comes from its daughters' dowries. Girls as a consequence are sometimes circumcised as infants to let there be no doubt among the local community that a family's females are of the required standard. A small number of Islamic villages across wide stretches of Africa also harbour a witch doctor as the local circumciser, usually female, who is a respected and revered figure, charges a high price for the operation – and whom it is socially unwise to offend. Additionally, any woman of a mind to make her own inquiries in the holy scriptures as to the practice's validity may be doubly frustrated; women in remote rural areas where circumcision persists (usually against the will of national governments) can rarely read, and even if they can, are never allowed access to the Koran.

The troublesome cult of FGM was always, its practitioners say, a way of protecting women's health, which, due to their sexually voracious nature, would suffer from too many orgasms. The practice has attracted unexpected admirers in the West, most notably the Victorian Arabist and erotic pioneer Richard F. Burton, who translated the *Kamasutra*. Burton also made known to the modern world the continuing practice of clitoridectomy, but saw merit in it. In his 1865 *Memoirs of the Anthropological Society of London*, he wrote: 'The reason for such mutilation is evident. Removal of the prepuce blunts the sensitiveness of the glans penis, and protracts the art of Venus, which Africans and Asiatics ever strive, even by charms and medicines, to lengthen.' The idea of the sexually insatiable woman who, by having her orgasmic desire 'damped down' somehow raises the erotic temperature for both a husband and his wife, manifestly remains a popular contention in parts of the African world today; 90 per cent of young women interviewed recently by the Cairo Family Planning Association had had some part of the clitoris and labia removed, which would suggest that their fathers, at the very least, thought it a good idea.

There are, it has to be said, some curious side issues to FGM when it comes to sexual pleasure. An American social psychologist and leading campaigner against FGM, Hanny Lightfoot-Klein, has looked specifically at the question of sexual pleasure as experienced by circumcised women in Sudan, and reported her rather extraordinary findings in the *Journal of Sex Research*.

Circumcision, Lightfoot-Klein found, is universal even among educated people. Far from being a furtive activity, it is celebrated with family festivities, a woman's circumcision day being more important than her wedding. It is seen not only as healthily dampening a young girl's sex drive (overt sexual enjoyment in women is regarded as disgusting and animal-like), but as a precaution additionally against the belief, held even among middle-class people, that if the clitoris is not cropped, it will grow to dangle between a woman's legs like a penis.

Against such a background, Lightfoot-Klein's discoveries in Sudan are all the more surprising, and throw a fascinating new light – if they are taken at face value – on our Western belief in the primacy of the clitoris in orgasm. For a larger proportion of circumcised Sudanese women – allegedly as great as 90 per cent – claim enthusiastically in interview with female researchers they enjoy sex *and* frequently or always have orgasms – even though for them to react outwardly to sex in any way other than by remaining totally inert is a strict taboo, for which their husbands could divorce them summarily. The minority who attest to not enjoying sex, according to Lightfoot-Klein, are those who still suffer intractable pain from their circumcision – or are in unhappy marriages.

'Some women,' Lightfoot-Klein found, 'said that they had intense, prolonged orgasms, and this was verified by their happy and highly animated demeanour as they described it.' A smaller study in Egypt in 1982 by Marie Assaad, a Cairo-based social reformer and campaigner against circumcision, also found a surprising level of perceived or claimed sexual enjoyment. Assaad revealed that 94 per cent of 54 circumcised women interviewed enjoyed sex and were happy with their husband.

'How is orgasm possible at all under such conditions?' Lightfoot-Klein asks reasonably. She draws on a number of sexual theorists who before her work in Sudan had suggested that while the clitoris may be the most erotically sensitive organ in uncircumcised women, in clitoridectomised females, other sensitive parts of the body, from the labia minora, to the cervix, breasts, anus and lips, plus the psychic factors of emotional involvement and spiritual connection, come into play in the clitoris' stead.

Clitoridectomised Sudanese women's descriptions of orgasm – and Lightfoot-Klein does not believe they lied – suggest as cogently as can be imagined that the modern, Western, mechanised view of female orgasm as a simple stimulus-response effect, by which orgasm happens as and when the clitoris is stimulated, is something of an over-simplification. It also seems to be almost *prima facie* evidence that, nerveless and inert as the vagina may be, there is still a distinct vaginal orgasm – not, as Freud and his acolytes, imagined a distinct and *superior* orgasm, but an orgasm all the same, and one that, surprisingly, can be experienced in the absence of a clitoris.

When women were questioned by Lightfoot-Klein, they never spontaneously mentioned their genitalia. 'Women tended to name their lips, neck, breasts, bellies, thighs or hips . . . This is due, at least in part, to the fact that a virtuous and modest Sudanese woman is required to never speak of that part of her body. When the genitalia were addressed directly by the question, "What about the area of your scar?" and following that, "What about inside?" erogeneity of one or the other (or of both areas) was admitted, or even glowingly described by many women.'

Circumcised women's recorded responses to the question of whether and how often they experience orgasm were remarkable. 'We have intercourse every two or three days. I never have orgasm during the first time, even though my husband maintains an erection for 45 minutes or an hour. When we have intercourse a second time, about an hour later, I am able to reach orgasm,' was one such.

Among the others: 'With my first husband, I almost never had any pleasure, and I had orgasm only a handful of times over the years. It was an arranged marriage, and although he was a kind man and good to me, I did not have any passion for him. My second marriage is a love match and I always have strong orgasm with him, except on rare occasions, when I am too tired or one of the children is sick.'

'When I was younger, I used to have it happen nine times out of ten. Now there are so many children and grandchildren in the house that we can have intercourse only every second or third week. We have so little privacy, and we have to be very quiet about it . . . I am able to come to orgasm once in a while now, perhaps one time in ten.'

The descriptions of (presumably) vaginal orgasm that Lightfoot-Klein collected could equally have come from the women's pages of a Western newspaper, apart from their being vastly more insightful and poetic: 'I feel as if I am trembling in my belly. It feels like shock going around my body, very sweet and pleasurable. When it finishes, I feel as if I would faint . . . All my body begins to tingle. Then I have a shock to my pelvis and my legs. It gets very tight in my vagina. I have a tremendous feeling of pleasure, and I cannot move at all. I seem to be flying far, far up. Then my whole body relaxes and I go completely limp . . . I feel as if I am losing all consciousness, and I love him most intensely at that moment. I tremble all over. My vagina contracts strongly and I have a feeling of great joy. Then I relax all over, and I am so happy to be alive and to be married to my husband . . . I feel shivery and want to swallow him inside my body. Then a very sweet feeling spreads all over my entire body, and I feel as if I am melting. I float higher and higher, far, far away. Then I drift off to sleep . . . I feel as if I am losing all consciousness. It is such a strong feeling. I hold my husband very, very tightly, and if the baby fell out of the bed, I would not be able to pick it up.'

Trying to explain this apparently paradoxical phenomenon, Sudanese psychiatrists quoted by Lightfoot-Klein theorise that the crippling effects of circumcision can be counteracted by an unusually strong bonding between marriage partners. 'In the opinion of most,' Lightfoot-Klein concludes, 'the sexual response of Sudanese women is largely nothing more than a kind of stereotypic response. They think that since orgasm entails both cerebral as well as muscular responses, and involves also respiratory and vascular reactions, the physiological phenomena are present but damaged or lessened in circumcised women. In compensation, they suggest that the cerebral component may be heightened.'

Whether the psychiatrists – all male, as Lightfoot-Klein wisely points out – are suffering from a spot of wishful thinking can only be speculated upon. But what is most striking about the findings from the clitoridectomised wives in Sudan is how these women, whose sex life is probably the same as their hundred times great-grandmothers in the Nile Valley in the time of the Pharaohs, are echoed curiously in part by American women interviewed in the 1970s by Shere Hite.

Hite, it may be remembered, discovered that the great majority of women were more interested in affection, intimacy and love in bed than in orgasm, and did not even rate orgasm as the most important bodily sensation of their sex life; the bulk of her questionnaire respondents cited the moment of penetration as the most satisfying in the whole gamut of sexual behaviours.

Can it be that in some way Sudanese Islamic women following ancient and, to us, repellent traditions, have nonetheless helped us to pinpoint a universal truth about orgasmic pleasure – a truth that, in spite of the appalling suffering of victims of FGM, could nevertheless be of some comfort to Western women who, thanks to the prevailing beliefs of our time and culture, regard themselves as chronically anorgasmic?

There are, obviously, many factors to consider before coming to the most tentative conclusion that FGM has anything positive to teach us about female orgasm. The doubts crowd in the more one considers Lightfoot-Klein or Assaad's research findings. It seems most odd that these, to us, abused women manage to have vaginal orgasms which, research such as that done in 1999 by QueenDom.com tells us, 26 per cent of liberated, educated, uncircumcised, Western women with access to *Cosmo*, vibrators *and* willing partners still seek in vain. One can only wonder if the circumcised women contrived their answers because they felt insecure and embarrassed about admitting they were non-orgasmic, were unwilling to admit to strangers that their marriages were not blissful, or were simply being polite to the nosey foreigners. And even if their accounts of non-clitoral orgasms are true, the question is begged as to whether their cultural tradition is worth bleeding to death and contracting septicaemia for – not to mention the accompanying package of economic and cultural suppression.

The strange case of circumcised women apparently having orgasms is paralleled, it might be noted (if only out of academic interest rather than practical), by the odder still, yet only passingly documented, phenomenon of male eunuchs having both erections and a form of ejaculation. In those Middle Eastern cultures where eunuchs – castrated servants – existed and in Ancient Greece, too, it was known that removing the testicles and penis did not remove sexual desire. Nor did castrates who retained their penis lose the ability to have an erection. That rare thing, a eunuch's wife, once revealed that her husband could ejaculate, albeit after a long bout of intense stimulation. The probability is that the ejaculate in such a case is fluid from the prostate gland.

It should also be noted that the more acceptable and sensually literate 'Eastern' take on sexuality and the sacrosanct Nature of both male and female orgasm (but especially female) is far more manifest and homogenous across ages and religions than isolated centres of Female Genital Mutilation. It was around 1500, over a thousand years after the *Kamasutra* appeared in India (of which much more later), that the Tunisian poet Sheikh Umar ibn Muhammad Nefzawi wrote *The Perfumed Garden*. But its tone and philosophy were to all intents identical to the earlier Indian work, and it may provide a flavour of the truer sensual Nature of Eastern and Islamic cultures.

> Long ago there lived a woman named Moarbeda, a noted philosopher who was once reputed to be the wisest person of her time. It is recorded that one day some questions were put to her, and these were some of her replies:
>
> 'In what part of woman's body does her mind reside?'
> 'Between her thighs.'
> 'And in what place does she experience her greatest pleasure?'
> 'The same.'
> 'And what is a woman's religion?'
> 'Her vulva.'
> 'And with what part of herself does she love and hate?'
> 'The same . . . We give our vulva to the man we love and refuse it to the man we hate.'

It is *The Perfumed Garden* that advises men, 'Do not unite with a woman until you have excited her with playful caresses and then the pleasure will be mutual', and famously recommends that choosing the right moment for penetration will awaken 'the sucking power of her vagina' – which in turn leads to great orgasmic pleasure for both. Choosing the perfect timing for penetration may be less of an undertaking than average for male readers of *The Perfumed Garden*, however. Nefzawi tells of a man named Abou el Keiloukh who 'remained erect for thirty days without a break' – by eating onions. (The *Kamasutra* had advocated another method of attaining Olympian levels of priapic virtuosity. It recommended that eating 'many eggs fried in butter, then immersed in honey, will make the member hard for the whole night'.)

As befits orgasmically sated societies, there was an equal acceptance at this time of homosexuality (male, at least) across North Africa, Turkey, Persia, all the Near East, India, and as far from the Middle East proper as Indonesia. Bypassing the sterner strictures of the Judaeo-Christian god's followers, gayness in the Eastern world flourished from ancient times to the present day. The Persian proverb 'A boy for pleasure; a woman for children' is still guiltlessly followed by many in Arab lands. Sodomy in Persia stood for 'higher' virtues, anal intercourse a way of reaching spiritual highs and teaching young men. It was equally respected in Egypt, by Moors, Arabs, Berbers, and in Afghanistan, where Pathans sang a gay love song, 'Wounded Heart': ('*There's a boy across the river with a postern like a peach, but alas! I cannot swim*!') Nomadic tribes in mid-Asia, too – Cossacks, Huns, Tartars, Mongols, Turkomans – were keen sodomisers.

When Islam held sway in Spain, Arabic poetry reflected its homosexual predilection. Muhammad ibn Malik, in twelfth-century Andalusia, wrote a poem called 'Facing Mecca':

> *Friday*
> *in the mosque*
> *my gaze fell upon a slim young man*
> *beautiful*
> *as the rising moon.*
> *When he bent forward in prayer*
> *my only thought was*
> *oh, to have him*
> *stretched out*
> *flat before me,*
> *butt-up,*
> *face-down.*

What legacy have such unexpectedly liberal and sensual ancient Middle Eastern sexual mores bequeathed to the present day? An eloquent answer is given by the British writer Yasmin Alibhai-Brown in an essay called 'Why East Beats West When it Comes to Sex' which appears in a 2001 book edited by Stephen Bayley and unambiguously entitled *Sex*.

She quotes from a letter she received from a British millionaire after an article of hers appeared on the ambiguous sexual message given by the *hijab*, the Muslim women's traditional head covering: 'I am writing this to you with some trepidation,' the letter read, 'but I can't stop looking at the pictures which accompanied your article. I find them a real turn on. I am sick of bodies on display everywhere. Women who cover themselves have real power over our fantasies. I know I cannot ever have one of these women, and I am jealous of men who can. It must be like opening up a beautifully wrapped birthday present.'

Many Muslim women, Alibhai-Brown explained, are wise and knowing sexually. 'In East Africa where I grew up, Zanzibari Muslim women were thought of as sophisticated, addictive lovers who could weave invisible bonds around a man and keep him intoxicated mostly by never giving him the whole of themselves. But you never saw them. They were always completely covered up in black robes, yet their eyes were animated in ways which cannot be described.' Alibhai-Brown confirmed in the essay that in Islamic texts, Allah gives a woman the right to physical gratification which she can demand of her husband.

'If one could step back from the ubiquity of sex in public spaces, and allow it to retreat into those small inner places where unfolding and disclosures become possible once more, we may yet discover the difference between a fuck and an experience,' Alibhai-Brown contends. She describes the modern anti-sex trends in Islamic countries and the Indian subcontinent as having 'marched in, like unwelcome storming soldiers, through history to deny; destroy or punish physical love and lust. The forces of suppression won ground in the twentieth-century; particularly in Islamic countries which responded to political powerlessness by killing all joy and earthly pleasures in their populations or by forcing them to live in a way that masks these. Hindu fundamentalism is displaying the same tendencies.'

Ironically, she concludes, 'the thrusting, commercialised sexuality which is sweeping across the globe is also destroying that genuine and deep sensuality which was once so carefully nurtured among these groups. Late-night sex shows on TV may work a treat for the beer-filled bloke who tumbles back at midnight and would like but can't have a shag (and many Asian men would be among them), but coarse, pretend sex cannot appeal to those with more refined tastes and traditions. So here we are, part and yet not part of the modern sexual revolution; more knowledgeable and at ease with physical love than many westerners (we have not until recently had to suffer any anguish about imperfect bodies), yet unable or unprepared to join in the scrum which passes for satisfying sex today.'

8

Sex and the City State

'She demands eight obols to give him a peck on his prick'
Hipponax of Ephesus (6th century BC)

The Ancient Greeks were the first culture we know to have applied method to the pursuit of pleasure. It is in Greek that we see the first sexual use of the word *orgasmus*, meaning, 'to swell as with moisture, to be excited or eager'. The pursuit of pleasure for its own sake, an overlooked, yet defining, feature of civilisation, is seen in the Greeks' championing of theatre, art, comedy, sport – and sex. 'Let there be lewd touching first and games before the work,' says an anonymous saying from the fifth century BC. One of the many roles of the Greek deity Mercury/Hermes was as the god of masturbation. Burgo Partridge, in a book entitled *A History of Orgies*, commented: 'The culture of the Greeks is entirely a song in praise of pleasure, and the Nature of that pleasure was an intense and ingenuous sensuality. At all intellectual levels the people recognised the essential part played by voluptuous materialism in human affairs.'

The Christian idea of 'sin', along with all the misery and self-flagellation that went with that, was a concept which had yet to be formulated. For the moment, guilt-free enjoyment of all the body and mind had to offer was the bonus prize for living in the most giddily advanced civilisation there had ever been. Yet it is easy, at the same time as noting the hedonism of the Greeks it, to exaggerate it. More fundamental was the emphasis in Greek life on balance, of guarding against the pursuit of one thing destroying the enjoyment of another. Marriage, for instance, afforded opportunities for sensual delight, but it equally concerned procreation, the preservation of bloodlines and the proper education of children. The Greeks recognised that unbounded sexual indulgence would lead to the sexual satisfaction of nobody, and that it might destroy the other things worth having.

On the other hand still, while it was understood that hedonism is not always compatible with momentary gratification, the concept of balance did not rule out the odd sexual excess either. It was appreciated that man is part human, part animal, and that the interests of the two can conflict. So the craving for sexual pleasure was an accepted addiction; orgies were seen as a kind of societal safety valve, and one Aristophanes comedy, *Lysistrata*, has the women of Athens organising a sex boycott to try to stop their husbands fighting wars.

The appetite for orgasm was not seen as exclusively male. Greek men were terrified by what they perceived as women's insatiability and sexual energy. Euripides' *The Bacchae* played on this Greek male fear that women would willingly tear men limb from limb in their lust for pleasure. Young virgins, especially around puberty, were seen as particularly wild. The abundant medical writings of the time ascribed women's drive for sexual activity to an inbred psychological urge rather than the desire for pleasure that tempted men. But whatever caused it was immaterial, so long as a civilised Greek city state kept this dangerous hunger for sex in check. This it did by promoting the necessity of marriage, preferably at puberty, after which pregnancy would completely cure any remaining sexual rampancy. As a double indemnity, women were cloistered at home, often surrounded by guard dogs.

Marriage, love and sex were not particularly inter-connected, though. The object of marriage was to produce a son and heir, but a Greek husband rarely looked to his wife for companionship or for sexual delight. Around 600 BC, an Athenian judge, Solon, established some sexual laws pertaining to marriage: an heiress was given the legal right to demand her husband fulfil his conjugal duties at least three times a month; a cuckolded husband was given leave to kill his adulterous rival. But in practice men and women had little to do with one another after marriage. The refined sexual pleasure appropriate to a gentleman was openly available via a number of extra-marital avenues. Prostitution was entirely accepted – the same Solon instituted fixed-price, state-run brothels outside the walls of cities, with prostitutes in gauzy garments openly advertising their attractions. Most were slaves or women taken as trophies of war. Additionally, the general expectation was for men to be bisexual, or 'ambidextrous', which opened up an even greater number of erotic opportunities for men.

These erotic opportunities included paedophilia. Parents routinely colluded in the sexual initiation of even young children by older men, and would express outrage if their children were not seduced. *The Birds* by Aristophanes, includes this denunciation by one character of another: 'Well, this is a fine state of affairs, you villain. You meet my son from the gymnasium, fresh from the bath, and you don't kiss him, you don't say a word to him, you don't hug him, you don't even feel his testicles. And you're supposed to be a friend of ours!'

The Greeks simply had no concept of sex as a sin or a forbidden fruit. They considered sexual relations to be a natural, everyday phenomenon. They had no sense of shame in connection with sex, and attached little stigma to any of its aspects save a little snigger at masturbation, which they did with gusto anyway. Unlike in Egypt, 'respectable' Greek women showed little by way of bare flesh, but the men were practically full-time nudists. The Greeks even fought their battles nearly naked. Male Greek culture was sexually charged at all levels.

The Greeks referred to various situations in which godliness seized a man, a state they knew as theolepsy. Hearing music, dancing and alcohol could cause this sense of divinity, and they also found it present at orgasm, when the bounds of personality particularly seemed to melt away and one's reality merge with the infinite.

Greek pottery provides us with something close to a pornographic record of the libidinous sex life of these ancient sybarites. A huge amount of it is overtly sexual; naked satyrs and nymphs romp away under olive trees; young lovers, heterosexual, homosexual, whatever, bathe together, dance and make love. We can only assume the whole jolly scene was drawn more or less from life, as artists saw it.

If anything, the Greeks were more suspicious of the one form of sexual desire that would be endorsed by the Christians – that occasioned by love. It was love, after all, that made a man copulate too often and lose strength through the over-draining of his semen; sexual pleasure was necessary to human well-being, and not copulating *enough* could cause a semen build-up and damage health. But sex had to be mastered, which expressly meant limited.

The incessant physical longing of love was regarded as a disturbance of the body's natural balance, a disease which deprived the mind of its control of the body. What we revere as the serotonin rush of post-orgasmic bliss, for the Greeks was a transitory dullness of intellectual power. Those who admired the intellectuality of the Greeks would still be echoing this belief thousands of years later. A nineteenth-century philosopher, Hartmann, wrote: 'Love causes more pain than pleasure. Pleasure is only illusory. Reason would command us to avoid love. If it were not for the fatal sexual impulse, therefore it would be best to be castrated.'

The tradition arose, then, that only the conscious intellectual or physical diversion of the mind would do the trick, a spot of hard work, whether it was fishing, poetry or just thinking. And like twentieth-century anti-pornography campaigners who became obsessed with pornography, anti-sex Greek philosophers thought a great deal – about sex.

In the fourth-century BC, Plato related in *The Republic* that someone once asked Sophocles in his old age, 'How do you feel now about sex? Are you still able to have a woman?' He replied, 'Hush, man; most gladly indeed am I rid of it all, as though I had escaped from a mad and savage master.' Socrates opined that satisfaction of the sexual instinct worked against moral perfection. Epicurus saw love as 'an impetuous appetite for sexual pleasure, accompanied by frenzy and torment'. He denounced carnal intercourse as the worst enemy of the serenity of the wise man: 'Sexual intercourse never did anybody any good and one can think oneself lucky if it doesn't do one harm. A wise man will neither marry nor have children. Nor will he yield to the passion of love.'

But a certain proportion of these thinkers' pronouncements was underwritten by hypocrisy, or by the sour grapes attitude of old men who could no longer take advantage of the gloriously licentious society in which they found themselves. As the Greek scholar Robert Flacelière wrote in his 1960 *Love in Ancient Greece*: 'In practise the Epicurean outlook closely resembled that of an egotistical old bachelor, valuing peace of mind above all things, and dignifying it with the high-sounding name of "philosophic wisdom".'

In Ancient Greece much as in other male-dominated societies through the ages, from the quadrangles of the great Western universities to the Muslim world, the exclusion of women from the mainstream of life in the interests of a supposedly greater intellectual or spiritual calling led to an extraordinary amount of homosexuality, especially between the sixth and the fourth century BC. According to Aristotle, it was initially employed in Crete as a method of birth control, and that island provides the earliest representation of homosexuality, in the form of a bronze plaque of around 650 BC. It was soon considered shameful in Crete for a well-born boy not to have an older man as lover.

Gay love on the Greek mainland was originally about male companionship and devotion to warrior buddies, shared bravery rather than sex, and certainly nothing to do with population control. Too many young Greeks died in their warrior years, having experienced no love life other than the closeness of military bonds. The military requirement for fitness segued into a homoerotic culture. The fitter a boy, the more orgasmic pleasure he seemed equipped to deliver to his comrades; the more beautiful he was, the better mind he was assumed to possess. The best-looking boys, therefore, attracted older male sexual lovers as tutors.

Anal sex was perfectly legal and the common sexual mode for men, being considered an enjoyable, healthy, and uplifting activity. One Greek physician explained that men enjoy anal intercourse because sexual enjoyment depends on friction of the part of the body where seminal fluid is secreted, and due to a birth defect, theirs happens to be in the rectum.

Vase paintings of anal intercourse usually show the participants as being members of the same age group. But paintings and illustrations on the male drinking cups used at evening *symposia* (drinking and intellectual discussion parties) tended to show the older men bringing themselves to orgasm between a naked teenage boy's thighs. The boy would be the guest of honour, attending with his father's permission, having been invited by another older man who fancied him – and who would commission special crockery celebrating his beauty.

Homosexuality was seen straightforwardly as morality in action, the unquestioned proper way to bring up a boy to be an upstanding citizen. A man was not really considered to be upholding community standards unless he practised sodomy. Lycurgus, the Spartan legislator, refused to consider any citizen to be a worthy man if he did not have a male lover. And any parental squeamishness about anal intercourse would damage a boy's education. The best teaching was conducted through the love between teacher and student, called *paiderastia*, and there was a common feeling that virtue could literally be implanted in a boy by receiving the anal ministrations of his teacher.

At the same time, however, heterosexual prostitution was an enormous and respected industry. While wives were breeding machines secluded at home, prostitutes, of the market place or the boudoir, were the liberated women with whom men could explore their sexual fantasies. The beauty and sexual skills of the upmarket and fabulously wealthy *hetairai* are illustrated on precious vases. One *hetairai* practice said to have survived in Greece to modern times is that of using the feet to masturbate a lover. Lower-ranking streetwalkers had their charms, too; some would wear a sandal whose sole imprinted on a dirt surface the words, 'Follow me'.

One of the most famous women of ancient Athens was a positively regal prostitute, Phryne, who held court with the leading men of her day in the late-fourth century BC. Her affair with Praxiteles, the greatest of the Attic sculptors and the most original artist of his day, was famous. She was his model for the first ever naked statue of a woman, Aphrodite. The statue became the masturbatory fantasy of every red-blooded Athenian man unable due to youth or lack of cash to aspire to a prostitute of Phryne's shimmering wonder. *Erotes*, by the satirist Lucian of Samosata, tells of a rather earnest young soldier who falls in love with the Phryne Aphrodite statue and achieves his ambition to spend a night in its company. 'The statue is a flawless work of Parian marble. The goddess's lips are slightly parted in a disdainful, ironic smile. No garment veils her charms, but one hand screens her modesty with a casual gesture,' Lucian writes.

'I need not,' he continues, 'be so indiscreet as to recount the details of the crime that he committed on that disgraceful occasion. When daylight returned, the goddess had a stain as a tribute to the traumas she had been through. After perpetrating this outrage, the young man threw himself into the sea.'

Despite this slightly disdainful tone, the Greeks regarded

male masturbation – which they called *cheiromania* or 'passion with the hand' – as wholly normal and a safety-valve substitute for men to whom sex was not available. The Ancient Greeks also knew that a man could 'see stars' and feel as if he was fainting when his prostrate gland was probed, and they were known to experiment with prostates in the manner of modern people and soft drugs. (The prostate gland is still used today by some as a kind of male G-spot.) Masturbation proper was talked about avidly, and frequently featured in comedy. In an Aristophanes play, a slave talks about his much-manipulated foreskin, saying that it is soon going to look like the back of a flayed slave.

Cynical philosophers encouraged masturbation as a defining act of self-sufficiency. Masturbating men were depicted on vases and terracottas; the Royal Museum in Brussels has a cup showing a garlanded youth performing the act. Plutarch records that Dio Chrysostum, a Stoic philosopher of the first century AD, praised the fifth-century BC philosopher Diogenes for his stance on masturbation. Diogenes was generally anxious to contravene social convention, especially in the matter of performing natural functions in public. He argued that sexual competitiveness was a destructive force in society, unnecessary since it was possible to find, 'Aphrodite everywhere, without expense'. When someone asked what he meant, Diogenes started to masturbate in front of his audience in the marketplace, saying to his surprised fans and critics: 'Would to Heaven that by rubbing my stomach in the same fashion, I could satisfy my hunger.' He attributed the origins of masturbation to Pan, who was distraught when Echo had left him bereft of sexual fulfilment; he was duly taught masturbation by Hermes/Mercury, his father, whose special subject it happened to be.

In a reversal of the custom at *symposia* for older men to perform frottage between a boy's thighs, vase paintings show young men masturbating by placing their penis between the thighs of an older, stooping man. This was delicately called, 'interfemoral connection'. Fears that masturbation wasted semen were not taken seriously, even though it was commonly believed that it took forty parts of blood to create one of semen.

As for female sexual pleasure, especially masturbation (of which more later), there is, as in so many matters Ancient Greek, a dichotomy. Greek mythology, with its chaste and demure goddesses – even Aphrodite was portrayed with a hand casually shielding her genitalia – displays women in an idealised form, minus the rampant sexuality believed to exist in mortal woman. But the evidence is that in reality, as in so many cultures where a formal decorousness surrounds female sexuality, Greek women, while not the nymphomaniacs of their men's fears, were not as obedient and modest as the men thought, either, even after their wilder sexual feelings had been strictly curbed by their closeted existence.

The freedoms of a woman such as Phryne were unbelievable to wives who were supposedly only allowed out of their house once a year for fertility ceremonies. Yet it is now known that there were secret women-only sanctuaries at which they could take breaks, get-away-from-it-all oases which had their own erotic life and ceremonial. Here, they told rude sexual jokes, reclined to eat from naughty pottery designed to amuse women, and conspired to win back a modicum of control over their own bodies, offering the gods plants like pomegranate that were thought to be contraceptives and abortifacients – the polar opposite of the fertility they were imagined by their husbands to be on holiday to improve.

The very existence of contraceptives, real or imagined, argues eloquently for Greek wives having had a more assertive, independent Nature than that with which they are generally credited, even if only a determination to enjoy a little sex without the encumbrance of pregnancy. Another popular method of birth control was 'misy', copper sulphate, which was thought, when drunk in solution, to ward off pregnancy for a year. The idea of heavy metals as contraceptives persisted for thousands of years. Women of other ancient cultures would drink water from blacksmiths' fire buckets, and English women near Birmingham as recently as 1914 drank, as a contraceptive, water boiled with copper coins.

Most literary genres in classical Greece depict women less like chaste goddesses or dangerous energy-sapping sex addicts, and more like the women of these sanctuaries – normal mortals suffering or enjoying the same sort of erotic desires as men. The reality of female sexuality was not as well represented, however, by the medicine of Ancient Greece, the famous Hippocratic Corpus, the body of so-called medical knowledge either garnered by a mysterious 'Hippocrates' (or, more likely, a group of practitioners around 400 BC who all called themselves Hippocrates) and celebrated in name to this day by the Hippocratic Oath.

It is self-evident that the ancients, free from such tedious requirements as scientific data, more often than not made up their medical pronouncements as they went along. Medical experts were, naturally, all men, and they unsurprisingly devised a model for sexuality that seems rather to justify male sexual acquisitiveness. They averred that women 'need' regular sex so as not to become ill – which conveniently gave men free range to 'treat' them without any issue of morality intervening. They were, after all, doing women a medical favour by having sex with them. Women therefore dare not for their own health's sake refuse their husbands sex, even when they felt no desire for them. In Ancient Egyptian and Hippocratic medicine conversely, an active female desire for sex, its symptoms including arousal, erotic fantasy, vaginal lubrication and generally melancholic or irrational behaviour, was known as an illness called *hysteria* – literally, a sickness caused by shifting of the uterus; the word *hysteria* has the same root as hysterectomy.

The Hippocratic author of a section of the corpus called *De Virginibus* asserts that the suicidal tendency of young virgins was due to lack of sex, rather than their being married off at the age of twelve to dirty old men. He thinks that the best cure is for women to marry young. Women, he contends, have a psychological need for sex yet no conscious desire for it or knowledge of what they really require physically. The satisfaction of their appetite has little to do with pleasure, apart from simply relieving the pressure of their blood on their heart. (It should be noted that the concept of hysteria as being almost interchangeable with orgasmic was only officially dropped by the American Psychiatric Association as late as 1952, and only then because of the confusion with the more modern meaning of hysteria.)

A connection is also seen in Ancient Greek medical hypothesising between female orgasm and the enhancement of fertility. For the Hippocratics, it has been said, a woman's enjoyment of sex is not proof that she will become pregnant; rather, becoming pregnant is evidence that she enjoyed intercourse. Some women apparently insisted accordingly that their husband bring them to orgasm as a means of improving the chance of begetting children. This, at least, is the way they put it.

With fertility in mind, then, if not shared adult pleasure, people were willing to go to great trouble to try to bring about simultaneous mutual orgasm, using, according to Thomas W. Laqueur, a physician and historian of sex at the University of California, Berkeley, both foreplay and an assortment of natural remedies. (*Tribulus terrestris* or 'puncture vine' from Bulgaria, oriental ginseng, the bark of the tropical African *Corynanthe yohimbe* tree, the Brazilian *Muira puama* ('potency wood') stem, the Mexican damiana leaf, wood betony from Europe, ashwagandha (winter cherry) root from India, Central American saw palmetto berries and *Avena sativa* – wild oats – have all been used at various times in the past as aphrodisiacs to enhance erection, sexual stamina and so on.)

Female sexual pleasure and orgasm were not necessarily synonymous in Greece, however. Women were thought to feel pleasure from the moment of penetration, and then experience a steady level of enjoyment from the friction of the penis in the vagina, rather than any kind of peak of excitement. 'Once intercourse has begun,' one Hippocratic writer states, 'the woman experiences pleasure throughout the whole time, until the man ejaculates. If her desire for intercourse is excited, she emits before the man, and for the remainder of the time she does not feel pleasure to the same extent; but if she is not in a state of excitement, then her pleasure terminates along with that of the man.'

The Hippocratic notion of the womb was of an independent entity within a woman's body that could override the woman's will with its own desires. The best way to fight the womb back was to sit the woman on some perfumes and burn mule-dung under her nose. The womb would flee from the bad smell around the brain and be attracted towards the good odour, closer to where it belongs, in the region of the pelvis. The woman's sexual desire – not surprisingly, given the mule dung under her nose – would then dissipate.

The Hippocratic Corpus was not the only repository of medical assertion and supposition. Everyone was allowed to chip in with a bit of homespun sexual advice. Aristotle, a man of extraordinary breadth of interest and wisdom but not known as a physician, felt free nevertheless to opine in *The Nicomachean Ethics* that, 'Erection is chiefly caused by parsnips, artichokes, turnips, asparagus, candied ginger, acorns bruised to powder and drunk in muscatel, scallion, sea shellfish, etc.' – and be taken seriously. (To be fair, Aristotle's body of opinion on sexual matters was generally more considered than this. He was often ahead of his time. He did not agree, for example, with those who believed that a woman's need for sex was caused by the displacement of the womb.) Plato, Aristotle's tutor, could similarly mention in passing (although quite sensibly, as it happens) that male and female sexual experience are 'owing to the same causes', and be taken, again, as an authority. This free-for-all in medical advice-mongering, although not unique to the intellectual hothouse of Athens, still seems alien and bogus compared to later cultures that valued what we recognise as expertise.

Despite the wide variety of great thoughts in circulation on sex, there was still no agreement in Ancient Greece as to how sexual reproduction took place. In Homer's day, around 800 BC, it was believed that the female became pregnant as a result of airborne '*animalculae*' that somehow found their way inside the woman. The Hippocratics believed in some form of meeting between male and female seed. Male seed was provided by a sudden, and highly pleasurable, ejaculation from the body, so it was thought that the mother must also contribute the same sort of fluid to help form the foetus This would explain why a child resembled its mother, but was not taken as an indication that women played an equal part in reproduction to the extent that husbands ought to be concerned that their wives' enjoyed sex. The Hippocratics connected the production of female seed with orgasm, too, but were not convinced that the wife had to enjoy her orgasm as the husband enjoyed his. Aristotle, who tended to dissent from Hippocratic conclusions, thought females produced semen from their ovaries and an egg was in the womb, this from a combination of menstrual blood and the female sperm. Four hundred years later, the physician Galen, whose influence dominated medicine in Europe and the Muslim Middle East until the seventeenth century, came close to a modern understanding of conception with his theory that female testes made semen which, when mixed with male semen in the womb, produced an embryo.

Galen also believed women could desire intercourse for its own, pleasurable sake. He was one of the first physicians to advocate therapeutic masturbation or finger 'stimulation' for frustrated women such as widows and nuns. Previously, in the section of the Hippocratic Corpus known as *De semine*, women's orgasm was seen as a reflection of men's. Aristotle thought both sexes' desire for intercourse could be prompted by the need to expel fluid after a period of celibacy. The

Hippocratics did not imagine orgasm could serve as an incitement to a woman to want intercourse; if a woman fails to orgasm, therefore, she would not be left frustrated. Aristotle, again, disagreed with this. Putting an interesting new spin on the question, he professed himself aware that both sexes could be incited beyond any immediate need for sexual intercourse by the memory of past pleasures. Therefore the desire for orgasm could be summoned up from the imagination in both sexes even when there was no pressing physical build-up of seminal material.

Despite the lack of consensus over so much about sex, the Greeks did at least come quite close to working out what and where the clitoris was, even though it would be thousands of years before this information was properly understood in Europe. The Hippocratics did not mention the clitoris (although neither did *Gray's Anatomy* as late as 1918), but literary figures such as the poet Hipponax of Ephesus and the playwright Aristophanes both seem to refer to it by the name *myrton* (myrtle berry), and some scholars believe Sappho, the sixth-century BC poetess and Lesbos's most celebrated inhabitant, used the word *nymphe* (bride) to refer to the clitoris. Aristotle mentioned it, too: 'An indication that the female emits no semen is afforded by the fact that in intercourse, the pleasure is produced in the same place as in the male by contact, yet it is not the place from which the liquid is emitted,' he wrote.

A peculiar feature of Greek medical thinking on sex and orgasm was an emphasis on the importance of heat to the process, almost as if it were a chemical reaction in a test tube that required help from a Bunsen burner. That is why in Greek writing on sex there is much imagery of fire and water, loosely applied as and when it could illustrate a particular theory. Some held, for instance, that the womb was a dry, painful area that could only be sated by men's sexual fluid. Others suggested that fertility required extra heat in the vaginal area.

In the Hippocractic Corpus, a writer contends: 'In the case of women, it is my contention that when, during intercourse, the vagina is rubbed and the womb is disturbed, an irritation is set up in the womb which produces pleasure and heat in the rest of the body . . . A woman also releases something from her body, sometimes in the womb, which then becomes moist, and sometimes externally as well, if the womb is open wider than normal.'

The Hippocratic belief that women's sexual pleasure ceases promptly when the male ejaculates is similarly explained in terms of heat and combustion: 'What happens is like this: if into boiling water you pour another quantity of water which is cold, the water stops boiling . . . In the same way, the man's sperm arriving in the womb extinguishes both the heat and the pleasure of the woman. Both the pleasure and the heat reach their peak simultaneously with the arrival of sperm in the womb, and then they cease. If, for example, you pour wine on a flame, first of all the flame flares up and increases for a short period when you pour the wine on. In the same way the woman's heat flares up in response to the man's sperm and then dies away. The pleasure experienced by the woman during intercourse is considerably less than the man's, although it lasts longer. The reason that the man feels more pleasure is that the secretion from the bodily fluid in his case occurs suddenly, and as the result of a more violent disturbance than in the woman's case.'

Another Hippocratic theorist avers that if a woman is too excited before intercourse, she will 'ejaculate prematurely', meaning that her womb will close and she will not conceive. 'Like a flame that flares when wine is sprinkled on it, the woman's heat blazes most brilliantly when the male sperm is sprayed on it . . . She shivers. The womb seals itself. And the combined elements for a new life are safely contained within.' Another Hippocratic argued that the need to have fluid inserted into the womb to quench its thirst ensured that a woman could not turn to her female companions to free her from sexual dependence on her husband.

For Galen, writing long after the Hippocratics, heat during

sexual climax is crucial to conception; simultaneous orgasm generates enough heat to 'commingle the seed, the animate matter, and create new life'. Later texts also recommend mutual simultaneous orgasm as the best way of dealing with the unfortunate consequences for fertility of a woman being too cold or too hot.

In her anxiety to conceive, or at least to convince her husband it was important for him to satisfy her so she might better do so, it is easy to forget female masturbation and the role – considerable in fact – that it played in the sex life of Ancient Greeks. Female masturbation was discounted by Greek medical writing. For the usual reasons of sexual politics, the Hippocratic authors would not countenance the notion of women being able to dispense with the need for their husband through masturbation. One writer announces, quite spuriously one would guess, that women do not get stones in the bladder as often as men do 'because they do not masturbate'. Even if a wife used a dildo on occasion, in Hippocratic medical thought it could not benefit her because of its inherent inability to introduce semen into her womb. Masturbation simply could not fulfil women's sexual appetite.

Female masturbation is equally glossed over by most Greek painting and literature, but this is largely because women were so rarely portrayed in Greek art. Hans Licht, a classics professor at Leipzig University, nevertheless came to the daring deduction in a 1932 book, *Sexual Life in Ancient Greece*, that Ancient Greek women were avid masturbators.

But even though Licht was writing in the licentious atmosphere of Weimar Germany, on the subject of Ancient Greek womanhood and masturbation he was still careful only to mention their 'mysterious conduct', and a characteristic form of Ancient Greek leather-based dildo, known as *olisbos*, as depicted in many vase paintings, as well as references in *Lysistrata* to women lusting after them. Women described, however elliptically, in Ancient Greek literature as using *olisbos* are often portrayed as wearing the things out in frustration.

Licht writes of a bowl in the British Museum in his time showing a contented-looking naked woman with two *olisboi* in her hand. Similar items were held by the Louvre and the Berlin Museum, which also had a vase showing a woman douching after use of the *olisbos*. Lucian and Plutarch write about a Lesbos full of *olisbos*-wielding lesbians, including Sappho, who in her own work describes one of her solo orgasms, obtained (after due prayer to Aphrodite) during a sexually fallow patch:

When I but glance at thee, no word from my dumb
lips is heard
My tongue is tied, a subtle flame
Leaps in a moment o'er my frame,
I see not with mine eyes, my ear can only murmurs hear,
Sweat dews my brow, quick tremors pass
Through every limb, more wan than grass
I blanch, and frenzied, nigh to death,
I gasp away my breath.

Dildo-making became a specialised commercial cottage industry around 500 BC in one particular city, Miletus, in Ionia. Miletus *olisboi*, which were exported across the Greek world, were manufactured by shoemakers from wood and padded leather, and were designed to be used with olive oil for lubrication. The shoemakers' skill was to sew the kid leather carefully so the stitches would not hurt women users.

The *olisbos* was used by both heterosexual women and lesbians, or *tribads* as they were known. The device was used either alone in privacy, or communally by two or more women together. A passage in Lucian's *Erotes* alludes to the latter use. A character called Charicles, who is outraged by the *olisbos*, complains about 'the invention of such shameless instruments, the monstrous imitation made for unfruitful love, lest a woman embrace another woman as a man would do; let that word which hitherto so rarely reaches the ear – I am ashamed to mention it – let tribadic obscenity celebrate its triumphs without shame.'

In other works, there is less implicit criticism of the practice. The third-century BC writer, Herondas, in *The Two Friends, or Confidential Talk*, describes how the friends discuss their *olisboi* without embarrassment. The play's complicated, gossipy and always risqué storyline would be far from out of place in a modern *Sex and the City* episode – or at an Ann Summers party.

Just as, in the modern world, we have bought a bit too readily into the idea of the dreamy, philosophical Greeks as out-and-out hedonists, we have cleaved a little too much to the notion of the technically savvy, sophisticated Romans as sexually a bit perverted and debauched. The situation is similar in many ways to how perceptions might be a few hundred years from now of twenty-first-century Britain and the US. From the evidence of contemporary popular culture, it could appear that old-fashioned Britain is rather refined in sexual matters, while technically advanced America, home of 90 per cent of Internet pornography, is brash, sexy and liberated. In practice, however, America is in many ways more modest and prudish about sex than the UK.

Rome had always been a city where prostitution flourished, fed by the ravenous orgasmic appetites of its men and women. According to a 1920s English scholar, W. C. Firebaugh, who is thought to have coined the expression that prostitution is 'the oldest profession', there was a complex grading system for prostitutes, and the women who practised this calling were by no means all from the lower levels of society.

The highest grade were the *Delicatae*, the kept women of the wealthy and prominent men, the direct equivalent of the Greek *hetairai*. Next down were *Famosae*, daughters and even wives of wealthy families who simply enjoyed sex for its own sake. Next came the well-respected *Meretrix*, the career, paid harlot; *Prosedae*, who waited in front of their brothel for passing trade, *Nonariae*, night walkers forbidden to appear before the ninth hour, *Mimae*, mime actresses, who were invariably prostitutes on the side, *Cymbalistriae*, cymbal players, *Ambubiae* (singers) and *Citharistriae* (harpists) who also moonlighted as prostitutes.

Further down still were the *Scortum*, whom Firebaugh defined merely as 'common strumpets', *Scorta erratica*, peripatetic, travelling strumpets who walked the street, *Busturiae*, who hung around funerals to service miserable mourners, *Copae*, bar maids who could be hired for the night by travellers as bedmates and from who we get the word 'copulation' as well as the still less delicate 'copping off', *Doris*, harlots of particular beauty who worked naked, *Lupae*, 'she wolves' known for the peculiar wolf-like cry they uttered when they came, or pretended to come, to orgasm, or alternatively for their tonguing skills (remember how a she-wolf licked Romulus and Remus better when she found and nursed them).

At the bottom of this substantial heap were *Aelicariae*, girls in bakers' shops, which were generally regarded as brothels, particularly where phallus-shaped cakes were sold (the *Aelicariae* would slip clients in when the bread was in the *fornix* – the oven – from whence comes our word 'fornicate'), *Noctiluae*, another grade of night walker, *Blitidae*, lower-class bar girls who got their name from a cheap drink sold in the tavern where they worked, *Forariae*, country girls who frequented highways, *Gallinae*, prostitutes prone to steal your wallet (so named after hens because of their propensity take anything and scatter everything), *Diobolares* – girls who would have sex for the bargain price of just two obols, *Amasiae*, enthusiastic amateurs whose reward was sexual pleasure alone, and finally, the bargain basement *Quadrantariae*, the lowest grade of all, whose charms were no longer merchantable.

So evidence of the Romans' debauchery is not exactly lacking, yet it is easily forgotten that same civilisation did also advance further than the Greeks ever did the cause of equitable marriage and the more equal distribution of orgasmic pleasure between the sexes. Musonius Rufus, a Roman Stoic of the first century AD, indeed, created the modern marital ideal seized upon later by Christianity. To paraphrase his words, Musonius's idea was that marriage should be a communion of souls with a view to producing children. With his strict ethical doctrine came a then rare sexual equality, with neither partner being allowed to have sex outside wedlock or before marriage. It was as a result of Musonius's beliefs that in the fourth-century AD, adultery was made punishable by a fine. On the other hand, the revenue from this taxation was so great that the state was able to build a temple to Venus with it.

But there are many examples too, as evidence of the novel Roman concept of Mr and Mrs Right, the perfect partners, of married couples putting up their own heterosexual erotic domestic sculptures. On the Roman-owned Aegean island of Delos there is a statue in a Roman house of a married couple locked together symbolically, part of the same piece of stone. What is most striking about this is that the statue was erected and paid for by the wife. More generally, too, in the Roman world, the matriarch was a commonplace figure, and one inconceivable to the Greeks. Middle-class and noble women seen in Pompeii frescoes give every impression of enjoying quite remarkable equality with their menfolk.

It is true to say that some of the old Greek fear of female sexuality pervaded Roman society also. Pliny believed women's sexual feelings such a threat that he advocated dousing them by applying to the labia mouse droppings, snail excrement and 'blood taken from the ticks on a wild black bull' – a combination that must have quenched both female and male lust. Juvenal in his *Satires* complained: 'And remember, there's nothing these women won't do to satisfy their evermoist groins: they have just one obsession – sex.'

But Roman sex manuals betray a more modern, less misogynist view of women. They portray the female orgasm as something all men should strive for, even if Ovid, in his first-century AD *Ars Amatoria*, does not quite strike the tone of a modern

Cosmopolitan article. 'Use force. Women like forceful men,' writes the poet. 'They often seem to surrender unwillingly when they are really anxious to give in. When you find the spot where a woman loves to be touched, don't be too shy to touch it . . . You'll see her eyes sparkle . . . She'll moan and whisper sweet nothings and sigh contentedly . . . But be careful that you don't gallop ahead, leaving her behind. And make sure that she doesn't reach the finish before you do.'

The fairest assessment of Roman sex is that it was a bit of a mixed bag. Venus, the goddess of love, appears in Roman life simultaneously as the guardian of honourable marriage, the matriarch of the Roman nation, the patroness of prostitutes – and the persuader-in-chief against licentiousness.

No appraisal of Roman society can be complete without a voyage round the more grotesque manifestations of the Roman male's lust for exotic orgasmic delights. The Romans' predilection for a high-octane sexual thrill may go back to the *Bacchanalia* tradition that originated in the far south of Italy and moved northwards. The *Bacchanalia* festivals, a debased form of the Greeks' *Dionysia*, were a distasteful hive of violence, high emotion and sex, in that order.

The immediate ancestors of the Romans celebrated marriage with a wedding orgy in which all the husband's friends had intercourse with the bride first, in the presence of witnesses. This is thought to be a survival of a so-called state of 'free prostitution' that preceded marriage in earlier times in which the idea of stable-coupledom was unknown. 'Natural and physical laws are alien and even opposed to the marriage-tie,' wrote Otto Kiefer in his 1934 *Sexual Life in Ancient Rome*. 'Accordingly, the woman who is entering marriage must atone to Mother Nature for violating her, and go through a period of free prostitution, in which she purchases the chastity of marriage by preliminary unchastity.'

Such practices may have co-existed with the *Bacchanalia*. But by the time of Rome's greatest influence, *Bacchanalia* was no longer regarded as a good thing. The historian Livy explained how more discerning later Romans thoroughly disapproved of the occasions: 'After the rites had become open to everybody, so that men attended as well as women, and their licentiousness increased with the darkness of night, there were no shameful or criminal deed from which they shrank. The men were guilty of more immoral acts among themselves than the women. Those who struggled against dishonour, or were slow to inflict it on others, were slaughtered in sacrifice like brute beasts. The holiest article of their faith was to think nothing a crime.'

Such orgiastic behaviour, according to the historian of the orgy, Burgo Partridge, 'became, not a purgative, but a habit-forming drug'. The corrosive Nature of *Bacchanalia* was not lost on Roman legislators, however. Some 7,000 people were prosecuted for taking part in them, many of whom were put to death. A decree issued by the Senate in 186 BC finally outlawed them. Nero's tutor Seneca was moved by such excesses to state that, 'pleasure is a vulgar thing, petty and unworthy of respect, common to dumb animals'. Yet the spirit of the *Bacchanalia* survived in the sex lives of many of the emperors, Nero notably included.

Of the Emperor Galba, we learn from Seutonius: 'He was very much given to the intercourse between men, and amongst such he preferred men of ripe age, *exoletes*.' (*Pedicon* was the Latin name for 'a man who exercises his member in the anus'. He was also called a pederast or drawk. The man 'who allows himself to be invaded in this way' was called the cinede *(cinaedus)*. If the cinede was 'worn out', and hence a looser fit, he was called an *exolete*. Domitian enjoyed heterosexual sex more. He described it as 'bed wrestling'. One of his favourite pre-bout activities was personally depilating his concubines to assist them achieve the pubic hairstyle of the time, which was for women to remove everything by either plucking or singeing.

The Emperor Augustus was the originator of strict and civilising Roman marriage laws, yet he, too, was pretty dissolute. He once instructed the wife of an ex-consul to attend him in his bedroom and sent her back to her husband's dining room, visibly flushed and with hair ruffled, leaving little to the imagination as to what had occurred. Augustus also got his wife to help him procure virgins.

Tiberius forbade the execution of virgins so that when such women were condemned to death, they would have to suffer the extra humiliation of being publicly deflowered by the executioner just before the sentence was carried out. Of Tiberius, Burgo Partridge, using Suetonius as principal source, writes: 'At his retreat in Capri he devised an apartment with seats and couches in it, and "adapted to the secret practice of abominable lewdness", where he entertained companies of girls and catamites who he called *Spintriae* and who defiled one another in his presence *triplici serie conexi* in order to arouse his flagging powers . . . In the Blue Grotto he swam like an old shark amongst a shoal of naked little boys, of an age when "they were already fairly strong but had not yet been weaned". These he called "his little fishes" and trained them to play between his thighs while he was bathing. The children were encouraged to suck on his penis as if it were a nipple.'

Caligula, according to Partridge again, would avail himself of any woman he fancied, regardless of whether her husband was there or not. After having sex, he would come back into the room and give a critical appraisal of his partner's charms or lack of them.

Nero would sail down the Tiber to Ostia with special debauchery booths erected on the shore in case he got caught short with the need for an orgasm. As he sailed, hopeful (or, more likely, nervous, given his unpredictability and boundless sadism) hostesses would stand in the booths and beg him to alight at their particular mobile boudoir.

Nero also had an incestuous relationship with his mother. It was rumoured that when he travelled with her in a litter he would have sex with her. At least, that was the usual explanation for the stains seen on his clothes when he ended a journey, but which were not there when he started it. Nero also invented a charming 'game' in which he put on one of a variety of wild animal skins and had himself 'freed' from a cage to launch himself on to the genitals of men and women who were trussed up in not-very-eager anticipation.

The Roman bourgeoisie had its own, slightly more discreet, charms. Archaeologists recently discovered a bathhouse in Pompeii with locker paintings depicting erotic sex scenes. Pompeii, and nearby Herculaneum too, had a classic Hamburg-type mix for an erotic city, a combination of wealthy merchants and visiting sailors which ensured a prosperous prostitution industry. Pornographic images from Pompeii and Herculaneum were stored in a secret room (*camera segretta*) in the Naples Archaeological Museum for several hundred years, until the public was recently allowed to see the artefacts by special request.

The material on show is creditable pornography, too, rather than mere graffiti. Among the well-executed paintings are one depicting a man having anal sex with a woman, several pictures of men with the kind of semi-erection favoured by modern porno magazines (it is achieved today by male models masturbating a few moments before the photo session), a half-man half-goat masturbating a (male) full-goat, women astride men, enormous penises framed by a temple's columns, a woman fondling another woman's breasts (the only lesbian scene found at Pompeii), an extremely graphic cunnilingus scene, and a curious threesome in which a male is having anal sex with a women, while receiving anal sex from another man.

Male masturbation was largely acceptable in the Roman world. The judgemental Juvenal deprecates the habit amongst schoolboys of mutually masturbating one another. The Latin poet Martial warns, 'What you are losing between your fingers, Ponticus, is a human being.' But it is Martial too who pronounces, '*Veneri servit amica manus*' – 'Thy hand serves as the mistress of thy pleasure'. He also describes Phrygian slaves masturbating themselves to overcome the lust occasioned by seeing their master having sex with his wife. In mythology, Mercury teaches his son Pan to masturbate when he is upset by the loss of his mistress, Echo. Pan later instructs the shepherds in the art of sex for one. The love poet Pacificus Maximus writes: 'Is there no boy nor girl to hear my prayers? No one comes? Then my right hand must perform the accustomed office.'

As for homosexuality, the Romans did not accord anal intercourse the same moral and institutional respect as the Greeks, but their enthusiasm for it appears to have been almost as fierce. Fellatio, curiously, was not greatly admired in Rome as it was seen as 'unnatural'. Some Romans, however, practised a form of fellatio in which the penetrating partner stayed still and declined to thrust his penis, leaving the receptive partner to do the work.

Poets meanwhile lauded anal sex, even while gently mocking well-known adherents, spreading rumours about who did and who didn't – and giving useful snippets of advice for the uninitiated. 'Stretch the foot and take your course, fly with soles in the air, with supple thighs, and nimble buttocks and libertine hands,' suggested Petronius in the first century AD. Curio the Elder alluded wittily to Caesar's apparent bisexuality by referring to him as 'the husband of all women, and the wife of all husbands'. Pacificus Maximus, in his 'Elegy II to Ptolemy', writes: 'For you, ungrateful boy, I keep my treasures all, and no one shall enjoy them but yourself; my penis is growing; while it used to measure seven inches, now it measures ten.'

One of the best measures of a society's progress towards sexual equality is contraception – whether it exists, whether it is legal, whether it is encouraged – and whether men and women, or just women, practise it. A British social historian, Keith Hopkins, suggested in a 1965 paper, 'Contraception in the Roman Empire', *Comparative Studies in Society and History*, that Roman men avoided large families principally by being unfaithful to their wives.

When wives felt under pressure to marry and have children,

and this made them feel worn out, Soranus, a first-century AD Greek physician who practised gynaecology in Rome, suggested that they abstain from sex periodically. But, as Hopkins points out, if your woman chose to abstain, the other way of continuing to have a sex life was to have sex with prostitutes, slaves – or men. Prostitutes tended to take steps such as interrupting coitus to prevent pregnancy but what might seem to be a worry with slaves – producing unwanted babies – was not a problem at all. The babies of slaves were raised like pets in Rome.

For women who did not want to give up their sex lives or have children, Soranus suggested the rhythm method, or the option of prolonging breastfeeding for longer than they might do otherwise. This has a contraceptive effect, and is an acknowledged birth-control method in some pre-industrial countries. In the ancient world, a woman was advised to continue feeding for around three years. To get round the practice, powerful Roman men often employed wet-nurses so their wives could start ovulating again and get back to the important business of producing heirs. Another Roman contraceptive method described by the poet and philosopher Lucretius was the custom of loose women to undulate their hips thus not only giving their partners maximum pleasure but also directing semen away from the uterus.

Lucretius, who lived from 96–55 BC, had another influential, and quintessentially Roman, belief that connected mutually orgasmic pleasure with successful conception. He argued that children resembled their parents because 'at their making the seeds . . . were dashed together by the collusion of mutual passion in which neither party was master or mastered'.

Heterosexual love? Mutual orgasmic passion? It was the kind of tender, romantic scenario that would have been distinctly alien to the Greeks. But by the second century BC, when the Romans invaded Greece, heterosexual, monogamous love was the modern, revolutionary way of living. Across a Mediterranean world about to be further revolutionised by

Christianity, the Roman legacy of the female citizen as joint head of a stable family was on its way to becoming a universal ideal.

9

Orgasm in the Orient

> 'Mr Simon Raven finds sex "an overrated sensation which lasts a bare ten seconds" – and then wonders why anyone should bother to translate the erotic textbooks of Medieval India. One good reason for doing so is that there are still people in our culture who find sex an overrated sensation lasting a bare ten seconds'
>
> Dr Alex Comfort, letter to the *New Statesman*

All over the modern Western world in the twenty-first century, men have begun looking for strategies to improve their lovemaking, both from their lover's point of view – and from their own. It is one thing to try to be a Casanova and please the ladies, another still to follow the lead of adventurous Victorian British sexual pioneers and study the Indian *Kamasutra* of 1,700 years ago, with its 529 sexual positions, or the Arabic *Perfumed Garden* of the 1500s, with a view to learning advanced techniques of stimulating your partner's clitoris and other erogenous zones.

But in the have-it-all twenty-first century, men started wanting orgasms as good as women's. All too aware that their sexuality tends to be focused on the ultimately disappointing goal of ejaculation, they wanted to be able to delay their climax for lengthy periods during sex, to enjoy more delicious and drawn-out orgasmic feelings when they did finally ejaculate – and to be able to repeat the performance. When a man becomes multi-orgasmic in this way, it is commonly held, he is able to satisfy himself more effectively, and also much more able fully to satisfy his partner. Becoming multi-orgasmic, twenty-first-century Western men now believe in large numbers, is one of the greatest gifts a man can give his partner.

The favoured source of how this goal might be achieved, of differentiating, as they believe possible, mere ejaculation from an altogether loftier form of 'dry' orgasm, and thereby enjoying more and better sex, is a society not everyone immediately associates with sexual virtuosity – not India, more commonly acknowledged as home of the famous 'Tantric sex', but Ancient China, where Tantric sexual practices traceable as far back as 6,000 years coalesced in a broad sweep of Chinese philosophy known as Taoism (aka Daoism).

Whereas in Indian sexual history little is documented before the third-century AD *Kamasutra*, there is documentary evidence that China was as sexually aware as any of the more westerly countries usually regarded as the cradle of civilisation. According to R. H. van Gulik, author of a 1961 book *Sexual Life in Ancient China*: '[sex] was never associated with a feeling of sin or moral guilt'. The defining feature of Ancient Chinese writings on sex is the emphasis on orgasmic pleasure for its own sake, irrespective of questions of reproduction. The subject of conception barely arises in the literature, so enraptured were the Chinese by the mechanical and emotional subtleties of lovemaking. 'The difference in subtlety between Ancient Chinese and modern Western views brings to mind the aphorism about the many different Inuit words for "snow",' comments the feminist author Naomi Wolf.

The Ancient Chinese divination book *Yi Jing* (I Ching) dates from the twelfth-century BC and contains a fire/water-based description of the orgasmic differences between men and women remarkably similar to the kind of concept the Greeks and Romans were musing on a thousand years on. 'Fire easily flares up,' says the *Yi Jing*, 'but it is easily extinguished by water; water takes a long time to heat over the fire, but cools down very slowly.' Thus the Chinese were yet another ancient people to make the observation, based equally on the time taken to reach the female orgasm, its ferocity and its near-infinite repeatability, that women are more carnal than men.

Five hundred years before Christ, and several hundred years before Hindu philosophy came up with their Tantric sex, Chinese devotees of the *Tao/Dao* ('The Way'), who studied human sexual response as part of a larger philosophical vision, claimed to have discovered a way for men to divert their orgasmic energy up the spine into the brain and back down again several times at will before triggering ejaculation, and also of 'retaining semen' while enjoying orgasms that did not culminate in ejaculation. A more elaborate version of withholding semen developed by Taoist masters was what is today termed 'injaculation' – a method of pulling semen up into the body in the reverse of the customary direction. The Taoists believe it is then absorbed into the blood.

According to Wolf, Ancient Chinese followers of the Tao were keen scholars of how to give their womenfolk as many orgasms as possible – but they had an ulterior motive. Their problem was that their male, yang energy, contained in semen, was finite and precious. In all Buddhist philosophy, to ejaculate without a purpose is a waste of spiritual energy. The yin energy women produce during sex, however, is boundless, and could in fact nourish a man's yang. Paying lavish attention to foreplay was additionally supposed to stir the yin, production of which peaked when women had an orgasm. It was ideal, furthermore, if the man remained inside the woman for as long as possible to absorb the maximum of yin. It is in this self-interested way, then, that the Chinese male came to view the female's orgasm as being as important as his own. (Some sages believed, in a rare acknowledgement of the reality of sex leading to conception, that the discharge of yang energy from the male is necessary in one circumstance only – successful insemination leading to pregnancy.)

As a *quid pro quo*, Chinese men explained to women that

sex with orgasm was beneficial to their health too, while unfulfilled sex could threaten a female's wellbeing. On the other hand, the Taoists also held that the more women with whom a man had intercourse, the greater would be the benefit *he* would derive from sex. But even if their culture was as male-dominated as that of any other ancient civilisation, the yin-yang concept ensured that Chinese women got a relatively good deal sexually. K'ung-fu-tzu, the fifth-century BC philosopher known in the West as Confucius, maintained that wives and concubines had sexual rights, and that it was a husband's duty to satisfy them. He suggested that intercourse once every five days should be enough to satisfy a woman under fifty.

As van Gulik implies, alternative sexual practices were not scorned. Fellatio was acceptable so long as not too much yang energy was lost, while cunnilingus was approved of both because it was pleasurable and because it accrued yin essence for the man. Their infinite supply of yin meant women were allowed to masturbate.

A kind of hyper-heterosexuality was the Ancient Chinese male ideal, however. To the middle-class Ancient Chinese male, the equivalent of playing nine consecutive eagles on the golf course was to attempt to have sex with nine women at the same time, making sure that all achieved orgasm, strictly controlling one's own ejaculation, and with lashings of cunnilingus all round. An Ancient Chinese husband had a solemn obligation to provide a regular sex life both for his four or five wives and even more concubines. With the latter, it was considered gentlemanly to have sex a minimum of once every five days. Husbands were allowed to retire from sex at sixty, but one sex manual of 79 AD states that if a man survives to the age of seventy, he should start again.

Anal sex was also regarded favourably and held in esteem as early as 500 BC, when it began to be referred to elliptically to as 'sharing the peach'. The tradition of gay male love in China continued to be strong until the early part of the twentieth century, when it began to be discouraged as part of the Westernisation of the culture. The new disavowal of the old ways survived both in Communist China and Taiwan, where it is, a little perversely, considered a heinous Western import and against traditional Chinese morals.

One of the Chinese culture's greatest claims to sexual fame is producing the world's earliest sex manuals – graphically detailed books which would be deemed pornographic by some today. The second-century pillow books, introduced for newlyweds, were just one example of this genre. They offered details of forty-eight different sexual positions, plus instructions for foreplay, oral and anal sex. Chinese erotica, which was really instructional material in the form of scrolls, novels and pictures, was also notable for its lack of insulting, misogynistic language or objectifying of women: the erect penis was the impressive-sounding 'Positive or Vigorous Peak', 'The Hammer,' the 'Heavenly Dragon Stem,' the 'Red Phoenix' and the 'Coral or Jade Stalk'.

The female pudenda were poetically named too. The clitoris was 'The Jewel Terrace', 'The Jade Pearl', 'The Golden Jewel of the Jade Palace'. The labia were 'The Examination Hall'; the vulva, 'The Golden Cleft and Jade Veins', 'The Open Peony Blossom', 'The Golden Lotus', 'The Jade Pavilion', 'The Palace', 'The Open Lotus Flower', 'The Receptive Vase' and 'The Cinnabar (or Vermilion) Gate'. Sex is 'Mist on the Mountains of Wu', 'The Meeting of the Dragon and the Unicorn', or 'The Clouds and the Rain'. An orgasm is 'The Bursting of the Clouds'.

Chinese sex manuals were expressly designed for both men and women. A poem by Chang, written around 100 AD, describes a young woman awakening her sexual desire on her wedding night by use of an erotic manual called *The Plain Girl*:

Let us now lock the double door with its golden lock
And light the lamp to fill our room with its brilliance.
I shed my robes and remove my paint and powder,
And roll out the picture scroll by the side of the pillow,

The Plain Girl, I shall take as my instructress,
So that we can practise all the variegated postures,
Those that an ordinary husband has but rarely seen,
No joy shall equal the delights of this first night,
These shall never be forgotten, however old we
may grow.

An erotic novel of the same period, *Jou P'u T'uan* (The Carnal Prayer Mat) by Li Yu, recounts the marriage of a scholar, Vesperus, to Jade Perfume, a beautiful and aristocratic but prudish young girl. She refuses to experiment sexually, or to make love other than in pitch darkness. The young husband persuades her to look at a sex manual, which she is soon perusing with enthusiasm. Her passion becomes 'greatly aroused' by the book, and her sexuality is awakened. *Jou P'u T'uan* has the distinction today of being banned by the Beijing authorities on grounds of indecency.

Naomi Wolf has written approvingly of both the language and female-friendliness of Chinese 'educational' pornography: 'When I read, as an adult, the Ancient Chinese erotic texts in translation, I felt oddly embarrassed,' she says in *Promiscuities*. 'The terms the Taoists used to describe women's genitals were metaphors of beauty, sweetness, artistry, rareness and fragrance . . . affection for women's genitals seemed, at first reading, hilarious, but also enchanting – like a life-enhancing comedy. Other Western women to whom I showed the Chinese translations had the same reaction.

'We should look at that response,' Wolf concludes. 'Just imagine how differently a young girl today might feel about her developing womanhood if every routine slang description she heard of female genitalia used metaphors of preciousness and beauty, and every account of sex was centred on her pleasure – pleasure on which the general harmony depended.'

Even in the highly sexed culture of Ancient China, sexuality had its special heyday. In the Han dynasty, from 206 BC to 221 AD, sexual desire, both male and female, was regarded as a powerful force of Nature, and female desire in particular, as Naomi Wolf puts it, 'was studied with the care that we now focus on the ecosystems which keep us alive and well'.

The most famous Chinese sex advice dates from the Han period, when Taoist philosophy was at its apogee. A modern Chinese sexologist, Jolan Chang, in a 1976 book, *The Tao of Love and Sex: the Ancient Chinese Way to Ecstasy*, distilled this ancient advice into a package accessible to Westerners. Much of it might be characterised as a distinctly feminine approach to sex. The point is made by Chang, as it is in another key modern Taoist sex guide, *The Multi-Orgasmic Man* by Mantak Chia, that the Ancient Chinese had no word for impotence. Taoism advocates a fail-safe technique that modern interpreters of the Tao call 'Soft Entry', by which a man can enter his partner when he does not have a bone-hard erection, or even when he is semi-flaccid.

Chang extracts from the Ancient Chinese texts specific instructions for better sex, four of which involve men learning how to arouse their female partner.

> He must know how to feel his woman's nine erotic zones.
>
> He must know how to appreciate his woman's five beautiful qualities.
>
> He must know how to arouse her so he can benefit from her flooding secretions.
>
> He should drink her saliva, and then his ching [semen] and her chi [breath] will be in harmony.

Chang also details ten signs of female desire that men ought to be able to recognise:

> She holds the man tight with both her hands. It indicates that she wishes closer body contact.

She raises her legs. It indicates that she wishes closer friction of her clitoris.

She extends her abdomen. It indicates that she wishes shallower thrusts.

Her thighs are moving. It indicates that she is greatly pleased.

She uses her feet . . . to pull the man. It indicates that she wishes deeper thrusts

She crosses her legs over his back. It indicates that she is anxious for more.

She is shaking from side to side. It indicates that she wishes deep thrusts on both the left and right.

She lifts her body, pressing him. It indicates that she is enjoying it extremely.

She relaxes her body. It indicates that the body and limbs are pacified.

Her vulva is flooding. Her tide of yin has come. The man can see for himself that his woman is happy.

The age of China's distinctive blend of sexual frankness and poetic lucidity did not come to a close with the end of the Han dynasty. In *Yufang Bijue* (Secret Instructions of the Jade Chamber), a Taoist text of several hundred years after Han, we can read of a technique for *coitus obstructus*: 'When, during the sexual act, the man feels he is about to ejaculate, he should quickly and firmly, using the fore and middle fingers of the left hand, put pressure on the spot between scrotum and anus, simultaneously inhaling deeply and gnashing his teeth scores of times, without holding his breath. Then the semen will be activated but not yet emitted. It returns from the Jade Stalk and enters the brain.' (Taoist masters would charge eager pupils vast sums for revealing the precise location of that key pressure point between the scrotum and the anus; it is sometimes described in translation as 'the million-dollar point'.)

Reay Tannahill, in her book *Sex in History*, paraphrases another typically juicy section from the seventh-century *Yek Tê-hui*, a sex manual by Master Tung-hsuan, a physician of the time: 'The Jade Stalk, he said, should hover lightly around the precious entrance of the Cinnabar Gate while its owner kissed the woman lovingly or allowed his eyes to linger over her body or look down to her Golden Cleft. He should stroke her stomach and breasts and caress her Jewel Terrace. As her desire increases, he should begin to move his Positive Peak more decisively, back and forward, bringing it now into direct contact with the Golden Cleft and the Jade Veins, playing from side to side of the Examination Hall, and finally bringing it to rest at one side of the Jewel Terrace. Then, when the Cinnabar Cleft was in flood, it was time for the Vigorous Peak to thrust inward.' Tung-hsuan, Tannahill adds, recommended the use of a penis ring, both to keep the Jade Stalk erect but also to stimulate the woman's Jewel Terrace during intercourse.

How sexually developed in antiquity, it may be wondered in the light of such advanced sexual material from China, was that civilisation's neighbour and great rival Japan? Given the more obscure, inward-looking and secretive Nature of ancient *Edo* compared to China, it is of little surprise that there is far less material on the sex life of the Ancient Japanese.

The most marked feature of the earliest religion of Japan, the Shinto cult, was animal worship, but evidence from various surviving erotic paintings show that even thousands of years before the geisha was heard of it, Shinto was profoundly hedonistic, as much so as the contemporary Greek culture. While early Chinese sexual manuals were mainly written for the health of male, old Japanese sexual texts centre more on the playful spirit required by both parties to enjoy sex.

Sexuality in Ancient Japan was celebrated through gods such as *Izanagi*, 'the male who invites'; and *Izanami*, 'the female who invites'. The deity *Kunado* was represented by a penis. Worship of the 'Heavenly Root' – the penis – was universal, with wooden and stone phallic symbols commonplace in town and country. They were thought to possess powerful healing and revivifying properties.

One custom was to dedicate beautiful young girls to the service of *Kwan-Non*, the Japanese Venus. The great temple of *Asakusa* was dedicated to her. Sacred prostitutes, in effect, they belonged to a religious order of nuns called *Bikuni*. It was a privilege to become a member of the *Bikuni*, girls being selected for their beauty and lovemaking prowess from all classes, including those employed in the more commercial brothels. Shinto temples, additionally, were the scenes of sexual orgies easily rivalling ancient Roman *Bacchanalia*.

Then there is the question of Japan's most famous contribution to the history of the orgasm, the egregious *Ben Wa* ('Joy') balls allegedly invented by a Japanese courtesan of unknown vintage, but called *Rino-Tama* and responsible for one of the staples of the modern sex shop. *Rino-Tama* it was who discovered that placing two marble-sized balls of suitable design inside the vagina could for a woman be like having a permanent portable vibrator in place. Hours spent moving about with *Ben Wa* balls installed could supposedly culminate in a subtle and discreet, yet impressive, mini-orgasm.

The earliest 'love beads', also known later as geisha balls, are believed to have been egg-shaped hollow balls carved from ivory. Subsequently, the casing would be made of gold or silver with a small weight – mercury in very ancient times – placed in the centre to roll around, creating rotating sensations within the vagina and sensitive surrounding tissue that could be orgasmic for practised women. The balls can be held low in the vagina or directly behind the G-spot. In the modern context, they are said to help to tighten and strengthen the pubococcygeal (PC) muscles, giving a better 'grip' during intercourse, as well as controlling the bladder and preventing incontinence with advancing age.

Some *Ben Wa* users swore by a specialised trick of the trade, which was to make just one of the balls hollow, the other ball solid. By their rubbing together in the vagina, a special kind of 'ringing' vibration would be set up, said to be still more conducive to a sly dose of orgasmic bliss on-the-hoof. Other women in Ancient Japan and later times have also enjoyed using *Ben Wa* balls for intercourse, and say their male partners enjoy encountering the smooth spheres during penetration.

Although the Chinese were the first Eastern people to bring advanced sexual knowledge into the public sphere with explicit sex manuals for the common man and woman, the original body of specialised, esoteric knowledge about sex, the Tantra, which later appears in Hinduism, Buddhism and Taoism, seems to have begun its development in Ancient India.

It is in the Hindu world, rather than in China, that the pursuit of the ultimate orgasm, irrespective of reproductive considerations, was revered as something close to a religious quest. The Ancient Hindus were more explicit than the hedonistic Greeks about pleasure. They considered earthly life had three distinct and equal purposes: religious piety (*dharma*), material prosperity (*artha*), and sexual pleasure (*kama*).

Every inquisitive schoolboy in the West in the twentieth century learned at some stage of the *Kamasutra*, and knew that in it the penis is called the *lingam* and the vulva, the *yoni*. But the *Kamasutra* is a relatively recent, third-century AD text – 200 years more modern, for instance, than Ovid's *Ars Amatoria*. The *Kamasutra's* most distant roots, however, can be traced to around 4,000 BC among the Harrapan tribe, which inhabited the area of present-day Sahiwal in the Pakistani Punjab.

The Harrapan worshipped femininity. Their goddess Shakti was represented in the form of a *yoni*. Sex was a way of combining male and female energies. Women were respected as well as revered, and rape was punishable by death. The growing sexual cult required males to do everything in their power to cherish and satisfy their wives. Impoliteness to women was banned, and neither could females be bought or sold.

The Tantric movement proper did not reach its height until about 700 AD, by which time the word had spread, suitably adapted by Tantric missionaries, to China. By this time Tantrism was embodied in a series of Tantra scriptures and had been colonised by Hindu sects, *Shaktis*, which venerated everything feminine. They were opposed by *Lingayatis*, who worshipped male gods. Tantrists came mostly from the middle castes, but their belief system was later appropriated by elite Brahmins. Isolated Tantric sects continued almost to the present day; one was studied in Bengal as late as 1980.

As in China, the most basic principle of Tantrism, drawn, it appears, from the folkloric belief in the super-carnal desires of women, was that men and women are like positive and negative in electricity – that energy flows via sexual intercourse from women to men. Shiva, the male god was thus often shown 'plugged in', in perpetual flagrante with the goddess Shakti.

Tantra defines orgasm as the blissful and indescribable result of interaction between the sexual potential of the two lovers, producing a polarisation of the bio-electric energies in the form of an ecstatic tension release similar to thunder. The orgasm produces in each of the lovers, separately or simultaneously, a profound feeling of contentment that has synchronous echoes in each plane of their being.

So fundamental to a healthy life and spiritual advancement was orgasm that some Tantric sub-sects required their monks and nuns to have sex as a religious duty. Temples dedicated to the goddess of love, Kama, were erected to celebrate Shiva and Shakti's lovemaking sessions. The inner sancta of such temples would be built to represent Shakti's *yoni* and kept permanently moist by a natural spring.

The femininity of the Tantra, as well as many of the sacred books of Hinduism, cannot be overstated; there is a real sense that the Hindu culture places as central to sex the woman's desire for safety, confidence, commitment and luxuriously extended physical fulfilment. Andre Van Lysebeth, a respected teacher of Tantric sex who spent decades studying in India, explains in his book *Tantra: The Cult of the Feminine*: 'For Tantra, each and every woman incarnates the Goddess, *is* the Goddess, Absolute Woman, the Cosmic Mother . . . A Tantrist worships the cosmic goddess Shakti in all women. In all women: fat or thin, young or old, able-bodied or disabled.'

The male Tantrist, for Van Lysebeth (and it is significant before getting too excited about the cult, that nobody mentions there being any female adherents), 'is able to "feminise" his sexual experience. To the ordinary male, sex is a convergent experience both in time and space – revolving around his sex organs and becoming progressively narrower in space and time. When the spasm is over . . . men's desire vanishes and men turn away from women, wounding their self-esteem.'

The Tantrist, however, 'does not make love to a vagina, but to a human being as a whole, i.e., the physical, psychic and cosmic woman, the incarnation of the cosmic Shakti . . . He intensely shares the Shakti's ultimate sexual emotion when she experiences a deep orgasm. This makes him aware of the sacred part of the woman, without trying to appropriate her body or her sex life. He does not think nor say, "This is my wife, her vagina is my property, I own her sensuality." He perceives sex as the manifestation of cosmic creative power, which is suprapersonal.'

The Hindu/Tantric tradition fostered an open attitude towards sex, as would ultimately be exemplified by the *Kamasutra*. But a great liberality, bordering on licentiousness, was in evidence long before that. Much of the continuing racy reputation of Tantric sex has its foundation in the practical eroticism of Ancient India. It is curious that Ancient China, with its equally sexy reality, has tended to be remembered in modern times more in terms of its martial arts and fireworks.

Precise directions are given by Tantric sex masters – part of the reason, perhaps, why male adherents in the modern era are often derided as the kind of men who like to read instruction books in bed. Sex must only take place when a woman is sexually excited. The goal is, for some, not ejaculating at all; for others, not ejaculating until the woman had one or more orgasms. As in Taoism, the longer the male can stay in some sense inside the woman, even if his penis is not erect, the more female energy he will absorb for his own benefit. This practice of *coitus reservatus* or *askanda* is sometimes portrayed in Hindu art by images of a flaccid or 'pendulous' penis.

As to what methods should be used to delay or block ejaculation altogether (*coitus obstructus*), Tantrists were advised to use meditation, self-discipline and manual intervention via 'the million-dollar point'. To avoid premature ejaculation, Tung-hsuan is at one with Indian thinking when he advises that at the last moment, 'the man closes his eyes and concentrates his thoughts: he presses his tongue against the roof of his mouth, bends his back, and stretches his neck. He opens his nostril wide and squares his shoulders, closes his mouth, and sucks in his breath. Then he will not ejaculate and the semen will ascend inward on its own account.'

It is not clear whether Tantrism was the tail or the dog in the sexual foment of Ancient India. A sect that utilises sex as a means to spiritual development sounds a little *avant-garde* by the standard of most societies, yet in India it is striking how almost every form of orgasmic attainment conceivable was venerated in antiquity by one sect or another. Fellatio, cunnilingus, prostitution, masturbation, anal sex, voyeurism, incest, transvesticism, masochism, coprophilia, bestiality and even, in rare cases, cannibalism, were acceptable to the holy men of some cult somewhere. Necrophilia was quite common; there are many images in Hindu art of the goddess Kali making love to the corpse of Shiva, and resuscitating him through orgasm. Tantric sex seems quite a tame interest against such a colourful background

It does not rule out gay sexuality, which is considered sacred and divinely ordained by many Tantric adherents. What harm, after all, if two males agree to share their sexual pleasure? They may not gain the essential feminine energy from the experience, but so long as they are bisexual they will not suffer lasting damage. The male's sacred spot in some Hindu scriptures lies between testicles and the anus, and is best invoked by a thrust of *lingum* in the anus. Many Tantrists aver that this gives a thousand times more pleasure than penetrating *i yoni.* Married women who indulged in lesbianism, however, were not tolerated by the early Tantric cultists. They were supposed to be punished by being shaved bald, having the two relevant fingers cut off and being led through their town on an ass.

Tantric sex is not necessarily synonymous with love and affection. In some ways Tantric sex is sex for its own sake. Removal of self is the central tenet of the Tantra, not devotion to any one other person. The important thing is that at the point of orgasm, the Tantric practitioner rises above 'self', the consciousness of one's own being. This is why ritual intercourse is an important part of Tantric sex for many practitioners. The term *Chakrapuja* (circle worship) refers to the basic religious ceremony of early Tantrism. The guru conducting the ceremony was charged with preventing it from becoming an orgy, but the event was still essentially about sex with strangers.

Such evenings were lubricated by wine or hashish, after which the small group of couples moved forward to feasting and then lengthy, promiscuous intercourse. For the truly dedicated Tantrist, the next step towards enlightenment was a ceremony that culminated in ritual intercourse with specially trained women known as *dakinis.* The ultimate orgasmic bliss on this continuum, the closest approximation to union with the divine, was described, as 'intercourse with oneself', in which the same feeling as sex with a woman could be reproduced as a soloist. This was the entrance to a new state of bliss in which the male Tantric practitioner, though a man and therefore weak and spineless, could be freed from dependence on women.

Heterosexual, mainstream and recommended for married

couples, the *Kamasutra* – it means, 'treatise on sexual pleasure' – was very much the late-to-market, mass-consumption version of India's lubricious culture. It is, nevertheless, the most famous sex book of all time. It is astonishing that the World Wide Web contained nearly a million references to the *Kamasutra* in 2003, as opposed to less than a hundred for the equally explicit Chinese pillow books of the same period.

The *Kamasutra* is an edited collection of Indian writings, both spiritual and practical, going back hundreds of years from the time of the author, one Mallanaga Vatsyayana, about whom nothing is known other than that he lived in the city of Benares on the Ganges, he was upper-middle-class, and assumed his readers similarly had servants – and took himself a little too seriously; he states at the end of the *Kamasutra* that he wrote the work 'in a state of mental concentration and chastity'.

The *Kamasutra* was composed in Sanskrit for precisely the same people as Dr Alex Comfort would aim at in *The Joy Of Sex* 1,700 years later – young, educated, broad-minded urbanites, known in Vatsyayana's India as *Nagaraka*. Despite the highly detailed illustrations with which the book is generally published, it was not a sex manual, unless you were a Yogic contortionist – athletic, often implausible, sexual positions are what the *Kamasutra* is most famous for. Neither was it a work of pornography. Graphic sexual paintings were common in Indian culture; wealthy Hindu husbands would commission pictures of themselves having sex with their wives as routinely as eighteenth-century Europeans had themselves painted in formal poses with their hunting hounds.

Unless one was a Sanskrit scholar, it has only been possible very recently to read the *Kamasutra* free from the androcentric language and prejudices of Victorian England. The standard translation has always been that published in 1883 by Sir Richard Burton, a daring progressive in his time, but very much of that time when, broadly speaking, only male sexual desire was acknowledged.

In 2002 a new translation appeared which had almost the

same clarifying effect on our view of the sexual life of long ago as when astronomers saw the first images from the Hubble space telescope. The new translation by Wendy Doniger, Professor of the History of Religions at the University of Chicago, and Sudhir Kakar, a leading Indian psychoanalyst and a senior fellow at Center for Study of World Religions at Harvard, both flatters the sexual democracy of Ancient India – and lays bare its sometimes shocking male prejudice.

It is now clearer that the *Kamasutra*, as an avowedly popular, mass-market book, did not attempt to place itself at the leading edge of sexual knowledge; it was more a conservative, accessible guidebook, a compilation of what all sensible, normal people should know about sexual pleasure from the accrued wisdom of the ages thus far. It is not state-of-the-art, certainly not experimental, and cannot have seemed particularly progressive to the real sexual gourmands of fourth-century India. In modern terms, the *Kamasutra* is more at the suburban level of the *Daily Mail* or *USA Today* market than the *Playboy* or even the *Cosmopolitan* reader – although Vatsyayana was far more liberal about both sexes having affairs outside marriage than would be acceptable in suburbia today. The view on fidelity from fourth-century India was that it is inevitable that couples tire of one another sexually, and so long as they are discreet, the odd sensuous fling with another man's wife or another woman's husband was acceptable.

The mixture of advanced positional gambits and geometric advice as to the relative coordinates of *lingam* and *yoni*, all of which excited twentieth-century young men in the West, is surprisingly bland stuff. Vatsyayana gave no credence to the reality that most women can never reach orgasm solely from intercourse, no matter what Tantric heroics the man goes to in prolonging his thrusting into the small hours. The *Kamasutra* also seems to disapprove of oral sex: 'It should not be done because it is opposed to the moral code,' he says, explaining that during intercourse, according to scripture, a woman's mouth is an exceptionally holy place, and should not be defiled by fellatio. If men must enjoy it, the *Kamasutra* says, it should be with 'loose women, servant girls, and masseuses' with whom a man 'does not bother with acts of civility'. Cunnilingus is also rather damned with faint praise by Vatsyayana: 'Sometimes men perform this act on women, transposing the procedure for kissing a mouth.' (He does go on nonetheless to give detailed and elaborate instructions for what he calls 'sucking the mango', and even notes without comment that some men even enjoy sucking each other's mangoes.

There is no unambiguous mention in the *Kamasutra* of the clitoris, and certainly nothing on the arcane matter of female ejaculation. Doniger and Kakar believe the clitoris, or even the G-spot, may have been alluded to, but only elliptically. Burton's *Kamasutra* refers vaguely to 'the part of the woman's body which should be touched while making love', specifying that the man, 'should always make a point of pressing those parts of her body on which she turns her eyes'. Doniger says this is wrong, but what the Sanskrit more accurately says is not a great deal clearer: 'What the text says is that when a man is inside a woman and touches her and when her eyes roll around, he should touch her more in that place,' Doniger has explained.

On female ejaculation, the sources Vatsyayana used were venerable to the point of being thoroughly out of date. For the final word on the female ejaculation question, for example, he turns to Svetaketu, a Vedic sage of over a thousand years earlier: 'Females do not emit as males do,' Svetaketu opined, and Vatsyayana echoed. 'The males simply remove their desire, while the females, from their consciousness of desire, feel a certain kind of pleasure, which gives them satisfaction, but it is impossible for them to tell you what kind of pleasure they feel. The fact from which this becomes evident is that males, when engaged in coition, cease of themselves after emission, and are satisfied, but it is not so good with females.' (It is probable that more progressive authorities were already aware of female ejaculation. It is clearly described in the seventh century in a work of the poet Amaru called the *Amarusataka*.)

Naomi Wolf, again, has reported positively on the *Kamasutra*'s esteem for women, and it is undeniable that part of the book is expressly written for them, which would be admirable if women of the time had been taught to read, which they rarely were. The bulk of the text, Wolf argues, shows respect for women's sexuality 'in ways that would be foreign to frat boy culture'. She quotes from Vatsyayana: 'For a man to be successful with women he must pay them marked attention . . . Do not unite with a woman until you have excited her with playful caresses, and then the pleasure will be mutual.' Women, in other words, Vatsyayana was saying, are as sexual as men, and men should strive to provide their partners with erotic pleasure, orgasms included.

Imaginative lovemaking, Wolf further observes from the text, is important because it generates 'love, friendship, and respect in the hearts of women'. Seduction of a virgin on a wedding night, 'includes the man's gently shampooing the woman's limbs, one by one, as she gives him her consent to do, and it involves not just talking to her but asking her questions and listening to her'. Sex toys and sex while bathing are encouraged. Vatsyayana also warns men against pressing their advantage too strongly: 'A girl forcibly enjoyed by one who does not understand the hearts of girls becomes nervous, uneasy and dejected, and suddenly begins to hate the man who has taken advantage of her.'

By Doniger and Kakar's account, the early Hindu sexual culture Vatsyayana describes was also surprisingly like our own. Young women of status are recognised as full, sensual participants in sex, free to date men and select their husbands. Vatsyayana deals with the skills required of mistresses, pragmatically suggesting they should look after their own future by a little harmless theft from their patrons. Married women, in his world, are also at liberty to take lovers. Vatsyayana tells virgins how to attract husbands and insists that men learn ejaculatory control: 'Women love the man whose sexual energy lasts a long time, but they resent a man whose energy ends quickly because he stops before they reach a climax.' He also instructs men to ensure women are treated by skilful use of foreplay in such a way, 'that she achieves her sexual climax first'.

But Wolf, who was writing before Doniger's revised translation, seems to pass over some of the less female-friendly material in the *Kamasutra*. Part of the book contains tactics to seduce a maiden, so long as it is done graciously, and precautions to be observed when seducing another man's wife. There is instruction on how two or more men can share one woman – men should take turns, one inside her, the other fondling her. (There is also, as a balance, advice on how two women can share one man at the same time; to satisfy both, he must fondle one while having intercourse with the other.)

But then there is an awkward question in the *Kamasutra* of the apparent legitimisation of rape. If a virgin is unwilling to have sex, men are counselled to have a brother ply her with alcohol. 'When the drink has made her unconscious, he takes her maidenhead.' Such rape is acceptable for other women, too: 'A man may take widows, women who have no man to protect them, wandering women ascetics, and women beggars . . . for he knows they are vulnerable.'

On safer ground, but only just, the *Kamasutra* encourages rough sex. Slapping, spanking, scratching and biting are all recommended orgasmic aids, and all resultant shrieks and moans are measures of increased passion rather than pain. Hitting and hissing are explicit ways in which men and women show how they are progressing towards orgasm. Females in particular should hit their partners' hands, shoulders, chest, thighs and sides of the body with their fists, palms and fingers during foreplay, but not intercourse. 'There are no keener means of increasing passion than acts inflicted by tooth and nail,' Vatsyayana says in Doniger's translation. 'Passion and respect arise in a man who sees from a distance a young girl with the marks of nails cut into her breasts.' He also advises that words spoken in the pre-orgasmic frenzy can mean the opposite of what they say: in this context, 'stop' and 'enough' mean 'don't stop' and 'more, please'.

Shaking of the woman's hands, according to Vatsyayana, is the first symptom of her reaching orgasm. He says she will feel a rush of profound love and will not want her partner to withdraw, even though he has ejaculated. After orgasm, both the partners should feel a joyful sense of fatigue and no sound other than heavy breathing should be heard. The partners should not immediately part. By the man staying in the woman and both feeling their partner's continuing presence proper intimacy is better achieved. Even after they separate, the partners should sit together, talk lovingly and eat.

If they return to bed to have sex again, they should reverse roles, the woman taking the initiative by starting the touching, stroking and kissing and other foreplay and finally mounting the man. This provides relief to the tired male and helps him in the harder task of regaining his desire for orgasm. It can also be adopted at other times for a change of sexual experience.

What kind of sexual climate, then, did the *Kamasutra* create in India? Was there any equivalent to the cold wind of asceticism and abstinence that blew through the parts of the world that adopted Christianity in the early part of the first millennium AD?

A Swiss Sanskrit authority, Professor Johann Jakob Meyer, examined sexuality in the India of around 500 AD in a 1930 book, *Sexual Life in Ancient India*. He attempted, albeit in a curiously tendentious style for an academic work, to give an account of the lives of women in Ancient India, based on the information on the relationship between the sexes that he found in the two great Hindu epics, the *MahaBharata* of around 350 AD and the *Ramayana*, an older text but one whose popularity, which persists today, was at its height at around the same time.

Without the delights of love (according to Meyer), the woman pines. 'The enjoyment of women is, countless times in Indian literature, praised as the most glorious thing in heaven and earth, as the one meaning and end of living, or anyhow of the years of youth,' Meyer observed. He quotes a proverb to the effect that: 'The power of eating in the woman is twice as great as the man's, her cunning four times as great, her decision six times as great, and her impetuosity on love and her delight in love's pleasures, eight times as great.'

'Eastern literature in general shows the woman as being resolute, full of fire, passionate in comparison with the often so slack and sinewless man,' Meyer writes. 'The characteristic comes out first of all in love. The fire of the senses is here also in the Indian doctrine far stronger in the woman, and it is not for nothing that the tender, timid sex so often takes the leadership in the Indian love tales, especially when it is a case of the fair one bringing about the tryst for the delights of the sexual union that they desire as soon as possible.

'The ascetic, too, often sees the love of many and lovely women, shining before him as the goal and reward in the world beyond, or in a future incarnation . . . For, to the woman, there is never anything higher than sexual union with the man; that is her highest reward. Driven on by love, women live after their own appetites.' Holy men, such as priests and kings, were accordingly allowed in some areas to relieve a young girl of her virginity.

Copulation, Meyer explains, was for the Indians one of the very highest things leading to perfection. 'In the Indian view the woman has also a far stronger erotic disposition, and her delight in the sexual act is far greater than the man's.' However: 'Means of heightening the sexual powers of the man, so much even that in one night he can satisfy a thousand women, are known in the old Indian literature in great abundance.'

Meyer includes many factual nuggets. 'The women in Bengal are known for beginning with their mouth the business of the vulva to excite the man's desires, owing to their excessive craving for the joys of love. Such a thing is, of course, strongly condemned by the law books, and punishable by death as "unnatural intercourse".' Mainstream Hinduism regarded oral-genital contact as a sin that could not be expiated in fewer than a hundred reincarnations. Yet some other Indian religions, according to Meyer, tolerated fellatio, while others still actually incorporated oral-genital contact into their rituals. Fakirs or monks in some parts of India had devotees fellate them. One erotic manual of the period around which Meyer studied included an eight-step plan to be used by eunuchs when performing fellatio. Other eunuchs, in areas where men maintained harems, were required to perform cunnilingus.

Masturbation was not encouraged in Ancient India, however. Even involuntary shedding of seed – nocturnal emission – had to be atoned for. As for what to do when overwhelmed by sexual feeling, Meyer records that Brahmins were advised: 'If passion arises in him, then let him undergo mortification. If he is in great erotic straits, then let him put himself in water. If he is overwhelmed in sleep, then let him whisper in his soul thrice the prayer that cleanses sin away.'

As for the legacy the Ancient Indian sexual mores have left for the modern-day populations of the subcontinent, there is much insight to be gained again from Yasmin Alibhai-Brown's essay 'Why East Beats West When it Comes to Sex'.

'As a British Asian,' Alibhai-Brown writes, 'I come from nations where the art of sexual pleasure was so central that for hundreds of years it was enshrined in religious iconography, in poetry, art and dance. Everything that the human imagination and body has ever craved or tried is represented in the ancient sculptures of India, delicate Indian miniature paintings and the *Kamasutra* of Vatsyayana, which was translated and published in Britain in the late-nineteenth-century by two members of a secret sex club known as the Karma Shastra Society. The translators, an ex-British Army officer and Arabist, Richard Burton, and a retired Indian civil servant,

Forster Fitzgerald Arbuthnot, saw their treatise as an essential education for English men who had only ever learned "the rough exercise of a husband's rights" without understanding the need for a wife to feel passion too and to participate as a willing part of a coupling rather than a victim of it.'

Alibhai-Brown explains how Eastern garments are designed to be modest, yet easily removable. The *sari* and *shalwar khameez*, according to an Indian Muslim cultural historian, was supposed to enable young married couples to have sex in the fields during lunch breaks. An eighteenth-century Persian lacquer box Alibhai-Brown saw in a Sotheby's catalogue illustrates this tradition well: 'Two apparently fully-clothed lovers are shown locked in a sexual embrace on a carpet in a wood. She appears to be the more energetic and determined partner and he is obviously enjoying his good luck. Two musicians play for them and a maid plies them with wine. Even as you spy upon this private act conducted in the open, their clothes lock you out and grab back the intimacy from your inquisitive eyes. It is this play between giving and withholding, between the explicit and the implicit that creates the excitement.'

Both Urdu poetry and Hindi songs, she writes, are obsessed with *ishq* (passion) and the devastation it can cause. 'My mother's favourite song is a tease about the red mark on the *chunni* (scarf) of the bride and the way she is glowing after her first night of sex. These traditions have long influenced sexual relationships and fantasies in the Indian subcontinent and the Middle East.'

10

Faith, Hope and Chastity: Orgasm in the Early Christian World

'Grant me chastity and continence, but not yet'
St Augustine of Hippo

At the same time as Vatsyayana was busy denying himself by the Ganges to clear his head for writing the *Kamasutra*, St John the Divine was on the island of Patmos working on the *Book of Revelations*. The chasm between the two works' aims, preconceptions and their view on the desirability of sexual pleasure could not have been wider. After the decadent sexual free-for-all of Ancient Egypt, Greece, Rome, India and China, the Christian church had spent the three centuries since Christ's death taking the Jewish codes and making them trebly harsh, with sex and orgasmic pleasure in particular always among the primary targets of the grave, new morally 'cleansed' world.

The Christians did this in the name of modernity, setting in train a kind of perpetual revolution in sexual mores that intensified through the centuries. Jesus Christ himself seemed to have no views one way or the other on sexual pleasure. When he entered the temple looking for sins that would damn the soul of man, sex was not among them. According to Mark, he spoke out against adultery and condemned divorce, characterising both it and remarriage as licentious actions. But this was virtually all. Yet after His early death, His followers tried to extrapolate what their master's thoughts *might* have been on the matter. Hedging their bets and possibly acting in error of Jesus's real beliefs, they led a large proportion of humanity forward into a brave new world of sexual repression, hypocrisy and guilt.

Nobody can say how sincere the new morality's originators were. St Paul, as an ethnic Jew of the sexually enthusiastic Hebraic tradition, may well privately have had a soft spot for sexual pleasure. But he made sure to develop, or at least to profess, a thoroughly modern fondness for celibacy. 'To the unmarried and the widows,' he wrote in *Corinthians*, 'I say that it is well for them to remain single as I do. But if they cannot exercise self-control, they should marry. For it is better to marry than to be aflame with passion.' Tertullian (second century) went a step further, calling sex shameful, while for Arnobius (third century) it was filthy and degrading, for St Jerome (fourth) unclean (he likened the human body to a darkened forest filled with roaring beasts only controllable by diet and avoidance of sexual attraction), St Ambrose (fourth) a defilement (he thought sexuality an ugly scar on the human condition), and St Methodius (nineth) unseemly.

So in the early Christendom after Jesus of Nazareth's execution, sexual abstinence for both men and women was conceptually anchored to God; sexual indulgence to sin. In the same way as prudishness would one day be the Victorian equivalent of political correctness, abstemiousness was the modernity of the new, post-classical era following the Crucifixion. Successive standard bearers of the modernistic Jesus cult would duly clamour to out-do one another in proclaiming the virulence of their distaste for the sinful, guilty phenomenon of sex, or – given that it was a disgusting job but one that someone had to do – their outright opposition at least towards anybody so spiritually unclean as to *enjoy* it. In an increasingly intellectual, ascetic Christian world, it seemed strange that God could not have invented a less disgusting and animalistic, a more *enlightened*, method of procreation.

The writings of these cheerleaders for self-denial drummed home the PC message for century after century that the sole purpose of sex is procreation; that intercourse is a tribulation necessary for the production of babies, but absolutely no cause for joy or pleasure, and best avoided by serious people. The consistency and cross-cultural spread of these ideas is quite amazing.

Here is Tertullian, a native of Carthage in North Africa, pontificating in his *A Treatise on the Soul* around 200 AD: 'I cannot help asking whether we do not, in that very heat of extreme gratification when the generative fluid is ejected, feel that somewhat – our soul has gone out from us? And do we not experience a faintness and prostration along with a dimness of sight? This, then, must be the soul-producing seed, which arises from the outdrip of the soul, just as that fluid is the body-producing seed which proceeds from the drainage of the flesh. In a single impact of both parties, the whole human frame is shaken and foams with semen, in which the damp humour of the body is joined to the hot substance of the soul.'

Then there was Origen, an early third-century theologian in Caesarea, Palestine, who was so taken with Matthew's mention of 'eunuchs, which have made themselves eunuchs for the kingdom of heaven's sake' (a metaphorical reference to those who choose celibacy for religious reasons) that he castrated himself as a young man. (Other self-styled divines developed a habit of burning off their own fingers in order to resist sexual temptation.) Abba Cyrus of Alexandria, a third-century Copt, wrote of sex: 'If you do not think about it you have no hope . . . he who does not fight against the sin and resist it in his spirit will commit the sin physically'. St John the Dwarf writes at the same time in Egypt's western desert: 'When one wants to take a town, one cuts off the supply of water and food. The same applies to the passions of the flesh.

If a man lives a life of fasting and hunger, the enemies of his soul are weakened.'

It is at this time that the First Ecumenical Council of the Catholic Church, the Council of Nicaea, ruled that celibacy was required of bishops. A hundred years later, Christianised Roman emperors such as the joint rulers Constantius and Constans were keen to follow a similar legislative path. In 342 they prohibited sexual relations between man and wife in any fashion that did not involve penetration of the vagina by the penis. Their intent was obviously to outlaw anal and oral sex. They had the backing of contemporary divines such as St Jerome, who opined that all sexual intercourse was impure, and advised young women to avoid hot bathing, because it 'stirs up passions better left alone'.

It was St Augustine of Hippo in Egypt, who lived from 354–430, who most coherently crystallised the belief that sex was fundamentally disgusting, and spread this teaching across the Christian world. As a young man, Augustine was eager to convert to Christianity, but he could not overcome his lust for his mistress and debauched lifestyle. Retrospectively appalled by his own experiences, he wrote his *Confessions*, the classic book of Christian mysticism and the story of his conversion. In it he confessed to having prayed regularly to God, saying famously, 'Grant me chastity and continence, but not yet.' Augustine's new-found loathing of earthy bodily reality was exemplified by his statement that 'we are born between faeces and urine'. He came to see sex as vile, lust as shameful, and all acts surrounding intercourse as unnatural. He regarded celibacy as the highest good. And adultery, by men as well as women, was naturally the Devil incarnate. Adam and Eve's fall, Augustine opined, was loss of control over the body, especially the penis. If sex had one purpose other than reproduction, Augustine believed, it was to combat male fornication.

After St Augustine came more detailed codification of self-denying lifestyles, as mankind's sexy, prelapsarian past faded into distant obscurity. John Cassian, a monk and ascetic writer from Southern Gaul, wrote in his *Collationes* around 420 that there were six degrees of chastity. The first consisted of not succumbing to the assaults of the flesh while conscious. 'In the second,' he specified, 'the monk rejects voluptuous thoughts. In the third, the sight of a woman does not move him. In the fourth, he no longer has erections while awake. In the fifth, no reference to the sexual act in the holy texts affects him any more than if he was in the process of making bricks, and in the sixth the seduction of female fantasies does not delude him while he sleeps. (Even though we do not believe this to be sin, it is nevertheless an indication that lust is still hiding in the marrow.)'

A modern psychoanalyst would doubtless have a field day with the early Christian sex-phobes. It is tempting to suggest that there was an element of refusal to lose control with these devoutly abstemious men. Among modern celibates, this seems almost expressly to be a motivation. Salvador Dali refused to have orgasms, either through sex or masturbation, arguing that to do so would deprive him of his artistic power. 'My brush is my penis and my paint my semen,' Dali would say. Balzac would say after every orgasm, 'There goes another novel.' And as late as 1982, Peruvian soccer fans would blame a World Cup defeat by Poland on players who broke a pre-match sex ban.

Oddly enough, celibacy was not imposed on the early Christian clergy, as opposed to monastic celibates, until the Second Lateran Council of 1139, and then only for financial reasons. The Church's idealisation of celibacy derived originally from the supposed virginity of Christ and his mother, but priestly celibacy was in fact introduced by Pope Innocent II principally to stop married priests handing down Church property to their heirs. The clergy themselves, used to having wives and mistresses and passing on both goods and priestly power within the family, still tended to resist the new view that they should become ascetics and encourage their flocks, who were also clearly copulating far more than the necessity of mere procreation demanded, to regard sex as disgusting.

The drive to enforce virginity on the entire clergy was not successful for another two hundred years.

The Church's more principled loathing for sex was not supplanted by business considerations although it was rarely more eloquently expressed than by Odon, Chancellor of Notre Dame in Paris, who in 1169 proclaimed sexuality to be no less than the principal means by which the Devil secured his hold on the world. But it should be noted that Saint Thomas Aquinas, whose theories and writings have become the cornerstone of the Roman Catholic Church's attitude to sex, arrived remarkably later on the sexual repression scene.

It was not until the thirteenth century that Aquinas expressed his robust opinion on sex, which he called 'lust' unless it was strictly procreational. His top four perversions were bestiality, homosexuality, using any sexual position other than the 'missionary', and masturbation, which he condemned, although only insofar as he considered it effeminate in men. He was surprisingly tolerant of prostitution, which he saw as a necessary adjunct of morality, an outlet for troublesome male libido. 'A cesspool is necessary to a palace if the whole place is not to smell,' he said. Arguing against prostitution for a churchman of the time was a little like a sailor complaining about the sea; it was a financial mainstay of the Church, as well as one of the best ways a priest's male parishioners had of ensuring that he would not sleep with their wives. Priests were wily when it came to withholding absolution to get induce wives to sleep with them, as well as compelling fornicators to name their partner, in order to get, so to speak, a hot tip.

The intense official anti-orgasm propaganda hit home, nonetheless. In a very real sense, these strictures remain the consensus today, and not just in the Christian world. Just as in the twentieth century the dominant American culture was adopted in part even by its enemies, the other major religions and cultures were also dragged into the fashion for prudishness. The extreme misogyny and abhorrence of sex in modern fundamentalist Judaism and Islam, a religion which, it is often forgotten, developed after Christianity, seem to have a curiously Christian momentum behind them.

Would modern Orthodox Jews, in their bizarre, wholly non-Hebraic medieval European garb, be so confused about sex (large numbers of their men are aggressive prudes, yet major users of prostitutes) if it were not for the Christian ethic influencing Judaism in the Middle Ages? Would modern fundamentalist Muslims (avowed woman-haters, yet desperate for female company when they visit the West) hold the views they do if Christianity had not poisoned the well of sexuality by associating it so strongly with guilt?

Even today's Indians and Chinese appear to have revolted against their sensual roots as if in sympathy with the ascetic momentum of Christianity, the notion that the renunciation of pleasure, self-flagellation and the pursuit of a monastic lifestyle are the surest ways of entering a state of grace. Some of the modern world's most abbreviated foreplay occurs in rural China, where 34 per cent of couples are now said to spend less than a minute on sexual preamble and oral sex is almost unknown, the modern Chinese, even in sophisticated cities, regarding it as 'too dirty'. Researchers have found a significant proportion of Chinese girls and women today say that they 'feel like vomiting' when questioned about sexual matters.

Communism may have been to blame here. China may be one of the last communist states remaining today, but it imported, via the USSR, an ideology with distinctly Christian Western roots which, amongst a lot else, redefined sexual pleasure as decadent. Accordingly, from the 1950s to the 1970s there was no sex education at all in China for a population of eight to nine hundred million people. The only sex manual, *Xing-di-zhi-shi – Knowledge of Sex* – was published in 1957. It was devoted almost entirely to love and marriage.

Whether India can blame such indirectly Christian influences as the British Raj is not clear. But just as in China, in India modernity has, perplexingly to Westerners, become congruent with increased coyness and reticence regarding sex. There has, to adapt an expression of Professor Lionel Tiger's, been a quite startling decline in the country's gross national eroticism: there is widespread disapproval of the erotic possibilities of married life. Sexual taboos are still so powerful in some Hindu communities that higher caste women do not have a name for their genitals. In modern India, the Mumbai film industry is only now reverting cultural norms to a more erotic – sometimes downright raunchy – model.

The orgasm's story from Aquinas onwards is to a large extent the struggle between a near-global orthodoxy that sexuality is a regrettable, animal characteristic (paradoxical since animals rarely and barely experience orgasm) – and occasional cultural blips when sex asserts its old power again and begins to re-establish its ancient primacy.

Given how physically pleasurable orgasm and its preamble are, the Christian achievement in bending the collective human mind away from thoughts of sex as anything beyond an occasionally necessary bodily function was no small feat. Sexual matters were far from ideal or liberal in the Ancient Mediterranean and Eastern worlds, especially when it came to the rights of women, but there is a huge gulf between the world view of Ancient Greek women comparing notes over who made the best masturbating sticks, or Ancient Hebrews believing sex is the holiest experience and undertaking known to man, and the subsequent generations of Christians who were successfully conditioned to be perfectly righteous in their perverse belief that sex and its entire hinterland of affectionate and pleasurable behaviours is a contravention of God's will, that only intercourse without a 'lustful appetite' was acceptable.

Neither, in the absence of any unambiguous sexual directive or taboo having been imposed by Jesus, was there was a clear prime mover for the establishment of the Christian sexual ethic, no inventor of abstinence, self-denial and censoriousness as a virtuous way of living. In the Roman world, before it became Christianised, there had been an anti-decadence voice in one Musonius Rufus, a Stoic of the first-century, who created the modern marriage ideal. Christianity appreciated his sentiments so much that it borrowed them wholesale. They meshed perfectly with the emerging Christian idea that human beings would sacrifice living in Paradise if they succumbed to weakness of the flesh. (Islam later followed this pragmatic approach more directly; Paradise was expressly promised for those who kept sex from getting out of control on Earth.)

Musonius said marriage should be no more nor less than communion of souls with a view to producing children. His ethical doctrine, for those who bought into it, imposed something that had not really been seen since Neolithic days – a measure of sexual equality. And therein may lie the root of Christianity's fervour for new sexual ways. Neither partner, in Musonius's vision, was allowed to have sex outside or before marriage. The rules were also strictly the same for rich and poor. The wealthy he sees as 'the most monstrous of all, some who do not even have poverty as an excuse'. Here then, we start to see an extension of the recently dead Christ's spiritual, if not economic, communism into the sexual area. Except that there is no notion from Musonius of equality of sexual gratification. For the first time, sex was only to be 'indulged in for the sake of begetting children', never for fun.

Given, as we have seen, the Old Testament's easy acceptance of a veritably throbbing carnality, plus the New Testament's blithe lack of concern with the matter, the Christian difficulty with sex is even odder. It extended, what was more, to all forms of pleasure: the third-century Lebanese philosopher Porphyry, regarded by St. Augustine as the father of Christian morality rather than Musonius Rufus, condemned not only sex, but horse racing, the theatre, dancing, marriage and mutton chops, averring of the latter that, 'those who indulged in them were servants not of God but of the Devil'. (In a bizarre footnote to the history of sexual abstinence, there exists to this day a vegan group in Christchurch, New Zealand, called Porphyry's People.)

In the shining, sexless new Christian world which Musonius ushered in, the obsolete, yet still extant, holy books of the old Hebrews became something of an embarrassment. In their drive to understand human beings as a finer, more cerebral, more moral form of wildlife than they were in the past, the writers of the Gospels, the radically-minded *post-mortem* historians of Jesus, had a problem with antique, but still undeniably holy, relics such as the 'Song Of Solomon.' It could hardly be struck out of the Bible without offending God, who must have had his reasons for allowing such muck to be published in his name. So instead it was reinterpreted as a metaphor of Christ's love of his Church, just as the more hardline Jews rebranded such verses as an explanation of God's love of Israel. The story of Onan was similarly adapted for the new creed's benefit by being used fallaciously to denounce masturbation, especially by the male, who has always, in line with Musonius's idea of a repolarisation of the sexual status quo, been condemned more vigorously for sexual sin than the female.

These modern spiritual regimes across the Christian world triggered a form of what can only properly be regarded as madness, spiralling down the generations, as bodily pleasure (or, at least any admission of indulging in bodily pleasure) became progressively more taboo. The prohibitions were so numerous – no sex (or, needless to say, masturbation) on Sundays, Wednesdays and Fridays, which effectively removed five months in the year. Then it was made illegal during Lent (the forty days before Easter), during any penance, during the forty days before Christmas, for the three days before attending Communion, during Saints' days and from the time of conception to forty days after giving birth, or until the end of breast feeding in some cases, which ruled sex out for at least a year. It seems the Church elders wanted ideally to turn all Christians into semi-professional monks. But even then, they needed to take care; if a man experienced a nocturnal emission, he was required to intone thirty-seven psalms on awakening. The absurdity of all this was ably mocked by the Italian poet Boccaccio in his *Decameron.*

The veto on masturbation was just as puzzling. Here was a harmless act that did not lower the value of a woman, did not break either her hymen or her heart, and did not produce unwanted, illegitimate children. All masturbation does is to produce an orgasm. If any proof were needed that it was the pleasure of orgasm more than, say, false intimacy or illegitimacy that was being targeted by the killjoy Christians, and would continue to be right up until the present day, it is surely the unfathomable obsession with orgasm.

Like all bad political policies, the drive to marginalise sex and those who enjoy it threw up conundrums for the policy makers. For example, absolutely all practices – diverse sexual positions, oral sex, anal sex – that a man and woman might discover they enjoyed were outlawed. But at the same time, a second cardinal rule in almost direct opposition to the first had to be instilled in the public.

Because marital sex may have been a shocking thing, but was not half as shocking as adultery, the Church found itself having to insist strenuously on the concept of 'marital debt'. Married people were told, confusingly, that they *must* grant sex to one another on demand. Other isolated incongruities in the monolithic Christian party line on sex can be found. In Christian Byzantium, for example, it was believed that a woman's erotic pleasure could positively determine her baby's health and temperament. But the overwhelming weight of belief was on the side of sex being highly regrettable, and the pursuit of satisfactory orgasm, especially in women, a near atrocity.

The communal drawing back from the lascivious delights of the old world and rejection of the bodily equipment their God had provided them with was a strange phenomenon. It was as if human beings had been given by God, Nature, evolution, whichever, a sports car to enjoy and chosen to use it as a tractor. Why should one of mankind's most intellectually advanced ideas to date – Christianity was revolutionary in so many other ways – have imploded on itself in this way? The historian of sex Reay Tannahill can only suggest that authoritarian societies had worked out that in disciplining sexual relations, it was possible to control the family and thus, critically the stability of the fragile new phenomenon that was the state. However, she says that even so, the legislators of matters sexual limited themselves to intervening only when sex impacted on areas of public concern such as legitimacy, inheritance and population control.

Setting aside its peculiar obsessional Nature, the new Christian morality ran perversely counter to human psychology in the way it denied any legitimate outlet for sexual feelings. The terrible distortions and corruptions of personality that this would create were not unknown to contemporary physicians. Soranus of Ephesus remarked in the second century, 'If the body feels no sexual desire it seems to suffer just as the spirit does.' Much later, Ambrose, the official poet of the Third Crusade, confirmed the medical belief that lack of sex was bad for the health. 'A hundred thousand men died there', he wrote of the Crusade, 'Because from women they abstained / They had not perished thus / Had they not been abstemious'. Furthermore, around the time of *Magna Carta* in England, the Church, whilst proscribing sex at practically all times, simultaneously set down detailed instructions on how husbands should have sex with their wives to best effect.

A sage known as Giles of Rome, according to a 2003 book, *1215: The Year of Magna Carta*, advocated Galen's venerable advice on raising the 'temperature' of women by foreplay so as conception might successfully take place. When the wife began 'to speak as if she were babbling', Giles said, it was time for the husband to make his grand entry.

The idea that missing orgasms was unhealthy was almost certainly the perception of the ordinary man and woman. This sentiment continues to be a commonplace in the contemporary West, where it co-exists a little uncomfortably with Christian abstemiousness and fear of sexual desire. To choose but one example of semi-scientific confirmation that denial of sex is harmful, a 1983 survey, *Sex and Self-Esteem, Medical Aspects of Human Sexuality*, noted in conclusion: 'Orgasm and other forms of sexual expression are such a source of self-affirmation that two-thirds of psychiatrists believe people "nearly always or often" lose self-esteem when deprived of a regular outlet for sexual gratification.'

Yet the madness traceable to one peculiar obsession of a few misguided, if idealistic, men 2,000 years ago did not abate. The damaging effects of lack of sex have continued to rear their ugly Hydra head for thousands of years now, in a range of unpleasant manifestations. The obscene paedophilic excesses of a handful of Catholic clergy the world over in the modern age is just one of the more dramatic and corrosive of these. Some of the most egregious examples of the twisted behaviour engendered down the centuries by the Christian cult's rejection of sex and, worse still, denial of the enjoyment of orgasm, occurred when Europeans attempted to export their 'civilised' Christian ideals to colonial subjects whose ideas on sex, to our current view, were rather advanced and sophisticated.

Spanish colonists in particular encountered an ancient tradition of sexual equality and reverence for the orgasm in South and Central America. They attempted to introduce the radical new idea to the 'primitives' that women's genitals were *partes vergonzozas*, 'the shameful parts,' and that female sexuality was an abomination.

The indigenous people sometimes argued the case for their own ways. There is a description of a cleric, Fray Tomas Carrasco, preaching to a crowd against their 'promiscuity' and urging them to embrace monogamy. A woman bravely stood up and spoke out against the new European ways of male domination and female sexual shame. Unfortunately, she was killed by lightning in mid-oration, which the friar interpreted as proof she was a witch, while her own people saw it as evidence that she was quite right.

As Naomi Wolf has written: 'Europeans who witnessed these native women's assertion of their sexuality saw not divinity but depravity. According to the colonisers, Pueblo women could not even conceive of modesty or shame in relation to their bodies. Since, in the Western tradition, human beings in Eden were redeemed by shame, and particularly, as Christian theology evolved, by feminine shame, Europeans were inclined to see Hell where the Pueblo saw everyday pleasure.'

One of the world's most grotesque examples of a community warped to near destruction by the belief that Christianity and shame about sex go hand in hand is provided by the primitive island of Inis Beag, off the west coast of Ireland, as it was when anthropologists discovered its bizarre sexual culture in the 1960s.

Nudity, the researchers found, was abhorred by the islanders, even among small children, and animals regarded as sinful for going about undressed. Dogs would be whipped for licking their genitals. Girls and boys were separated at all times. Bathing was unknown, dressing was done only under bedcovers, and breastfeeding was highly uncommon. Any type of sexual expression needless to say, masturbation even to open urination, was severely punished by beatings. Parents believed that after marriage, 'Nature would take its own course'. As a result, there were many childless couples due to neither spouse knowing what was expected of them. Marriages were arranged and forced on couples, their average ages at marriage being thirty-six for men and twenty-five for women. A man was still considered a 'boy' until he was forty. The Church taught women that sex was a duty to be endured and that to refuse sex with their husband was a mortal sin. Underwear was kept on during sex and menopause regarded as an inevitable madness that afflicts women, some of whom confined themselves to bed at forty and lived as invalids until old age.

Psychologists studying the island found that its inhabitants sought escape from sexual frustration by masturbation, drinking, and alcohol-fuelled fights.

There is a popular view of the human psyche as a mattress; if our desires are thwarted in one area, they will simply prompt something else to pop up like a broken spring elsewhere in our behaviour. The mattress theory is famously illustrated in antiquity by Teresa of Avila, the fourteenth-century Spanish saint, who experienced rapturous visions of 'angelic visitation' which sound suspiciously like nothing more or less Godly than a rather spectacular orgasm.

As St Teresa put it: 'In his [the angel's] hands I saw a long golden spear and at the end of the iron tip I seemed to see a point of fire. With this he seemed to pierce my heart several times so that it penetrated to my entrails. When he drew it out, I thought he was drawing them out with it and he left me completely afire with a great love of God. The pain was so sharp that it made me utter several sharp moans; and so excessive was the sweetness caused me by this intense pain that one can never wish to lose it.'

Orgasmic-style, rapturous bodily sensations may be more common in those of a religious disposition than has been generally acknowledged. We spoke earlier of the oddly copulatory rocking of Orthodox Jews praying and reading the holy scriptures. Burgo Partridge, in his history of orgies, wrote wisely in reference to the early days of Christianity: 'Abstinence from sexual activity leads to an almost total mental preoccupation with the subject and psychoneurotic symptoms and sexual hallucinations were developed on a really astonishing scale. A terrific outburst of "incubi" and "succubi" swept the bedrooms of Europe. These were nocturnal visitors, connected in the minds of the Christians with witchcraft and devilry, who indulged in liberties with the afflicted person, always of a sexual Nature. They were particularly common in nunneries, and seemed also to be highly infectious.'

Common sense prevailed widely, though; many medical men

were aware that incubi were delusions, and it was frequently said at the time that, 'incubi infest cloisters'. More telltale still was the fact that these nocturnal 'spirit visitors' often left the nuns with a phantom pregnancy; a more eloquent example of the subconscious trying to impose itself through bodily processes would be hard to find.

It took Geoffrey Chaucer to point out satirically that incubi became much less spoken of after 'limitours' – wandering friars notorious for sleeping with women while their husbands were absent – appeared on the medieval scene. R.C. Zaehner in his 1957 book *Mysticism Sacred and Profane* further noted: 'There is no point at all in blinkering the fact that the raptures of the theistic mystic are closely akin to the transports of sexual union, the soul playing the part of the female and God appearing as the male. The close parallel between the sexual act and the mystical union with God may seem blasphemous today. Yet the blasphemy is not in the comparison, but in the degrading of the one act of which man is capable that makes him like God both in the intensity of his union with his partner and in the fact that by this union he is co-creator with God.'

Amongst women, nuns unsurprisingly seem to have been particularly unhinged by voluntary orgasmic deprivation. In 1565, an epidemic of erotic convulsions reportedly affected a convent in Cologne. According to a Dr De Weier, who was called in to investigate, the nuns would throw themselves on their backs, shut their eyes, raise their abdomens erotically and thrust forward their pudenda. Partridge mentions imaginary night-time visitations among the sex-starved by 'witches' and their demons, but 'witches' themselves who testified at various times in the Middle Ages to 'night-flying' were probably (broomsticks notwithstanding) using hallucinogenic drugs and ointments and enjoying LSD-like firework displays. But there were accounts too of self-proclaimed witches, often sexually inactive crones, swearing that they had flown when they outwardly appeared to be asleep; it is very possible that they were experiencing self-induced orgasm.

The mattress hypothesis was best exemplified, however, by a one-time priestly initiate in England, now a novelist, Paul Crawford, who has written on the dangers of celibacy in the modern age, and how this 'destructive force in the life of the Church' has distorted priests' behaviour. 'My experience training for the priesthood at Oscott College during the 1980s gave me first-hand experience of the unhealthy development of human sexuality among its clergy,' Crawford wrote in the *Guardian*. 'The life of the Catholic priest, with its marked isolation, loneliness and sexual denial, cannot fail to frustrate individuals and deform otherwise natural urges and desires into more bizarre, or simply counterproductive and pathetic appetites.

'Even masturbation was outlawed. I will never forget the burning faces of men being told that this was sinful, when in all truth it was an absolute necessity as far as most of us were concerned . . . To sublimate direct sexual expression, many of the men in the seminary entered a strange twilight world of glib affection, camply addressing one another as "Mother" or "Dearest". More humorous than in any way disturbing, this behaviour did seem to signpost sexual frustration. On one occasion, when a male relative of mine visited me, a student forced himself on him in the toilets and tried to kiss him.' Celibacy continues in a variety of cultures to be regarded as a source of power.

Nothing in history quite approximates the ferocity with which Christianity pursued its case versus the orgasm. There is, as Desmond Morris has observed, no one quite so obsessed by sex as a fanatical puritan. What is most puzzling about the continuing phenomenon is that the rejection of sexual enjoyment has at no stage been seen by Christian revisionists as in any sense a grievous insult to God, who must, surely, have created the parts of the body responsible for orgasm, which would suggest that the process is therefore free from sin.

A certain logic can be seen in parts of the Christian sexual tradition. For women, a practical reason for celibacy is that it was the only 100 per cent-sure way of avoiding, and particularly dying in, childbirth. It is also possible to contrive an argument against masturbation on the grounds that it is taking liberties, if only a little, with God's creation. Then again, if masturbation were a sin, surely God would have been bold enough to speak outright about it, rather than hide coded, ambiguous references to it in the Bible for uneducated preachers to interpret for him?

Self-denial and censorious zeal, it has to be concluded, seem to speak to something deeper in the human psyche than any professed religious belief. Even if Jesus had demanded free love and promiscuity of his adherents, the suspicion remains that the textual references would have been removed in mysterious ways – or re-interpreted to within an inch of their life. What it is that motivates the religious obsessive's hatred of sex could, for a psychoanalyst, be any number of childhood traumas and dysfunctions. However, an economic explanation has to be high on the list of possible reasons for the Christian church's obsession with sex.

If there is one thing, after all, the Church is more fixated on than sex, it is money. Mandatory priestly celibacy, as we have seen, was a mechanism for keeping Church property from leaking away by inheritance to clerics' children. Sex as a joyous activity for laity, it might also be argued, could equally damage the Church's fiscal interests, as it could lead to a surfeit of bastard children who would effectively be heathen – and ineligible to be part of the customer base of the Church.

Many other facets of human history have been founded on such obliquely commercial foundations; it is quite likely that the mass enjoyment of orgasm, too, has been seen as ultimately bad for religious business.

11

Orgasm in the Middle Ages

'O Venus, that art goddesse of plesaunce!
Syn that thy servant was this Chauntecleer,
And in thy servyce dide al his poweer,
Moore for delit than world to multiplye.'
Geoffrey Chaucer, '*The Nun's Priest's Tale*'

Throughout much of the world, the Medieval era and the Renaissance were the heyday of unfettered, guiltless, uncomplicated enjoyment of orgasm. China, India, Japan, the Middle East and Central America (before Christian values were fully hammered home by Spanish colonists) were all centres of orgasmic excellence. Even in the Christian world, as we shall see, there was widespread flouting of the stern official strictures against the enjoyment of sex; clerical marriages were officially allowed so long as husband and wife did not have sex, a prohibition that was manifestly no more than a formality when Popes routinely had children and Henry III, Bishop of Liège, was known to have sired sixty-five illegitimate offspring.

The dead hand of the anti-sex movement did not properly descend until the emerging Protestant Reformation in Europe began to threaten the formal dismantling of priestly celibacy, along with a more pragmatic all-round approach to sexual matters. In 1563, the Council of Trent reacted on behalf of the traditionalists to the reforming trend by pronouncing uncompromisingly, '. . . that it is more blessed to remain in virginity or in celibacy than to be joined in marriage'.

No such state of sexless, pleasure-free grace troubled the medieval Chinese. Huang O, a female poet of the sixteenth century, wrote as sensitively and yet erotically as any woman in antiquity in *The Orchid Boat:*

I will allow only
My lord to possess my sacred
Lotus pond, and every night
You can make blossom in me
Flowers of fire.

Huang O was almost certainly referring to her vagina in this verse – an obvious statement, one might say, were it not for the strange historical quirk that from the eleventh-century onwards the Chinese were collectively fixated on the female foot as being configurable as a kind of extra vagina. Their fetishistic passion was, more correctly, for the tiny 'golden lotus flower' foot that high-class Chinese parents produced in their girl children by binding their feet from the age of six.

A typical lotus foot in an adult was small to the point of deformity – some four inches long by a thumb's width across, according to Xiao Jiao, a writer on the history of sex in China, although one suspects this may be a downwards exaggeration. Encased in a tiny silken slipper, it was considered the most private area of a woman's body, touching it being the ultimate act of intimacy between her and her husband or lover. The toe of this deformed, bud-like extremity was exaggerated to simulate a small pseudo-penis, while the fleshy, soft area under the arches, was regarded, and used, both as a pair of secondary vaginas and as a female analogue for the penis.

An American China specialist, Howard Seymour Levy, wrote of the antique custom of footbinding in a 1966 book on the subject that: 'The ways of grasping the foot in one's palms were both profuse and varied, ascending the heights of ecstasy when the lover transferred the foot from palm to mouth. Play included kissing, sucking and inserting the foot in the mouth until it filled both cheeks, either nibbling at it or chewing vigorously, and adoringly placing it against one's cheeks, chest, knees or virile member.'

Some women, according to Levy, developed this erotic artistry to the extent that they could deftly grasp a partner's penis between their feet and guide it into the vagina. A few could also masturbate by rubbing or stroking their own lotus feet. For others, orgasm was said to be enhanced if a lover grabbed their feet during it. Chinese women, both lesbian and straight, would also reputedly practise simulated intercourse by mutually inserting their big toes into one another's vagina.

Footbinding and its accoutrements titillated many foreigners in Imperial China. Marco Polo, travelling there from Venice in the thirteenth century, admittedly managed to miss seeing the custom (he also failed to notice acupuncture and the Great Wall). But a Dr Matignon, attaché to the French diplomatic mission to China in the 1890s was, like most visitors in the previous thousand years, more observant, writing how: 'Touching of the genital organs by the tiny feet provokes, in the male, thrills of an indescribable voluptuousness. And the great lovers know that in order to awaken the ardour of especially their older clients, an infallible method is to take die rod [penis] between their two feet, which is worth more than all the aphrodisiacs of the Chinese pharmacopoeia and kitchen.'

Footbinding was banned in China in 1902, and appears to have been an extreme attempt by cruel, possessive men to keep women captive, immobile and helpless. There must also be a suspicion that the tales of women being able to masturbate by manipulating their own deformed feet long be more in the realms of male fantasy than reality. The rheumy modern eye, additionally, will not be slow to detect not a little inherent paedophilia in the notion of men being masturbated by feet the size of a six-year-old girl's.

Yet it is worth nothing that there were allegedly payoffs for the woman beyond the thrill a few achieved by having their feet fiddled with. Women reported that having their feet bound altered the vagina, causing 'a supernatural exaltation' during sex. It was also claimed by some that the teetering, swaying 'willow walk' of women with bound feet caused an upward flow of blood producing better enjoyment of intercourse.

India, where a little discreet foot fetishism was also known, continued to be China's main rival in promotion of the sexual arts. The Indian corpus of sexological texts, known as the *Karma Shastra* and dating from the eleventh century, acknowledge the area later termed the 'Gräfenberg zone' and then the G-spot in Europe, and also seem to be well acquainted with female ejaculation.

In the twelfth century another compilation of erotic greatest hits appeared, the *Ananga Ranga*, a self-help book which included section from the *Kamasutra* and was designed by its author, Kalyana Malla, to help prevent marital problems brought on by sexual boredom. Kalyana Malla was a court sage and poet commissioned by a Mogul emperor, King Ahmad, who wanted to improve his lovemaking skills to live up to his handsome appearance. Kalyana Malla stated: 'The chief reason for the separation between the married couple and the cause which drives the husband to the embraces of strange women, and the wife to the embraces of strange men, is the want of varied pleasures, and the monotony which follows possession.'

Imagination was the key to staying sexually fulfilled, he argued, as many sex therapists would still be saying a thousand years later. By thinking up new positions, he wrote, you could fantasise that you had dozens of sexual partners; each position, additionally, could open up another hitherto un-felt sexual delight. Kalyana Malla advocated a variety of what might be called gym equipment to facilitate adventurous positional gambits; the most popular of these was identical to a modern children's swing, but put to sexual use with the woman sitting on it with her legs up, presenting her vulva to her partner at convenient penis-height, and with the benefit of the effortless motion of the 'love swing' to ease herself up and down his stationary erection.

From sexual anatomy to subtleties of the emotions, the *Ananga Ranga*'s understanding of sex was not bettered in any Western culture until modern times. The *Madana-chatra* – or clitoris – the text specifies, is the upper part of the *yoni*. 'Before proceeding to the various acts of congress, the symptoms of the orgasm in women must be laid down. As soon as she commences to enjoy pleasure, the eyes are half-closed and watery; the body waxes cold; the breath, after being hard and jerky, is expired in sobs or sighs, the lower limbs are simply stretched out after a period of rigidity; a rising and outflow of love and affection appears, with kisses and sportive gestures; and, finally, she seems as if about to swoon. At such time, a distaste for further embraces and blandishments become manifest; then the wise know that, the paroxysm having taken place, the woman has enjoyed plenary satisfaction; consequently, they refrain from further congress.

'The woman who possesses *Chanda-vega*, or what is known as furious appetite, needs carnal enjoyment frequently and will not be satisfied with a single orgasm . . . And, moreover, let it be noted, that the desires of the woman being colder and slower to rouse than those of the man, she is not easily satisfied by a single act of congress; her slower powers of excitement demand prolonged embraces and if these be denied her, she feels aggrieved. At the second act, however, her passions being thoroughly aroused, she finds the orgasm more violent, and then she is thoroughly contented.'

Although we know less about its isolated culture, Japan in the Middle Ages was also both sophisticated sexually and of a mind to enjoy the pleasures of the flesh. *The Tale of Genji*, an eleventh-century Japanese sex-romp, is based around the practice of *Yobai*, or 'night crawling', a process of stealing in (by previous appointment, hopefully) to a girlfriend's bedroom for a little illicit lovemaking: 'With the lights dimmed and nineteen-year-old Saki feigning sleep on her *futon*, the reporter creeps into her dimly lit room. He parts the folds of her *yukata* [sleeping gown] and allows his fingers to creep up her reposing thighs towards the promised land. Already damp with lust, she makes purring noises and, in short order, the two are coupling in erotic *samurai* combat.'

The Japanese later designated part of the capital, Edo (modern Tokyo), as 'the floating city' – the equivalent of London's Soho, after the city was rebuilt following a fire. There was strict zoning, so pleasure palaces were all erected together in an area with over 3,000 registered prostitutes.

An overwhelmingly positive view of sexual desire, especially female desire, can be traced on the other side of the world, too. Peruvian drinking vessels, made by women potters, have been found in the shape of vulvas, which allowed the drinker to simulate cunnilingus. The Zuñi Indians of New Mexico and Arizona blessed baby girls in a formal ceremony that celebrated their luck in being female. The men would strap on enormous false penises, and support their womenfolk's equality, singing songs about how happy their sexual organs could make women. Sexual practices among Zuñi women, according to Naomi Wolf, 'spanned the spectrum from heterosexual to homosexual, and age was no barrier to eroticism'.

Among the Hopi Pueblos of North Arizona, too, before the arrival of Europeans, Wolf has discovered, 'female desire was the engine that, in a yearly woman-led rite celebrating female fecundity, sexuality and reproduction, would bring about the symbolic re-creation of cosmic harmony'. Pubescent Hopi girls would spend two days in an initiation ceremony dancing naked with other women, caressing clay penises, singing erotic songs and displaying their genitals to the men to incite them to lovemaking.

If gay sex, on the other hand, is the ultimate meter by which enjoyment of sex for sex's sake can be unambiguously gauged, the orgasm was set even fairer in the Medieval world. Out of range of the uncompromising Judaeo-Christian god, male gayness very nearly rivalled heterosexual values across the Middle East, North Africa, Turkey and as far to the east as Indonesia. In both China and Japan, male prostitution was widespread. China had a god of sodomy and boy prostitutes, Tcheou-wang, while Japan boasted male geisha houses. In both cultures, male prostitutes could be found on designated streets as late as the Second World War.

China was the source of one of the world's classic homo-erotic masterpieces. In the sixteenth-century *Jin P'ing Mei*, we read how the hero, Ximen Qing: '. . . opened the boy's robe, pulled down his pants, and gently stroked his penis . . . while the boy surrendered his bottom to a mighty warrior, Ximen stroked his stiff penis . . . Said the boy: "He pushed his poker so violently between my buttocks that today they are swollen with great pain. When I asked him to stop, he pushed his poker in and out all the more."'

That Christopher Columbus was engaged in the rather more macho business of exploring the globe as the *Ananga Ranga* was being written provides a fitting illustration of the apparent austerity that was the norm under the Christian sphere of influence in the Middle Ages.

There is certainly no scarcity of examples of this. An eleventh-century doctor of medicine in Germany, Albertus Magnus, suggested a method for the essential business of removing female sexual desire. It was, of course, to burn the penis, eyelids and beard of a wolf, 'and then make the woman drink the results without her knowing anything about it'. The twelfth-century Italian theologian Peter Lombard maintained in his *De excusatione coitus* that for a man to love his wife too ardently is a sin worse than adultery. Some Catholic men at this time wore a *chemise cagoule* in bed, a heavy nightshirt with a hole in the front to allow the erect penis access to the vagina. Like the bed sheet with a hole mythically used by

Orthodox Jews to have intercourse through, the *chemise cagoule* ensured that neither party would derive unnecessary tactile pleasure while procreating.

In the thirteenth century, Saint Francis was so deeply troubled by the thought of having an erection that every time he found himself sporting the beginnings of one, he felt obliged to throw himself into a thorn bush to deal with it. Masturbation was also deeply frowned on, but by Medieval times people liked there to be a scientific reason for a restriction; a variety of Medieval medical texts duly stated that semen is a precious rare fluid, and that it takes forty parts of blood to make one of semen. Better, therefore, not to waste it. (The same 40:1 proportion, curiously, appears in Hindu writings.)

Reducing sexual intercourse in the interest of modernism to a harsh, ugly, unromantic act had the paradoxical effect of decimating any semblance of female rights – woman being the sex which, be it biologically or as a matter of socialisation, invests the most emotion in sex. It was as a direct result of the Christian debasement of sex that women became more than ever the property of men. As far back as 585, Catholic theologians were arguing that women did not have a mortal soul. By the ninth century the *droit de seigneur*, that virtually gave noblemen the right to ravish any peasant woman on the road and to deflower the brides of their subjects, was widely established across Christendom.

Hypocrisy was also an inescapable element in the official Christian party line on sex. Tomas Sanchez, a prominent churchman and author, suggested in 1621 that any person who feels an orgasm coming on outside of marital intercourse should lie still, avoid touching the genitals, make the sign of the Cross, and pray fervently for God not to allow him to slip into orgasmic pleasure. Yet Sanchez also argues that 'if, when engaged in sex with a whore, a man withdraws before ejaculation, he is considered to have repented and not sinned against God's laws'.

And yet the first 1,500 years of Christianity actually encompassed more diversity in attitudes to sex than they are usually credited with. The older, more sexually 'progressive' Judaism was proportionately becoming less sensible by the day; Orthodox Jewish elders by the Middle Ages were declaring masturbation to be so heinous a crime that a man should not even touch his penis when urinating. According to a 1987 book, *Bizarre Sex*, by a London psychotherapist, Dr Roy Eskapa, men at this time were required to 'aim' when urinating by lifting the scrotum. Women, records Eskapa, were permitted to examine their own genitals, but only to check if they were menstruating.

Even in the supposedly abstemous Christian countries there was a black market, so to speak, in sex. In fact, there will have been far more people – both outside the Church and within it – enjoying their orgasms than trying their hardest to hate them. Many physicians held that celibacy was unhealthy and prescribed their patients more frequent sex. In everyday life, a quite overt sexuality prevailed. Archdeacons' courts heard constant sex cases – adultery, incest, fornication and homosexuality prevailed. Fornication was described as 'the vice of everyone and excused by many'. In the eighth century, Boniface, the Exeter monk and chronicler, characterised the English as a people who 'utterly despise matrimony, utterly refuse to have legitimate wives, and continue to live in lechery and adultery after the manner of a neighing horses and braying asses'.

Accusations of sodomy were bandied about almost casually in Medieval England. Anselm, a twelfth-century Archbishop of Canterbury, writing to Archdeacon William in 1102 to suggest a court crackdown on this sexual offence, says: 'This sin had been so public that hardly anyone has blushed for it, and many, therefore, have plunged into it without realising its gravity.' A serious crime it may have been, but the Knights Templar, famous during the Crusades, were accused of it simply to undermine them politically. The accusation was taken no more seriously than the words of a modern political spin doctor.

Reforming Christians in mainland Europe too, furthermore, were frequently less than high-minded when it came to enjoying a bit of sexual pleasure. Martin Luther (1483–1546) was ceaseless in his struggle against the Catholic loathing of enjoyment. In his own way, he was quite earthy and lusty. He believed sexual impulses were natural and irrepressible, that celibacy was invented by the Devil and that priests should marry. He himself did, arguing that marriage was not even a holy sacrament but a civil matter, and that Christ committed adultery with Mary Magdalene among others so he could know whereof he spoke. The forbidding-sounding John Calvin also accepted that such pleasure could be legitimate, although he also held that too much passion even in the marital bed was sinful, and believed in the death penalty for adultery in his brutal Geneva theocracy.

No account of the history of sexual pleasure at this time can ignore the odd development, from the eleventh century onwards right across Western Europe, of a peculiar, sex-free form of sexual indulgence, namely the famous courtly or 'true' love. The creation of a section of the aristocracy in Southern France, this romantic ideal, with its emphasis on character ennoblement through love, is often presented as a new model for loving, male-female relationships, previously unknown in Western civilisation. It was, however, similar in some degree to the Ancient Greeks' belief that gay relationships were more moral than 'straight'.

The two defining characteristics of courtly love were that it could only take place between unmarried couples, and that no intercourse – and certainly no orgasm – was involved. Such covert, frustrating, painful, forever-foreplay liaisons were considered morally uplifting, enchanting and exciting, making the knight involved a better man and warrior, the lady a purer, yet more delectable, soul. Intercourse was believed to be false, unspiritual love, the kind of banal and mundane transaction that took place only between man and wife. Troubadour poets begged their ladies not to grant them sexual favours. 'True love' meant kissing, touching and fondling, a modicum of naked contact, but never sex.

Most royal courts in Europe subscribed on and off to the courtly love ideal, which was in many ways the motor power of the Renaissance, and the concept still has some attraction today, although one is thinking more Jimmy Carter, who admitted to owning a lustful heart, than Bill Clinton, who enjoyed lust in the flesh. In 1122, the cultured and radical Queen Eleanor of France and England, William the Conqueror's granddaughter, presided over the most courtly of courts. She believed that love should be an equal relationship, based on mutual respect, admiration and the free interplay of mutual emotions. It was designed to elevate the status of women and inspire progress. A chaplain in Eleanor's court known only as André assisted the queen's daughter, Marie of Champagne, in writing a love manual called *Tractatus de Amore et de Amoris Remedio* (A Treatise on Love and Its Remedy). A poet, Chrétien de Troyes, on Eleanor's instruction, produced the original story of Sir Lancelot and Guinevere.

Among those for whom neither the strictest interpretation of Christianity, nor the curiously abstemious-yet-indulgent cult of courtly love, held much appeal, meanwhile, aphrodisiacs were popular in the Middle Ages. The root of the orchis plant, roughly the shape of a scrotum loaded with testicles, was thought especially potent at reviving a flagging non-courtly love life. Fashions were also devised to exaggerate the genitals, or simply put them on display. In 610, the Queen of Ulster and the ladies of her court came to meet Cuchulainn, Ireland's most famous warrior hero, in topless dresses, and flashed their vaginas for good measure to show due respect. Revealing trends in fashion were still not sufficiently saucy for Leonardo da Vinci however, who publicly declared that men should celebrate their genitals by drawing attention to them via their dress. A fashion soon followed for wearing lightly coloured tights and showing off the loins to their best effect with coloured ribbons tied round the groin and waist.

Popular culture was also stiff with bawdy verses, some of which suggest that mutual orgasm was far from an unknown concept, even if a slightly *avant-garde* one. One anonymous fourteenth-century English poem, 'Sexual Intercourse', reads:

That girl was skilled at thrusting under the trees;
she gave me, we were brave, a blow for every blow,
and after my masculinity drops came in her.

Another snippet, an anonymous fourteenth-century ditty included in a collection called *Loose Songs* published in the late-eighteenth century by Bishop Thomas Percy, gives a glimpse of the timeless male shame of premature ejaculation.

Then off he came
And blusht for shame
Soe soone that he had endit.

Male gayness in Medieval Europe does not seem to have been substantially curtailed by Christianity. It was rampant before the new religion took off ('The Celts [of France] take more pleasure in pederasty then any other nation, to such a degree that amongst them is no rarity to find a man lying between two minions,' according to the Greek writer, Athenaeus) and still going strong among the Normans when they invaded England in 1066. William the Conqueror's son, William Rufus, (King William II), was so overtly homosexual that the Church refused to bury him in consecrated ground after he was killed by a hunting arrow in 1100. And, as we saw from Anselm's letter to Archdeacon William, the native English were not at all averse to gayness. The manner of Edward II's murder in 1327 suggests his love for his favourite Piers Gaveston was less than Platonic: he was, it was recorded, 'sleyne with a hoote broche putte thro the secret place posteriale' – the red-hot-poker-up-the-bottom story so beloved by generations of bloodthirsty British schoolboys.

Officially, sodomy was condemned everywhere, but it never really abated for long. It was enormously popular in Spain and Holland, while a Scottish traveller and writer, William Lithgow, wrote in 1610 of Padua: 'For beastly Sodomy, it is rife here as in Rome. Naples, Florence, Bullogna, Venice, Ferrara, Genoa, Parma not being exempted, nor yet the smallest Village of Italy: A monstrous filthinesse, and yet to them a pleasant pastime, making songs and singing sonets of the beauty and pleasure of their *bardassi*, or bugger'd boyes.'

Throughout the Middle Ages, the Church was keen to fight the subversive enlightenment of the emerging Renaissance and reassert the darkness of religion. St Thomas Aquinas's reaction to courtly love was to state that kissing and touching a woman, even without thought of sex, was a mortal sin. Wound up to near madness by the continuing popularity of sex and sexual thoughts, priests and religious fanatics took to mutual flagellation, going from town to town to pray ostentatiously and whip each other like the finest modern S&M enthusiast.

Women as temptation incarnate became targeted as inherently sinful. Any physically desirable woman was viewed by elements in the Church as evil – a superstition that still persisted in Ireland into the late-twentieth century when the Magdalen Homes finally ceased incarcerating girls and condemning them to a lifetime's servitude, in some cases merely for being pretty. In 1450, official Catholic dogma began to assert that witches existed and flew by night. Jacob Sprenger and Henry Kramer, Dominican brothers and Professors of Sacred Theology at the University of Cologne, tortured 'confessions' from women and burned to death over 30,000 'witches' on charges of fornicating with the Devil, whose penis, the Church claimed, was covered with fish scales.

Yet despite such outrages, sex continued its slow but inexorable journey towards its place in the sun. If the celebration of masturbation, especially by women, is one of the best indicators of a sexually liberated society, this example from *The Delights of Venus* by Johannes Meursius, a seventeenth-century

Dutch poet, is suggestive of a culture not overly burdened by Christian strait-lacedness; Meursius's lines were in praise of dildos.

> *Just at sixteen her breasts began to heave*
> *Yet scarce knows what the titillation means.*
> *All night she thinks on Man, both toils and sweats,*
> *And dreaming frigs, and spends upon the sheets;*
> *But never knew the more substantial bliss,*
> *And scarce e'er touched a man, but by a kiss.*
> *Her virgin cunt ne'er knew the joys of love*
> *Beyond what dildoes or her finger gave . . .*
> *Come this way, I've a pretty engine here,*
> *Which us'd to ease the torments of the fair;*
> *This dildo 'tis, with which I oft was wont*
> *T'assuage the raging of my lustful cunt.*
> *For when cunts swell, and glow with strong desire,*
> *'Tis only pricks can quench the lustful fire.*
> *And when that's wanting, dildoes must supply*
> *The place of pricks upon necessity.*

Contraception, another key indicator, along with masturbation and gay sex, of the existence of sex for fun, was officially unacceptable but widely practised from the third century onwards, at least by prostitutes. Chaucer mentions contraceptive sponges and tampons used by those who have sex 'moore for delit than world to multiplye'. As far as male contraception was concerned, *coitus interruptus* was described in a 1375 *Book of Vices and Virtues* as a sin 'agens [against] kynde and agens the ordre of wedloke'. But it was still used routinely by husbands to separate orgasmic pleasure from child-begetting.

Shakespeare, inevitably, is our barometer of how far an understanding of sex was ingrained in the popular culture of post-Medieval 'Merrie England'. As the mainstream BBC1 or HBO dramatist of his day, he trod a skilful path between rudeness and respectability; hidden reference was his method of getting sexy bits in. Thus he mentions dildos, if a little obscurely, in 1611 in *A Winter's Tale* ('He has the prettiest Loue-songs for Maids and with such delicate burthens of dildos and fadings'), although Ben Jonson did so the previous year in *The Alchemist* ('Here I find the seeling fill'd with poesies of the candle: And Madame, with a Dildo, writ o' the walls'). The lines, 'Graze on my lips; and if those hills be dry, Stray lower, where the pleasant fountains lie', are often attributed to Shakespeare but alternatively believed to be Alexander Pope's. But scattered throughout confirmed Shakespeare material there are also numerous less-than-subtly concealed terms for genitals, both male ('cod', 'thing') and female ('quaint', 'count', 'sheath'). 'Arise' or 'stand' often refers to an erection; in *Macbeth*, the porter describes how alcohol can make a man 'stand to or not stand to', illustrating his theory by pointing and then dropping a key.

The Elizabethans were in thrall to the rather melodramatic notion of every orgasm shortening a person's life by one minute. This was more likely a seduction ruse by men to convince the ladies that each selfish orgasm 'hurts me more than it will hurt you', but Shakespeare exploited the idea for all it was worth. Thus the word 'die' is often interchangeable with 'orgasm' (the French idea of *le petit mort* was curiously pervasive), although it is only fair to add that a love-death connection is also common in love poetry, as well as to connect romance with concepts of transience, impermanence and mortality.

But one can be fairly sure that patrons of the cheap seats at the Globe would have known precisely what Shakespeare was getting at when Cleopatra kills herself after Antony's death, holds the asp to her breast and calls, 'Husband, I come!' *Romeo and Juliet*, similarly, is stiff with death-orgasm images. There is a recurring motif in the play depicting death as Juliet's bridegroom including a passage where Romeo dreams he is dead and anticipates that he will 'lie' with the dead Juliet. There is also what seems to be a distinctly sexual image where Juliet kills herself with Romeo's dagger.

It is in the Sonnets that Shakespeare loosens his ruff a little more. But while Sonnet 116 ('Let me not to the marriage of true minds / Admit impediments . . .' is the staple of wedding services, extolling as it does the correctly chaste ingredients for a seemly relationship, few of us can quote so freely from Sonnet 129, which is on the trials and tribulations of the orgasm. Nicknamed 'the Lust Sonnet', it deals with sentiments quite different from the sort of meaningful relationship alluded to in 116. In the Lust Sonnet, the word 'love' does not appear at all.

Shakespeare elucidates instead upon the problems surrounding sex and the pitfalls and disappointments of orgasm that are still keeping modern-day sex therapists in business. So neurotic is 129 on the matter of sex – it names and shames orgasm as the source of lust, which is characterised as 'perjur'd, murd'rous, bloody, full of blame Savage, extreme, rude, cruel, not to trust' – that one scholar, John Robertson, suggested in his 1926 book *The Problems of the Shakespeare Sonnets* that Shakespeare did not write it; but he may merely have been protecting the Bard's reputation in a Britain still suffused with Victorian sexual attitudes.

Dr Marvin Krims, a psychoanalyst and lecturer in psychiatry at Harvard Medical School, gives particular attention to the Lust Sonnet in his essay 'A Psychoanalytic Exploration of Shakespeare's Sonnet 129'. 'From the initial pejorative portrait of lust as emotional exhaustion and orgiastic offal ('Th' expense of spirit in a waste of shame / Is lust in action') . . . the sonnet delivers a dismal account of what the words also endorse as an exceedingly enjoyable and profoundly satisfying experience,' writes Krims. ('Spirit' here means semen, so its waste or expulsion is ejaculation. 'Shame' refers to genitalia, since the Latin 'pudenda' (not actually used by the Romans) derives from *pudere* – 'to be ashamed' and translates as 'things to be ashamed of'.

'Sex,' Krims deduces of Shakespeare's attitude, 'is both desirable and dreadful, from seduction to final satiation. And the concluding couplet, in a tone of supreme irony, assures us that there is absolutely nothing we can do, "To shun the heaven that leads men to this hell".'

Another academic in the psychiatric field, Brett Kahr, a psychotherapist at the School of Psychotherapy and Counselling, Regent's College, London, makes the point that Shakespeare, reflecting the culture around him, was no early advocate of women's liberation. 'In *Measure for Measure*,' Kahr wrote in a 1999 essay in the *Digital Archive of Psychohistory*, 'Shakespeare has described men as "great doers", and as creatures who go "groping for trouts in a peculiar river" [i.e. having sex recreationally]. These fragments offer some idea of how Renaissance men continually failed to understand women, regarding them instead as objects of obscurity who must be probed and prodded by a "rapier and dagger man", another one of Shakespeare's references to male sexual behaviour which teems with violence of an extreme Nature.'

The defining sexual advance of the Medieval Christian world (achieved in spite of Christian values rather than as a result of them, it must be stressed) was made in 1559 by a scientist in Venice. That his name, in this age of exploration, was Columbus – more properly Mateo Renaldo Colombo – is one of history's sweeter ironies because his achievement was to plant the flag, you might say, on a piece of uncharted territory which was arguably as unexpected and important a discovery as the New World found by Christopher Columbus. For Mateo Columbo it was who identified a small organ which, he believed, was 'pre-eminently the seat of woman's delight'.

Like a penis, Columbus reported in his book *De Re Anatomica*, 'if you touch it, you will find it rendered a little harder and oblong to such a degree that it shows itself as a sort of male member . . . If you rub it vigorously with a penis, or touch it even with a little finger, semen swifter than air flies this way and that on account of the pleasure . . . Without these protuberances, women would neither experience delight in venereal embraces nor conceive any foetuses.' We know this 'protuberance' as the clitoris, from the Greek for 'little hill'.

There are more than a few sub-ironies in the Mateo Columbus story, as wonderfully unravelled by an Argentinian psychotherapist Federico Andahazi in his poetic novelised version of it, *The Anatomist*, which was published in 1997. Christopher Columbus, Andahazi pointed out, 'discovered' America when the Native Americans knew where it was all the time; Mateo Columbo revealed what half the world's population – women – had a pretty shrewd idea about, so far as location and function were concerned, and none more so than on the pre-Christian American continent.

'Mateo Colombo searched, travelled and finally found the "sweet land" he longed for,' wrote Andahazi, 'the organ that governs the love of women.' The *Amor Veneris* (such is the name the anatomist gave it, "if I may be allowed to give a name to the things by me discovered") was the true source of power over the slippery, shadowy free will of women.'

'What would happen,' Andahazi speculated in his preface, 'if the daughters of Eve were to discover that, between their legs, they carried they keys to both Heaven and Hell?' He went on to tell us precisely what did happen: the Dean of the Medical School in Padua, it turns out, was not best pleased by the other Columbus's New World of orgasmic pleasure for women: 'In the eyes of the Dean, the anatomist's newest findings had exceeded all limits of tolerance. The *Amor Veneris*, Mateo Colombo's America, went far beyond what was deemed permissible for science. For more than one reason, the mere mention of a certain "pleasure of Venus" made the Dean's gorge rise.'

Colombo, we learn from Andahazi, discovered in his anatomical researches a woman from Spain, Inés de Torremolinos, who had what appeared to Colombo to be a tiny penis, and to us, presumably, would have been an unusually well-developed clitoris. The mysterious organ, Colombo noted, was 'inflamed, throbbing and moist'. (It is probable that, by happenstance, Inés was suffering from clitoromegaly, an enlargement of the clitoris that causes it to appear like a small penis; the condition is caused by excess androgen and is usually accompanied by heavy hair growth on the body.)

Colombo obviously thought at first that he was examining a hermaphrodite, but by the medical understanding of the day in these cases both sets of sexual organs are withered and reproduction is impossible. Yet Inés was a mother of three.

> 'Intuitively, the anatomist took hold of the strange organ between his thumb and index fingers, and with the index finger of his other hand he began gently caressing the red and engorged gland. He then observed that every muscle in the patient's body, up to then completely relaxed, tensed suddenly and involuntarily, while the organ grew somewhat in size and throbbed with brief contractions.'

The second Columbus went on to examine over a hundred other women, both living and cadaver. To his considerable shock, he realized that the Inés de Torremolinos 'penis' existed, 'small and hidden behind the fleshy labia', in all women. As a scientist, he was delighted to find that Inés's odd orgasmic behaviour, too, was repeatable experimentally. The anatomist, who admitted he set out on his quest to try better to understand women from a romantic point of view, had discovered the key to love and pleasure. 'He was unable to explain,' Andahazi wrote, 'how this "sweet treasure" had remained undetected for centuries, and how generations of scholars, anatomists from the West and from the East, had never seen that diamond that could be observed with the naked eye simply by parting the flesh of the vulva.'

Colombo reported his momentous findings to the Dean of his faculty in March 1558. Whereas today he might have won a Nobel Prize for Medicine, in the sixteenth-century his reward for 'discovering' the clitoris was being arrested in his classroom within days, accused of heresy, blasphemy witchcraft and Satanism, put on trial and imprisoned. His manuscripts were confiscated, and his 'America' was never permitted to be mentioned again until centuries after his death.

Another sexual pioneer working at precisely the same time in Venice had better luck. In the wake of history's first recorded syphilis epidemic, Gabriel Fallopius – whose name was given to Fallopian tubes – invented the condom, or more correctly re-invented it since the Ancient Egyptians and others had used rudimentary sheaths. Fallopius's condom was a linen sheath, designed ostensibly for protection not as a contraceptive. Fallopius advocated the very unusual and progressive notion of hygiene in a book entitled *De Morbo Gallico* – The French Disease. 'As often as a man has intercourse, he should, if possible, wash the genitals, or wipe them with a cloth,' Fallopius advised. 'Afterwards he should use a small linen cloth made to fit the glans, and draw forward the prepuce over the glans; if he can do so, it is as well to moisten it with saliva or with a lotion.'

So far as is known, the pre-rubber Fallopius sheath was not at first appreciated as a contraceptive. 'For the rake cared little whether he left his victim with child or not,' explained Gordon Rattray Taylor in his 1953 *Sex in History*. 'Women, however, were beginning to equip themselves with effective contraceptive devices, and Casanova relates how he once stole a supply of the devices, which are so necessary (as he puts it) to those who wish to make sacrifices to love, leaving a poem in their place.'

The Casanova reference indicates how the invention did not catch on for sexual purposes until a couple of centuries later. It was only named in print a century after Fallopius, in 1665, in *A Panegyric upon Cundum* by a notorious sexual swordsman, John Wilmot, Earl of Rochester. Some believe the word was derived from a Dr Condom, who procured contraceptives for the libidinous Charles II, but this may be a myth. More likely, it comes from the Latin *cunnus* (for the vagina) and *dum* (in its sense of 'able to be fooled'). Equally likely, especially given the Restoration timing of Rochester's reference, it was a pun alluding scurrilously to Charles's supplier.

Either way, Fallopius's condom, before it was perfected in the twentieth century by the invention of rubber, acquired a reputation as a not always effective, or honourable, part of the armoury of lovemaking. The French wit Madame de Sévigné (she who famously did not have time to write a short letter, so wrote a long one instead) later described the contraceptive sheath as 'gossamer against infection, steel against love', while the glib Casanova seems to have used them as a method of seduction by referring to them as a device 'to put the fair sex under shelter from all fear'. James Boswell, meanwhile, attested to their failure even to hold back the male orgasm, which was always one of their supposed benefits. He recorded in his diary in 1764: 'Quite agitated. Put on condom; entered. Heart beat; fell. Quite sorry, but said, "A true sign of passion."'

The mid- to late-seventeenth century was a time of steady, if unsure, progress towards a better understanding of sex. For one thing, the completely new concept of young married couples living together in a dwelling exclusively their own began to develop in Europe as the norm. It cannot be a coincidence that the word 'orgasm' begins at the same time to be used in its present Oxford English Dictionary sense of 'the height of venereal excitement in coition'. The English physician Nathaniel Highmore re-coined the term '*orgasmum*' from the Greek around 1660. The first manifestation in English of plain 'orgasm' is in the translation of a 1684 book by a Swiss physician, Théophile Bonet: 'When there appears an orgasm of the humours, we rather fly to bleeding as more safe.' It was not until 1771, however, according to the OED, that a writer called T. Percival, mentions something more unambiguously sexual, 'a kind of nervous orgasm, or spasm on the vitals'.

Even before orgasm had its own name, it was seen as a problem in women. In 1653, Pieter van Foreest, a prominent Dutch doctor, published a medical compendium with a chapter on the diseases of women. For the affliction commonly called hysteria ('womb disease') he advised that a midwife massage the genitalia with one finger inside the vagina, using oil of lilies or similar as lubrication. In this way, van Foreest said, the afflicted woman can be aroused to 'the paroxysm'. This form of therapeutic masturbation for frustrated women was not a new 'treatment', but a revival of the finger 'stimulation' first recommended, as we saw earlier, by Galen for widows, the chaste and nuns, and periodically mentioned again every few hundred years. In 1660, however, Highmore summed up the problem of the considerate male lover through the ages when he likened genital massage to 'that game of boys in which they try to rub their stomachs with one hand and pat their heads with the other'.

A midwife, Jane Sharp, in her 1671 *The Midwives Book*, one of dozens of supposed midwifery guides (more likely pornography) published in the full flush of liberated, Restoration London, wrote of the clitoris: 'It will stand and fall as the yard [penis] doth and makes women lustful and take delight in copulation.' A Danish physician, Caspar Bartholin, similarly explained in his *Anatomy* of the same period that the clitoris is 'the female yard or prick . . . [which] resembles a man's yard in situation, substance, composition, repletion with spirits, and erection'. The Dutch anatomist Regnier de Graaf, writing again in the same decade, argued: 'If those parts of the pudendum [the clitoris and labia] had not been supplied with such delightful sensation of pleasure and of such great love, no woman would be willing to undertake for herself such a troublesome pregnancy of nine months.' The English surgeon, William Cowper, in his *Anatomy of Humane Bodies* (1698), shows the clitoris for the first time as a distinct organ – a reference that was mysteriously missing again in early-twentieth-century editions of *Gray's Anatomy*.

In 1684, the first practical sex manual was published anonymously in London, entitled *Aristotle's Masterpiece, or the Secrets of Generation Displayed in All Parts Thereof*. Neither a masterpiece nor anything to do with Aristotle, it was a best-seller for over a century. Besides advice on pregnancy and so on, it also strongly advocated foreplay and promoted male stimulation of the clitoris, stating that, 'blowing the coals of these amorous fires' pleased women. *Aristotle's Masterpiece* remained the definitive 'dirty book' of the next hundred years or more, a volume young men with only limited interest in gynaecology and obstetrics would study in private out of sheer academic interest.

A tide of basic smut too, not all of it necessarily progressive, accompanied the Restoration. Visual pornography from Europe began to appear for the first time – the home-grown product did not generally start to be produced until the next century. There are exceptions: an example of engraved English porn from the 1660s shows a plume of female public hair being worn by a man in his hat. For the moment, however, most British porn was literary. The Earl of Rochester in a 1680 collection, *Poems on Several Occasions*, wrote colourfully of '. . . the common fucking post / On whom each whore, relieves her tingling cunt / As hogs on gates do rub themselves and grunt'. In his *The Imperfect Enjoyment* of the same year, he attested to what appears to be his own premature ejaculation:

But whilst her busy hand would guide that part
Which should convey my soul up to her heart,
In liquid raptures I dissolve all o'er,
Melt into sperm, and spend at every pore.
A touch from any part of her had done't:
Her hand, her foot, her very look's a cunt

And here is John Donne, from a rather more elevated literary stance, in 'To His Mistress Going to Bed':

Licence my roving hands, and let them go
Before, behind, between, above, below.
O my America, my new found land,
My kingdom, safeliest when with one man manned,
My mine of precious stones, my empery,
How blessed am I in this discovering thee!
To enter in these bonds, is to be free;
Then where my hand is set, my seal shall be.

One of the most interesting literary contributions to the history of the orgasm in Restoration England comes from an extraordinary woman named Aphra Behn. Before becoming the first female professional writer, Behn was an English spy, code-named Agent 160, and is also believed later to have been James II's mistress. Not surprisingly, given her background, Behn's great fascination was with the interdependence between sex and power, of both of which she had her fill in her intriguing life. So Behn it was who wrote the first female complaint against impotence in her poem 'The Disappointment'.

. . . The willing Garments by he laid,
And Heav'n all open to his view;
Mad to possess, himself he threw
On the defenceless lovely Maid.
But oh! what envious Gods conspire
To snatch his Pow'r, yet leave him the Desire!

He Curst his Birth, his Fate, his Stars,
But more the Shepherdesses Charms;
Whose soft bewitching influence,
Had Damn'd him to the Hell of Impotence.

Despite such advances for the cause of female sexuality, male masturbation remained a thoroughly awkward subject during the Restoration period, even in prematurely liberated Holland. In 1677 a Dutch microscopist, Anton van Leeuwenhoek, discovered sperm while examining a human semen sample. But how was he to explain where he obtained the guilty material? Leeuwenhoek carefully noted: 'What I describe was not obtained by any sinful contrivance . . . but the observations were made upon the excess with which nature provided me in my conjugal relations.' It was not for another century that Lazzaro Spallanzani, a priest and scientist worked out the role of 'spermatic worms'. He made oilskin 'trousers' for male frogs which allowed them to mount females in time-honoured fashion, but not for their sperm to escape. The females remained unfertilised. When Spallanzani took some of the sperm collected in the trousers and mixed them with the females' eggs, they were fertilised.

Less helpfully in this same generally progressive era, scientists were discovering (or rather substantiating their pre-existing prejudices) that women did not need an orgasm to conceive and concluding that it was best after all if the female sex remained passive during intercourse. As Thomas Laqueur has pointed out of this movement, Westerners, 'no longer linked the loci of pleasure with the mysterious infusing of life into Nature'. Or as Emma Dickens put it in her book *Immaculate Contraception: The Extraordinary Story of Birth Control*: 'Out, for scientists, went the Medieval idea of women as lusty equals in sexual congress, and in came a limited and boring role for women. This is not a world of which Chaucer's Wife of Bath would have wanted to be a part at all.'

12

The Foundations of Victorian Prudery: The Orgasm from the Late Restoration to 1840

'None of our wares e'er found a flaw, Self preservation's Nature's law'

From an advertising jingle by Mrs Phillips,
an eighteenth-century condom wholesaler in London.

News did not travel fast in the middle of the second millennium. In 1740, nearly two hundred years after Mateo Colombo, the clitoris was 'discovered' yet again, this time by a Swiss biologist, Albrecht von Halter. Von Halter observed how sexual feelings in women were focused on '... the entrance of the pudendum ... When a woman, invited either by moral love, or a lustful desire of pleasure, admits the embraces of the male, it excites a convulsive constriction and attrition of the very sensitive and tender parts, which lie within the contiguity of the external opening of the vagina, after the same manner as we observed before of the male. When the clitoris grows erect and the blood is flushed into the woman's external and internal genitalia, the purpose is to raise the pleasure to the highest pitch.'

The eighteenth century in Europe, in terms of sheer crudity, was the bawdiest century of all, twentieth included. A mass response to the austere days of Oliver Cromwell, just a few decades earlier, the 1700s were in many ways the first 1960s. The eighteenth century saw the profoundest schism yet between those who regarded orgasm as a pleasure to be enjoyed whether or not one was attempting to reproduce – and those who subscribed to the old Christian morality, which continued to teach that orgasm was only morally acceptable within marriage, and then only if experienced with the intention of bringing about conception.

Christianity, too, was deeply divided on the issue of enjoyment of sex. The religion was so segmented into competing brands by now that it is impossible to speak for Christianity as a whole. But a comparison between the Catholics and the Puritans should suffice to show the variety of views now available. The Roman Catholic Church continued on its schizophrenic way; a suitable vignette to demonstrate how it was developing can be found in 1714 when the Church ended the confessional requirement that men name the women with whom they had fornicated. This was not exactly a liberalising move; the reason as we have seen was that it had been discovered that priests were using the information they were given as hot tips for partners with whom they could quietly commit their own sin of fornication.

The Puritans, surprisingly, in spite of Cromwell's sober legacy, were not especially anti-sex. They were pious and severe, but also highly sexed and quite sentimentally romantic. According to one scholar, D. Daniel, in a 1966 essay 'Christian Perspectives on Sexuality and Gender', 'Married sex was not only legitimate in the Puritan view; it was meant to be exuberant', and the Puritans 'were not squeamish about it'. Even so, they still believed excessive desire was animal, and prayed before sex.

With the notable exception of the notorious seventeenth-century 'Blue Laws' in New Haven, Connecticut, which outlawed every form of pleasure and promoted public whippings, branding or execution for adulterers, and the Salem 'witch' killings, the Puritans rejected the heartless joylessness of European Calvinism and condemned the Catholic ideal of virginity. Puritan sternness was often only a seemly mask for a mischievous, playful culture.

Most Puritans were, it is believed, good lovers. John Milton typified the Puritan world view. He was showily virtuous, but had a healthy, idealistic and romantic view of sex. His epic poem *Paradise Lost* Depicts Adam and Eve as romantic lovers; he despised St Augustine's woman-hating and miserable-ism, and even lobbied Parliament for modern, easy divorce. But the Amish people, whose frank sexuality is to this day expressed in the practice of *rumspringa*, the Pennsylvania Dutch word for 'running around' (a period when teenagers are allowed to test to the limit the 'Devil's Playground' outside their closed communities), provide the best example of the Puritans' dualistic, but generally permissive, attitude to sex.

After the Restoration of the monarchy in 1660, however, the England of the seventeenth and eighteenth centuries, unlike the United States, was not remotely in the thrall of Puritanism. Despite having equally rejected the misogynies of Catholicism, it had, in most ways, become a bad place for women to live. The feminist writer Joan Smith explains that even respectable men were permitted to seduce servants and use prostitutes, and until 1774 in England it was legal for a man to place his wife in a lunatic asylum for any reason he chose. A legal decision of 1782 established that a husband could beat his wife with a stick so long as it was no thicker than his thumb. And a woman who at that time committed adultery would rarely or never see her children again. Even when the 1774 reform was passed by Parliament, Smith observes, locking up a wife (like Mr Rochester's 'mad' first wife in *Jane Eyre*) was legal, and as late as 1840 a judge ruled that a man was still entitled to lock his wife up to prevent her running away with a lover.

It is the eighteenth-century 'medicalisation of sex', as it has been called, that underpinned this period's polarisation between a clinical and a burlesque view of orgasm. It was widely believed once again (a view still in play in the twentieth century) that the female orgasm, or 'hysterical paroxysm' ('the most common of all diseases except fevers,' according to one contemporary doctor) helped women to conceive. This led to another revival of the idea of the therapeutic orgasm, but administered to women less as an expedient way of calming them down than as a course of medical treatment. Doctors in the eighteenth century were not as keen as they would be in the nineteenth to perform manually this laborious task, which could take up to an hour before the Victorian genius for mechanisation reduced it to ten minutes or less. So physicians used midwives and even husbands as their masturbating proxies.

Clitoral stimulation by a medical index finger was not the only method of rousing in recalcitrant women a healthy, conception-aiding orgasm. Other methods advocated were vigorous horse-back riding or simply sitting in a rocking chair. Bernard Mandeville, author of a 1711 book on hysterical paroxysm, *Treatise of Hypochondriack and Hysteric Passions*, recommended a stiff ride out followed by three hours of 'massage'. The reasons he put forward for the hysteria were a little contradictory; it was caused either by sexual frustration or masturbation.

Another pseudo-medical pioneer of the eighteenth century was James Graham, a moralist who, a little ironically, found himself imprisoned in 1783 for giving obscene public lectures on sex. He travelled round England and the American colonies spreading the word for healthy sex, within a strict marital context. He was also one of the world's first sex therapists, helping couples with difficulties conceiving. His patented sex aid was a magnetic, vibrating bed. When he was imprisoned, he was trying to raise £20,000 for an updated musical version of the bed.

In Europe, sexual advice was being given to the great and good by a prominent doctor Dutch physician, Gerard van Swieten, who was famous as a health reformer and founder of the Vienna Medical School. In 1740, van Swieten was consulted on a delicate matter by the Empress Maria Theresa of Austria. Maria Theresa was unhappy, unfulfilled and confused by her inability after three years of marriage not only to conceive, but even to manage a moment of penetrative sex with her shy young husband, the Holy Roman Emperor Francis I. There was no question of blaming the Emperor, so gossip in the Viennese court had it that the Empress was 'sterile' as it was called then, or sexually dysfunctional as we would say today.

Van Swieten's pronouncement on the matter does not look all that remarkable in the Latin in which he cautiously clothed it: '*Praeteria censeo, vulvam Sacratissimae Majestatis, ante coitum, diutius esse titillandam.*' Had van Swieten delivered it in modern language, the conclusion of his prescription might have caused a stir, however. The diagnosis translates as: 'Furthermore, I am of the opinion that the sexual organs of Her Most Sacred Majesty should be titillated for some length of time before coitus.'

And it would seem that van Swieten's advice, that sex worked better if it was pleasurable for both parties, hit home too. After three years of sexual drought, Maria Theresa almost immediately became pregnant and went on to have sixteen children, the eleventh of whom was Marie Antoinette, later Queen Consort of King Louis XVI of France. The couple also began to enjoy a reputation for being passionate and ardent lovers. All, it would seem, because the Empress was emboldened to ask her husband to tease her clitoris a little, to treat sex as a pleasure – and thus to turn their lovemaking overnight into something more than a clinical attempt to provide a successor.

Maria Theresa went on to become a positive advocate of sexual pleasure, according to Marie Antoinette's biographer, Antonia Fraser. Marie Antoinette was married at fourteen in 1770 to Louis Auguste, the fifteen-year-old heir to the French throne, but history repeated itself with the newly marrieds unable to consummate their relationship for another two years.

The Empress, who became quite domineering in middle age, saw this entirely as the soon-to-be Queen's fault, for failing to inspire sexual passion in Louis XVI. Marie Antoinette, according to Fraser, was interested in intimacy based on sentiment, not sex, which she regarded as a disagreeable duty. In the ribald popular pamphlets of the day, which were used to whip up Revolutionary crowds in 1789, Marie Antoinette was portrayed as both a lesbian and an unfaithful wife, since if the King was not satisfying her, somebody surely had to be; the Princesse de Lamballe, Superintendent of the Household, was said by the gossips to be working on the queen's sexual frustration 'with her little fingers'.

The level of disrespect to the Royals betrayed by this scurrilous pamphleteering about their sex lives is thought to have been one of the catalysts of the French Revolution. Yet Louis XVI and Marie Antoinette's disastrous sexual relationship had less to do with the King's foppishness and Queen's neurotic instability and more with a simple physical cause, according to research unveiled in France in 2002. The historian Simone Bertière showed that the King was the possessor of a '*bracquemart assez considérable*' (an over-large penis), and the Queen of a condition politely known then as '*l'étroisse du chemin*', an unusually narrow vagina.

With the Austrian-French alliance that was designed to be cemented by the marriage rendered rocky by the sexual problems between King and Queen and their resultant childlessness, Marie Antoinette's older brother Joseph II of Austria, according to Bertière, wrote to their brother Leopold with an interesting view on the best way of producing orgasm in sexually shy males. The French King, Joseph reported, was able to have 'well-conditioned, strong erections and introduce his member, stay there for two minutes without moving, withdraw without ejaculation and then, still erect, wish [Marie Antoinette] a good evening'.

He should, concluded Joseph, 'be whipped like a donkey to make him discharge in anger'. It was after more such extreme

Agony Uncle advice from Joseph that the unhappy couple finally conceived a daughter in 1778, four years after their succession to the throne, followed by three other children before both of them were executed in 1793.

So much, then, for the medicalisation of the orgasm in the eighteenth-century, a phenomenon that was going to be a key ingredient of the next stage in the evolution of sex: Victorian prudery. The endemic rakishness of Georgian London tells us something of the other end of the sexual spectrum that in due course contributed, by its conspicuous excess, to the Victorian mindset.

In 1780 an extraordinary kind of Good Whore Guide – *Harris's List of Covent Garden Ladies: Or Man of Pleasure's Kalender, Containing the Histories and Some Curious Anecdotes of the Most Celebrated Ladies Now on the Town* – began to be published annually. *Harris's* detailed the physical charms, attractions and prices of all the known prostitutes in the area. So successful did it prove that a companion guide to the ladies of Piccadilly was soon brought out.

The entries in *Harris's* appear today to be rather comical; it is easy to spot the kind of raw material that must have inspired *Blackadder Goes Forth*:

> Miss B., Titchfield-street: This child of love looks very well when drest. She is rather subject to fits, alias counterfits, very partial to a Pantomime Player at Covent Garden Theatre. She may be about nineteen, very genteel, with a beautiful neck and chest, and most elegantly moulded breasts, her eyes are wonderfully piercing and expressive. She is always lively, merry, and cheerful, and in bed, will give you such convincing proofs of her attachment to love's game, that if you leave one guinea behind, you will certainly be tempted to renew your visits.

A book of 1709, *The Secret History of Clubs of All Descriptions*, by a journalist, Ned Ward, gives a slightly less saccharine impression of the reality of eighteenth-century

London than does *Harris's* glowing review of Miss B. of Titchfield Street. 'Once drunk enough,' Ward reports of London 'gentlemen', they would, 'attack the mask'd ladies who hand about the theatre in their secondhand furbiloes to open the wicket of love's bear garden to any bold sportsman who has a venturesome mind to give a run to his puppy'.

Behind the bawdy, however, lay desperate social problems that would not be recognised until the next century. One view of the Victorians is that they were more reformers than mere prudes. Miss B., it is clear from reading *Harris's*, was very much carriage trade. On Drury Lane, a quickie with a lesser-starred whore could be had for a shilling or a cheap bottle of wine.

Unmarried or widowed women faced a choice between making their living as servants or shop girls, which was as underpaid a calling as ever, or taking their chances catering to the market for instant orgasmic satisfaction for the carbuncled, black-toothed and foul-breathed self-styled Georgian rakes. When arrested, the women faced being sent to a disease-ridden jail, or even transportation. And if eighteenth-century prostitutes were lucky enough to escape prosecution, they were more likely than not to contract syphilis or gonorrhoea. The *Times* reported in 1785 that 5,000 prostitutes a year in London were dying of venereal diseases.

England and Holland (where Mathijs van Mordechay Cohen had a thriving Amsterdam business selling 'condons' made of lambs' bladders) were a little more liberal sexually at this time than France. In 1723, police in Montpellier swooped on a meeting of an orgy cult, the Multiplicants. The sect's members held sham temporary 'marriages', which would be consummated publicly. The French authorities' reaction was surprisingly savage compared to the sexual anarchy in contemporary London. The leaders of the cult were hanged, the male followers were sentenced for life to the galleys, the women shaved bald and condemned to live out their lives in nunneries.

In the technology of contraception, nonetheless, the French

were stealing a march on their European rivals. The earliest bidets, a contraceptive aid that to this day confuses American tourists in Europe, came into use in France around 1710. The bidet was invented by Parisian furniture makers, according to a 1997 study by the domestic historians Fanny Beaupré and Roger-Henri Guerrand. It was known early in its existence as 'the ladies' confidant' – and named after a French term for a pony, since it was thought that using it resembled mounting a small horse.

The bidet's prime purpose was as unclear as it remains for some today; it was partially introduced for hygienic reasons since, at a time when bathing was still a once-weekly treat, sex was a smelly business. Added cleanliness was also seen as a protection against VD. It was, additionally, a method of contraception, one of a wide variety in use in France which was the first country on record systematically to reduce its birth rate; it fell between 1750 and 1800 and carried on declining into the nineteenth-century. The official, pre-Revolutionary French view of this trend was not favourable. The author of the 1778 *Récherches et considérations sur la population de la France* complained: 'Rich women, for whom pleasure is the chief interest and sole occupation, are not the only ones who regard the propagation of the species as a deception of bygone days; already these pernicious secrets, unknown to all animals save man, have found their way into the countryside; they are cheating Nature even in the villages. If these licentious practices, these homicidal tastes, continue to spread, they will be no less deadly to the State than the plagues, which used to ravage it.'

Contraception in Britain was an earthier and manifestly less successful business. It also became caught up late in the century in the unpleasant business of eugenics, when a British economist, Thomas Malthus of Dorking, became famous as an advocate of family planning – as a method of keeping down the numbers of the poor. In the mid-eighteenth-century, the condom market was a duopoly for two women – a Mrs Perkins and a Mrs Phillips, who had a wholesale business on Half

Moon Street in London. Mrs Phillips even had an advertising jingle:

> *To guard yourself from shame or fear,*
> *Votaries to Venus, hasten here;*
> *None of our wares e'er found a flaw*
> *Self-preservation's Nature's law.*

The bidet did not take off in Britain, but post-coital douching by other means was later promoted by an 1832 book, Dr Charles Knowlton's *The Fruits of Philosophy*. He explained the importance of the douche, claiming to have invented it, and advocated not water but a solution of alum (an industrial chemical used medically as an astringent) mixed with green tea or raspberry leaves. Dr Knowlton was arrested on a sales trip to Cambridge, Massachusetts, and jailed for three months for trying to sell his book there.

While the bidet was revolutionising French plumbing, Paris became the scene of a strange and rare cult of what seems to have been communal female masturbation. The deeply peculiar early hypnotist (or 'magnetiser'), the Austrian quack Friedrich Anton Mesmer (1733–1815), attracted a retinue of women and girls to sessions in the city where they would sit in a circle around a basin of 'magnetised water', holding hands and touching knees. Assistant magnetisers, generally handsome young men according to one account, 'embraced the patient between the knees', massaging her breasts and torso as they gazed into her eyes. A few jangling notes would then be sounded on a piano, whereupon the women would flush redder and redder, until they descended en masse into convulsive fits, reportedly laughing, shrieking, sobbing and tearing their hair, After the 'crisis' was over, Mesmer (later immortalised in the word 'mesmerism') would make his appearance and stroke the women on their faces, breasts, spines and abdomens so as to restore the 'insensible' to 'consciousness'. It has never been established what really went on in Mesmer's sessions, but they seem to bear the hallmarks of a masturbatory experience for the women.

More traditional male masturbation, however, continued to be the perennial bone of contention in the eighteenth century. 'Self-pollution', as it was now known, blossomed into the subject *du jour*. A prominent English surgeon, John Hunter, came close to approving masturbation, saying it was better for the constitution than sexual intercourse as the emotions were not roused. He was soundly criticised by professional colleagues for this view.

Yet the pendulum was swinging in the direction of approval of masturbation (and the related 'problem' of spontaneous ejaculation), which only a century before had been considered a serious sin. Previously, sounding off against the practice was a Church obsession. Now, in keeping with the medicalisation of the age, it was starting to be viewed on its merits. Edward Shorter, author of *The Making of the Modern Family*, explained that condemnation of masturbation, far from being a taboo of any importance, was effectively unknown before the Industrial Revolution, and was barely mentioned at all in serious medical texts before the eighteenth century; even pornographic literature seldom bothered with it. During the extended period of the Revolution, furthermore, rather than being suppressed, Shorter argued that masturbation came into its own as part of a 'premarital sexual revolution' that did much to 'modernise' relationships.

A wide range of experiential material on masturbation was published across the Western world. In Switzerland, Jean-Jacques Rousseau, in his 1792 *Confessions*, 'frankly and sincerely' told all. He admitted being addicted to masturbation, which he explained he did while thinking about beatings he received as a child from his foster mother. He also said his first sexual encounter left him feeling he had committed incest, leaving him unwilling to have sex with women.

But Rousseau was not the first celebrity masturbator. Samuel Pepys, at the height of the Restoration, recorded a supposedly involuntary orgasm (but more likely obtained by subtle frottage) whilst watching the King's mistress, Lady Castlemaine, in church. The scholar John Aubrey at the same time recorded how he finally worked out why his tutee, the Duke of Buckingham, lacked concentration in lessons – the young milord was masturbating. And in the colonies, Joseph Moody, the Harvard-educated schoolmaster of York, Maine, noted in a diary of 1720–4 fifty-five instances of his own masturbation, according to research by Brian Carroll of the University of Connecticut for an article in the *William And Mary Quarterly*.

> Thurs. [July] 19 [1722]. This morning I got up pretty late. I defiled myself, though wide awake. Where will my unbridled lust lead me? I have promised myself now for a year and a half that I would seek after God, but now I am perhaps farther away from him than ever before.

> Mon. [April] 13 [1724]. Pretty cold; wind from N. W. to S. fine weather . . . I dined with the doctor and schoolmaster Abbott. Then with the doctor I called on Captain and Ensign Allen. I stayed up with my love not without pleasure, but I indulged my desire too freely, and at night the semen flowed from me abundantly.

Thomas W. Laqueur's landmark 2003 book *Solitary Sex: A History of Masturbation* pinpoints a specific moment when, he contends, masturbation became a serious moral issue as opposed to something for country parsons to bang on about. Laqueur identifies the fulcrum point as the release in Boston in 1724 of an anonymous tract called *Onania: Or the Heinous Sin of Self-Pollution, and all its Frightful Consequences, in both Sexes, Considered*. It set out that masturbation caused many illnesses, including 'palsies, distempers, consumptions, gleets, fluxes, ulcers, fits, madness, childlessness' and even death. Laqueur argues that masturbation began now to worry not conservatives, but progressives. What could bother a progressive thinker of the Enlightenment, the Age of Reason, about a private, non-exploitative pleasure like masturbation? One might as well ask a twenty-first-century liberal what could possibly bother him about a private, non-exploitative pleasure like smoking; for sketchy, unscientific reasons, masturbation had become a matter of public morality and, very nearly, public policy.

The Enlightenment was all about learning to think of the self as regulated by one's own moral compass – not the dictates of religion. And that is why masturbation came to be perceived as a threat; it endangered self-discipline and self-care. Laqueur identifies three aspects of masturbation that troubled Enlightenment thinkers: 'It was secret in a world in which transparency was of a premium; it was prone to excess as no other kind of venery was, the crack cocaine of sexuality; and it had no bounds in reality, because it was the creature of the imagination.'

In eighteenth-century France, too, according to Michel Foucault in another milestone, if abstruse, sexual studies text, *The History of Sexuality*, masturbation and sexual thought in general came to be seen as a huge social problem. Not just the discipline, but the very architecture – the classrooms, the shape of the tables, the dormitories – 'all this referred, in the most prolix manner, to the sexuality of children. What one might call the internal discourse of the institution – the one it employed to address itself, and which circulated among those who made it function – was largely based on the assumption that this sexuality existed, that it was precocious, active, and ever present . . . Around the schoolboy and his sex there proliferated a whole literature of precepts, opinions, observations, medical advice, clinical cases, outlines for reform, and plans for ideal institutions.'

Another student of sexual history, Brett Kahr, regards André Tissot, the Swiss author of a 1760 anti-masturbation tract, as responsible for 'the greatest advance of the eighteenth century'. This physician, explains Kahr, urged generations of children to refrain from touching themselves. 'It may seem extremely strange to regard Tissot as a progressive thinker,' writes Kahr, 'because we all know about the psychological suffering that millions have endured in subscribing to the misguided belief that masturbation causes blindness and insanity. But Tissot not only instructed youngsters to refrain from fondling themselves, he also advised the boys and girls to forbid the adults from molesting them as well . . . his widely disseminated tract also laid the foundations for the crusade against child abuse, a cruel institution which damages one's ability to enjoy sexual relationships in adulthood.'

Masturbation was less problematic in parts of eighteenth-century Scotland, or so it seems. Here, ritualised 'self-pollution' was, a little bizarrely, the founding and practically sole purpose of the Beggar's Benison, a gentlemen's club of the time. That and another startling institution devoted to the worship of female pubic hair were the subject of a Scots-published 2001 book, *The Beggar's Benison: Sex Clubs of Enlightenment Scotland and Their Rituals*, by David Stevenson, Emeritus Professor in the Department of Scottish History at St Andrews University.

The Beggar's Benison and the Wig Club were ostensibly organised around 'the joys of libertine sex' and frolicking in the 'Colony of Merryland' (in 1652, a London journal had referred to a brothel-ship on the Thames as being 'bound for Merryland'). But the bourgeois and upper-class members were mostly interested in masturbation, many being past the age of being immediately attractive even to hired women. A prominent 'Beggar' was a gentleman called David Low. The club's activities must have provided an interesting diversion from his work; he was the Minister of an Episcopalian congregation and, says Stevenson, 'the nominal Bishop of Ross, Moray, Argyll and the Isles as well as'.

The Beggar's Benison was founded, unlikely though it seems, in 1732 in the herring fishing village of Anstruther in Fife, on the Firth of Forth. An extreme reaction to the dour

Calvinist moralism of these parts, it took its name from a (probably apocryphal) incident in Scottish history when the famously libidinous King James V, travelling incognito, was carried across a river at Dreel Burn by an old peasant woman of the kind that stationed themselves at river crossings to assist gentlefolk keen to avoid getting their feet wet. He paid the woman, possibly (Stevenson imagines) had sex with her, whereupon she thanked him with the blessing, or benison, 'May your purse ne'er be toom [empty] and your horn aye [always] in bloom.' Roughly translated as, 'May you always have plenty money and a fine erection.' This was adopted as the Beggar's Benison motto. The Beggars' emblem, reflecting their seaside origins, was a crossed anchor and penis.

Quoting from the Beggar's Benison's records, Stevenson described the club's initiation ceremonies: 'The novice was "prepared" in a closet by the recorder and two remembrancers "causing him to propel his Penis until full erection". He then came out of the closet, a fanfare being provided by "four puffs of the Breath Horn", and placed "his Genitals on the Testing-Platter", which was covered with a folded white napkin. "The Members and Knights two and two came round in a state of erection and touched the novice Penis to Penis." A special glass with the order's insignia on it was then filled with port and a toast drunk to the new member, and, in a brief parody of a church service, he had to read aloud an "amorous" passage from the Song of Solomon and comment on it. Investment with sash and medal followed, as the sovereign and other members intoned the benison or blessing: "May Prick and Purse never fail you!"'

The club record is oddly coy about spelling it out, but it was considered good form for the novice, if possible, to masturbate himself to ejaculation at the ceremony. 'The Beggars dined and drank in an atmosphere of happy obscenity,' Stevenson says, 'delighting themselves with how shocking they were being. The air was full of sexual innuendoes ranging from basic crudity to some quite ingenious and witty efforts.' Two of their toasts were, 'to girls lecherous, kind and willing' and 'to the mouth that never had a toothache'.

As befits a Rotary-style club dedicated to masturbation, the members (so to speak) also revelled in an early form of lap-dancing. 'Posture girls' were hired to pose and show their genitals. At the Candlemas meeting of 1734, we learn from Stevenson: 'One Feminine Gender, 17, was hired for One Sovereign, fat and well-developed. She was stripped in the Closet, nude; and was allowed to come in with her face half-covered. None was permitted to speak to or touch her. She spread wide upon a Seat, first before and then behind; every Knight passed in turn and surveyed the Secrets of Nature.' At another life class of 1737: 'Two girls, 16 and 17, posed, exhibited, and danced nude.' Appearing at the club was not always an advisable career move, however. One girl who displayed her body was shamed on her wedding day when a well-wisher shouted in the church, 'That's the bitch that showed her hairy cunt and arse to the gentlemen of the Beggar's Benison for five shillings.'

Just as *Playboy* magazine carries serious articles, the Beggar's Benison hosted occasional lectures on worthy-ish topics. One such, given in 1753 by James Lumsdaine, was on 'The Act of Generation'. Lumsdaine considered the Biblical 'Be fruitful and multiply'. This cannot have been Nature's only intention in equipping man with sexuality, Lumsdaine argued, for the sexual urge continued even after one had done his multiplication. Indeed, if men tried to stop being sexually active, they would be left with 'living plagues' – strong, unfulfilled sexual urges, causing anguish and wasting a capacity for great enjoyment. Nature, he argued, has given us 'this exquisite endowment', and it becomes us to enjoy the great gift. He did, however, urge the use of contraception. Ensuring that unwanted conception does not take place was, for Lumsdaine, central to love-making; men should not be thoughtless brutes or beasts in satisfying their desires. The 'sexual embrace should be independent of the dread of a conception which blasts the prospects of the female'.

The Beggar's Benison never quite became a national institution, but it prospered modestly in the surprisingly fertile soil of Nonconformist Scotland. It expanded to Edinburgh and Glasgow, then across Scotland. The Benison even reached as far as Russia. A Benison medal was identified in the 1980s in the Hermitage Museum in St Petersburg. Russia also developed its own sex clubs in imitation; Peter the Great established one called the 'Most Drunken Synod'. Lodges created over fifty job titles for members, from 'orator' to 'prick farrier' and 'cunt-peeper'. The punishment specified for wrong-doing members was to have their penis smeared with egg yolk and oats and then presented to a pair of hungry ducks who would gobble it clean. The club had a British branch especially for the Moscow expat community. It was called 'the British Monastery'.)

The Wig Club was founded in 1775 in Edinburgh, and named after its most venerated item of regalia – 'one renowned Wig worn by the Sovereign composed of the Privy-hairs of Royal courtezans' – a hairpiece, that is, made of the pubic hair of Charles II's mistresses. The Wig Club's members ranged from the ages of 21 to 68 with the average admission age being in the late-30s. 'Each member of the Wig Club had to provide locks of public hair suitably harvested to be added to the wig, as proof of having triumphed sexually,' recounts Stevenson. 'Whether this proof of ability on matters sexual is more or less distasteful than the Benison's masturbation may be debated.'

The Wig Club was more ribald than even the Beggars, who inclined a little towards solemnity. Stevenson tells of one Wig Club meeting where '. . . it was ordered that all members were to drink "to the Wig out of the Prick Glass", which, just in case things weren't achingly jocular enough, may have been a trick glass "designed so that part of the contents would be released suddenly when the glass was raised to a certain angle, so drenching the drinker".'

Almost wholesome by comparison with the Scottish sex clubs was another marker buoy of eighteenth-century sexuality, John Cleland's erotic 1748 novel *Fanny Hill, or the Memoirs of a Woman of Pleasure*. This tale of young Fanny's arrival in London as an innocent orphan from Liverpool, her swift descent into prostitution and subsequent experiences in a brothel run by Mrs Brown, takes the form of confessional letters bursting with all the gynaecological detail the most demanding male reader could want.

The first work of pornography to be written in English, it was an immediate bestseller, even though full, unexpurgated editions were not available until after the renowned indecency trial in 1963 of the publishers of *Lady Chatterley's Lover*. In the same way as that twentieth-century trial provided a huge publicity boost for an obscure pornographic text, the legal hoohah that accompanied publication of the first part of *Fanny Hill* in 1748 was a splendid advertisement for it.

John Cleland was a Westminster School boy in the 1720s and for some years worked for the East India Company in Bombay. He started writing *Fanny Hill* around 1730 in India. He returned to London in 1741 and soon fell into debt – he actually completed *Fanny Hill* in the Fleet debtors' jail. Cleland's financial problems combined with a legendary grumpiness isolated him from literary circles, but his father had been a friend of Alexander Pope, and he was known to the day's luminaries, Boswell, Garrick, Pope and Smollett. He survived to the impressive age of seventy-eight, his earnings from *Fanny Hill* only just keeping him going through a barren subsequent literary career. He wrote little else – three more erotic, but long-forgotten, novels and some hack journalism.

The moment *Fanny Hill* came out, there were the expected complaints from the Church, but Cleland was not prosecuted for indecency. He was, however, arrested for obscenity and summoned before the Privy Council. There he pleaded poverty; the Privy Council gave him a warning but accepted his mitigation that he was only doing it for the money and awarded him, puzzlingly enough, more money still in the form of an annual pension of a hundred pounds. Ralph Griffiths, the St Paul's Churchyard bookseller who stumped up a hefty twenty-guinea advance for the book, went on to make a reputed profit of £10,000 from sales of *Fanny Hill.*

What stands out about Cleland's work is that, in the global tradition of pornography for men, it mostly consists of wishful male thinking – to wit, that almost anything a man does will have the ladies simpering in delight. The orgasms are described from a female perspective, in keeping with the convention that even male writers never attempt to describe the male orgasm, preferring instead to imagine the female. Although it is ripe with voyeuristic, masturbatory fodder, it is surprisingly prim, light on rude words, heavy on euphemism and with just one fleeting allusion to oral sex; cunnilingus and fellatio were simply too smelly to consider in the unwashed eighteenth century.

Fanny Hill is, nonetheless, creditable pornography that succeeds in still being erotic today, as well as packed with sex action. Finding the sexual scenes, indeed, is like looking for hay in a haystack: 'My young sportsman wantonly takes my hand and carries it to that enormous machine of his that stood with a stiffness! a hardness! an upward bent of erection! and which, together with the inestimable bulge of lady's jewels, formed a grand show out of goods indeed! . . . Chiming then to me, with exquisite consent, as I melted away, his oily balsamic injection, mixing deliciously with the sluices in flow from me, sheath'd and blunted all the stings of pleasure and flung us into an ecstasy that extended us fainting, breathless, entranced . . . I no sooner felt the warm spray darted up my innards from him, but I was punctually on flow to share the momentary ecstasy.'

Then there is Fanny's first attempt at masturbation, again through the rose-tinted monocle of Mr Cleland: '. . . guided by nature only, I stole my hand up my petticoats, and with fingers all on fire, seized, and yet more inflamed that centre of all my senses; my heart palpitated, as if it would force its way through my bosom; I breath'd with pain; I twisted my thighs, squeezed, and compressed the lips of that virgin slit, and following mechanically the example of Phoebe's manual operation on it, as far as I could find admission, brought on at last the critical ecstasy, the melting flow, into which nature, spent with excess of pleasure, dissolves and dies away.'

And here is Cleland's male perspective, steering scrupulously clear of giving any hint of what orgasm is like for a man and concentrating instead on the visual – plus the male experience of the woman's orgasm: 'What firm, smooth, white flesh is here! . . . How delicately shaped! . . . Then this delicious down! Oh! let me view the small, dear, tender cleft! . . . This is too much, I cannot bear it! . . . I must . . . I must . . . Here she took my hand, and in a transport carried it where you will easily guess. But what a difference in the state of the same thing! . . . A spreading thicket of bushy curls marked the full-grown, complete woman. Then the cavity to which she guided my hand easily received it; and as soon as she felt it within her, she moved herself to and fro, with so rapid a friction, that I presently withdrew it, wet and clammy, when instantly Phoebe grew more composed, after two or three sighs, and heart-fetched Ohs! and giving me a kiss that seemed to exhale her Soul through her lips, she replaced the bedclothes over us.'

Fanny Hill nevertheless has an unexpectedly suburban dénouement, though one appropriate to a time in which, in spite of rampant bawdiness, Josiah Wedgwood took care to stick spurious fig leaves to the relevant bits of his classical china nudes. Fanny meets – well, is rented by – a young buck called Charles, falls in love with him and becomes his kept woman. They live in sin for a while before he vanishes abroad, leaving her to rekindle her old career in a new brothel. She becomes the kept woman of an older man, who then dies, leaving her his fortune. She is reunited with Charles, and they marry.

The era of *Fanny Hill* (and of the Venetian sexual adventurer Casanova, who lived from 1725–98, for that matter) begins to peter out towards the end of the eighteenth-century as the early signs of Victorianism – 'The Great Forgetting', as Naomi Wolf has called it – start to appear.

There were several last-gasp libertine carry-overs into the new century. In 1828, the by-then veteran political and sexual radicals, Richard and Jane Carlile, published an attack on conventional sexual morality, *Every Woman's Book, or, What is Love, containing Most Important Instructions for the Prudent Regulation of the Principle of Love and the Number of a Family*. The Carliles argued that passion, subject to birth control, should be given free rein. They believed that all kinds of illnesses were caused by sexual frustration, and wanted to build a Temple of Venus where men and woman could enjoy sex purely for pleasure and outside the bounds of marriage. In accord with the new trend for anti-masturbation sentiment, they maintained that masturbation led to physical and mental disease. It was wrong, they said, because it diverted sexual desire and pleasure away from the family The Carliles' free love vision, on the other hand, was for them the solution to the plague of prostitution and paedophilia.

The Germany of Goethe and Schiller proved highly influential in orchestrating the sexual tempo of the opening years of the nineteenth-century. Weimar, the German Oxford, became a hotbed of both heterosexual and homosexual passion, which were extolled with solemn eloquence – and practised with erotic verve. Weimar intellectuals from Schlegel to Liszt reconciled their outward Christianity with a sturdy defiance of its restrictions, demanding and grasping a freedom in sexual matters which scandalised burghers and Church alike.

But arguably darker forces were already at work across the world. Sex before marriage was so rife in eighteenth-century Britain that in the second half of the century more than 40 per cent of children were conceived as a result of sex before wedlock. Few societies can sustain such potentially destabilising non-compliance with consensus values, and the Marriage Act of 1753 stipulated that all sexual relations were to be banned until after marriage. Four years later, a censorious Society for the Reformation of Manners had been officially endorsed by George III, who issued a Proclamation against Vice to back it up. In 1802 a still more severe Society for the Suppression of Vice was founded in London by William Wilberforce, the anti-slavery campaigner. The society was designed, '. . . to check the spread of open vice and immorality, and more especially to preserve the minds of the young from contamination by exposure to the corrupting influence of impure and licentious books, prints, and other publications.'

There was now a perceptible swing away from what had been acceptable by default in the previous century. Medical texts became dismissive of sexual pleasure, witness the *Medical Journal*'s 1802 statement that, 'Many viviparous animals are subject to periodical venereal orgasm', meaning, one assumes, that humans are above that sort of thing. Middle-class, turn-of-the-nineteenth-century European men began to look to wife and home for relief from the morally tarnished life of coffee house, whores and scandals. In the same way as the early Christians tried to forge ahead with a modern approach to life, by which they meant ditching their licentious, sexy pre-Jesus ways, the coming of the new century in Europe prompted a bout of anti-sex modernism. A new middle class 'erotophobia', as it has been called by some historians, was formulated on an ostensible reverence for women's superior spiritual being. And just as a franker, brasher take on sex would spread from West to East in the twentieth century, in the early nineteenth, the 'cleaned-up' Western view of sexuality reproduced virally through many other parts of the planet. In the Islamic Ottoman Empire, for example, homosexuality, which had been happily tolerated since the fourteenth-century, became taboo with the adoption of a Western influenced culture. The same thing happened in India, where a hitherto fluid view of sexuality was more general until British influence began to establish itself.

The new Western worship of supposed womanly temperance came, however, at a cost for women themselves, as it precluded them from fulfilment of their sexual needs. Yet even women willingly bought into the myth. Even Mary Wollstonecraft, the virtual founder of feminism, scolded her peers in the 1790s for what she saw as their selfish pursuit of pleasure and argued, as did many others of her stripe, for the control of sexual passion by religion and reason. Women, for Wollstonecraft, were figures who 'civilise men and raise children up to virtue'. A century later the Suffragettes also willingly embraced the concept of female 'nobility', untainted by lust for orgasmic relief.

Literature was subjected to a revisionist fervour that would put the twentieth-century Taliban to shame. Jane Austen was condemned for allowing Lydia in *Pride and Prejudice* to elope without disaster befalling her. Dr Thomas Bowdler, a Swansea medical man and co-founder with Wilberforce of his anti-vice society, gave up medicine to practise a crude form of censoring surgery on the plays of William Shakespeare, the Old Testament and Gibbon's *Decline and Fall of the Roman Empire*. He removed from all these lewd works 'words and expressions which cannot with propriety be read aloud in a family', and gave the English language a new word – 'bowdlerising', meaning literary emasculation.

Public concern over masturbation reached hyperbolic extremes. In 1834, the Reverend Sylvester Graham, a Philadelphia Presbyterian minister and proselytiser for vegetarianism, published *A Lecture to Young Men on Chastity Intended Also for the Serious Consideration of Parents and Guardian*. Graham, who *inter alia* invented the ubiquitous American staple Graham flour and believed humans should be more like orang utans, set the pattern for a century of hysterical condemnations of masturbation and sex in general, predicated on no discernible research or thought process at all.

Decades before the notorious John Kellogg, another eccentric cereal founder portrayed in the film *The Road To Wellville*,

Graham believed sexual desire was heightened by 'high seasoned food, rich dishes, the free use of flesh', that masturbation reduced life expectancy and led to insanity, and that over-sexed married couples would be struck down with 'languor, lassitude, muscular relaxation, general debility and heaviness, depression of spirits, loss of appetite, indigestion, faintness and sinking at the pit of the stomach, increased susceptibilities of the skin and lungs to all the atmospheric changes, feebleness of circulation, chilliness, headache, melancholy hypochondria, hysterics, feebleness of all the senses, impaired vision, loss of sight, weakness of the lungs, nervous cough, pulmonary consumption, disorders of the liver and kidneys, urinary difficulties, disorders of the genital organs, spinal diseases, weakness of the brain, loss of memory epilepsy insanity apoplexy abortions, premature births, and extreme feebleness, morbid predispositions, and an early death of offspring'.

Graham's prescription for the good life was an overweening blandness in food and sex. But a general onset of mildness in reaction to the sexual excesses of the eighteenth century was already well in train. The idea that women's sexuality is inherently milder than that of men dates from this time. The 'modern' medicalised view was that women were more angelic than animal, their desires focused not on lust but on affection and domesticity.

There are many convoluted theories of what, other than the kind of crudeness illustrated by Hogarth's engravings, brought about this attitudinal shift. Democratisation – which certainly was the political drift of the period, increased literacy, belief in science above superstition, and changing patterns in economics have all been argued as causes for the new blandness. The most plausible theory is that women were recognised, biologically speaking, as being better equipped than men to control the impulsive, destructive drive of carnal desire. Libido was increasingly ascribed purely to masculinity. So middle-class women as never before, seeing their less fortunate sisters floundering in a morass of sexual vice and disease, took on sexual 'purity' as their defining virtue. For poor women prostitution continued to boom, but strictly as a service industry to cater – quite usefully – to men's base lust.

As for hope that a more sexually mature society might emerge in the nineteenth century from the anarchy of the eighteenth, there was precious little. In 1821, a Massachusetts court found a Boston printer, Peter Holmes, who tried to publish an American edition of *Fanny Hill*, guilty of smut-peddling. It was the United States' first obscenity trial. The judge had refused to see the book, or let the jury see it, or to enter passages from it into the court record. To do so would 'require that the public itself should give permanency and notoriety to indecency, in order to punish it'. Holmes appealed to the Massachusetts Supreme Court, pointing out that the judge had not seen the book. The superior court was not swayed. The Chief Justice wrote that Holmes was, 'a scandalous and evil disposed person' contriving to 'debauch and corrupt' the citizens of Massachusetts and 'to raise and create in their minds inordinate and lustful desires'.

In Britain too the retreat was being sounded for sexual liberation. The Wig Club closed for business in 1827. The Beggar's Benison died out in 1836, although rumour had it that as late as 1861, covert Beggars' groups were still meeting. But as Victoria was being groomed for the throne, it was simply no longer acceptable, even within the confines of a private gentlemen's club, to be overtly sexual.

13

A Tale of Two Sexes? The Orgasm From Victoria to Health & Efficiency

'Literally every woman who yields to her passions and loses her virtue is a prostitute'

Bracebridge Hemyng, quoted in Henry Mayhew's *London Labour and the London Poor*

Victorian prudery was not Queen Victoria's fault. It started before her time, and is argued by many historians actually to have been on the ebb by the beginning of her reign. The pious hope that the nineteenth century should be more 'civilised' than the seventeenth and eighteenth was already forming during the eighteenth. It was in 1791 that the *Gentleman's Magazine* boasted, with no irony intended: 'We are every day becoming more delicate, and, without doubt, at the same time more virtuous; and shall, I am confident, become the most refined and polite people in the world.' Neither was the disparaging of sex that is associated with Victorian Britain just a British phenomenon. As we will see in this chapter, some of the most enthusiastic 'Victorians' were in Europe and the newly liberated American colonies.

Just as the Christian disavowal of sexual pleasure early in the first millennium was made in the name of progress and modernity, the nineteenth-century trend towards a more sexually covert and sober society was informed by a mélange of the blossoming belief in science plus notions of moral 'purity'. There was a specifically new twist in this anti-sexual 'revolution', however – the placing of women on a pedestal, if a rather rickety one.

A new idea of women as pure and sexless began to be fostered. Very broadly speaking, the Victorians wanted to differentiate themselves from morally laxer days and to equate sexual abstinence or control with moral and religious purity – to assure the ascendancy, that is, of restrained 'female' sexuality over dissolute 'male' ways. The Queen herself, in so far as it matters, was by some accounts quite a bawdy soul, as exemplified by her flirtatious relationship with her servant John Brown.

One sexual legend about her is that she failed to give the Royal assent to an Act of Parliament outlawing lesbianism because she refused to believe such acts could occur between women. In fact, the Criminal Law Amendment Act of 1885 as it was sent for Queen Victoria's assent did not specify lesbian practices; whether this was because the government withheld mention of such things because it *might* offend the Queen is not known.

Nevertheless, history being the inexact study it is, the very idea of the sexually pure Victorian woman and her lascivious, cynical, whoring male counterpart is subject to frequent revisionism. No one scholar can alight on a single eternal truth; the best we can do in assessing the status of the orgasm throughout this period is to rely on informed generalisation and anecdotal detail – and to be suspicious of every new theory in case it is the scholarly equivalent of masturbation, the spinning of a new idea for the sheer pleasure of it.

As for the kind of material that could be regarded as evidence, some of the best-known is obscure in origin, while the clearest seems plain enough but needs to be understood in the context of a time when hypocrisy was as much in the ascendant as iron smelting. Take, for instance, the most totemic of supposedly Victorian beliefs about sex, the idea that for women it was a matter of 'lying back and thinking of England'. It is unclear if anyone ever said this, let alone believed it. Some sources suggest it was 'a Victorian mother' instructing her daughter on her wedding night about the birds and the bees. Others say the phrase was coined by Queen Victoria on being asked how best to endure the pain of childbirth.

There is an abundance of source evidence for Victorian women being encouraged to adopt what has been called 'the clinging-vine personality' – the art of appearing weak, anxious to lean on strong men and be dominated by them – to be very different, in fact, from Jane Austen's feisty women. The *Encyclopaedia Britannica* of 1842 stated how a woman being courted should develop a sweet, modest exterior; she ought to convey her feelings only by a 'timid blush' or the 'faintest of smiles'.

Mrs Sarah Ellis, who wrote a guidebook in the same year called *The Women of England, Their Social Duties and Domestic Habits*, emphasised the importance for women of recognising 'the superiority of your husband simply as a man'. Clergymen, teachers and journalists supported this consensus view.

Among the upper and middle classes, sexual desire was swept under the hearthrug (although not necessarily out of existence) by such a movement. By the mid-nineteenth century, the passionate, orgasmic Nature of women, which had been a given – regarded, indeed, as a threat – for thousands of years was missing, presumed dead. Not even marriage was permitted to liberate female desire; decent wives were expected to be willing but passionless sexual partners; ladies did not move during sex, and only consented to it at all to please their husbands and have children. It was not that women were passionless, but that their passion manifested itself in the superior maternal instinct rather than their sex drive. A French author, Auguste Debay, did suggest nonetheless that women should fake orgasms since 'man likes to have his happiness shared'.

There were other cultural influences, too, dictating that women's desire for orgasmic pleasure was muted. Staying indoors, out of the sun, and preferably reclining, as was the custom, certainly depleted well-to-do Victorian women's health and can only have nullified any sex drive that survived the propaganda onslaught against such unrefined thoughts. Some young women drank vinegar or forced themselves to stay up all night in the belief that one or the other would make them appear pale and interesting. The 'ladylike' approach to pregnancy, which viewed it as a form of disability requiring 'confinement', must also have played havoc with any desire by a woman to express herself as a sensual being.

The fashions of the time, too, seem designed to deplete female libido, even while increasing male desire. Corsets laced tightly to emphasise the (highly sexual) egg-timer figure frequently caused internal injury, lowering still further any chance of a woman consciously wanting sex. Women, according to some accounts, were also expected to wear bathing suits in the bath and remain mostly clothed during sex, which extinguished any last chance of their becoming aroused.

Ignorance of the female form on the part of men was endemic. The art critic John Ruskin, despite having seen the classical statuary of Paris and Venice, was allegedly so shocked when he saw his young wife Effie's pubic hair for the first time that he went into apoplectic spasms and later confided in the Queen herself that there was something wrong with his wife. It is hard to imagine Effie Ruskin's self-esteem, as we would term it now, let alone sexual desire, surviving her husband's horrified reaction to her body; he never consummated the marriage. Robert and Elizabeth Barrett-Browning short-circuited any such marital awkwardness by never seeing one another naked – a very common occurrence in nineteenth-century marriages. Theodore Watts Dunton, Algernon Charles Swinburne's literary agent, was treated to his first glimpse of female breasts on his deathbed. Watts Dunton was married, but had never seen a woman naked; his nurse bared all at a kindly friend's request. (It is interesting, considering how visual men's orgasmic experience is, to speculate on how very frustrated Victorian men must have been when sex was entirely about fumbling in the dark.)

The Victorian anti-sex fashion reached ludicrous excesses. With female legs never seen in public and rarely in private, it became a mark of indelicacy to offer a lady a leg of chicken ('dark meat'). She would automatically be given breast, or rather 'white meat'. 'Victorian' Americans took this level of refinement further still; just as they later invented 'English muffins' that are unknown in England, they reportedly took the English habit of modesty regarding legs a stage further by fitting the piano legs in some homes with elaborate crinolines. Perhaps they were being ironic, or it was simply an amusing fashion statement. Perhaps not.

There was one important group of women, however, in Victorian cities who were considered by the intelligentsia to possess boundless sexual desire. This was the poor who despite bad food and living conditions and consequent ruinous health were regarded as being an indefatigable repository of sexuality, forever producing illegitimate children by their ceaseless demand for sex.

Here, we cut into the still darker meat of Victorian sexuality. For the middle-class male mind, sexually starved as a result of its own imposition of sexlessness on 'respectable' women, could not stop dreaming of the terrifying but tempting forbidden treat that prostitution might provide. Poor urban women, rosy-cheeked country girls and cheap, plentiful female servants all appeared to the Victorian man as potential prostitutes.

They were the nineteenth-century equivalent of the witches that frightened and fascinated the Medieval mind. Using prostitutes, fantasising about them, or simply harrumphing about their supposed ubiquity were all ways in which Victorian men seemed to get in touch with their libidinous self. Street girls had been plentiful enough a hundred years earlier, but now pretty much any attractive single woman was regarded as a prostitute. Respectable working- or clerical-class courting couples in places like Clapham Common in London would be mistaken by outraged middle-class residents for prostitutes with their clients and reported to the police.

This moral panic, a rank mixture of obsession and repulsion, led to hugely exaggerated figures for the real numbers of prostitutes operating in cities (50,000 according to one contemporary estimate in London in 1850) as well as a disastrously confused notion of what genuine street women's motivation was. Whereas today prostitutes and other 'sex workers' happily – proudly, often – acknowledge that they are motivated primarily by economics and are no more or less likely than the average woman to enjoy sex, in the Victorian world view, if a woman enjoyed sex, it was a sure sign that she was a 'prostitute'.

Any amount of speculation can be entered into as to the below-the-line machinations in the male mind that led to this bizarre conclusion. It would not be far off the mark to suggest that men found themselves caught between the expectation that they should be strong, dominant and initiate sexual activity, and the often embarrassing reality that they were rather incompetent at it. In such a bind, it would be all too easy, rather than try to satisfy a critical wife, to buy sex from a seller who was not bothered how competent the purchaser was, so long as he paid up. It is tempting, but untrue nonetheless, to assert that all middle-class Victorian men used prostitutes. Equally, to argue that many women colluded in their respectable husbands' obsession with prostitution is bound to be a distortion of reality; yet we have Sarah Ellis arguing in her book that a woman should consider herself lucky if her husband saw prostitutes, because it meant she herself would not have to endure endless pregnancies.

The really damning charge that can be levelled against the Victorian man is that of prurience rather than the use of prostitutes. The prohibitions and restrictions of the age fostered a spirit of unhealthy, even grotesque, obsession over precisely what was going on under ladies' (or, in America, even pianos') crinolines. For instance, a single mother hoping to leave her baby at London's Foundling Hospital had to give the male admissions panel a detailed record of her sexual relationship with the child's father. The panel would want to know every detail – where the trysts had taken place, the Nature and intensity of her feelings and desire. She would also have to give names and details of those who could corroborate her story such as doctors, relatives, employers, and even provide 'proof' such as love letters. The pious hope was that the woman would be able to show her good faith had been betrayed, that she had succumbed to sexual passion only after a promise of marriage had been made, or else been taken advantage of totally against her will. Her conduct since falling pregnant had to be unimpeachable, her means non-existent, and the child under the age of one.

An 'official' myth of generalised male degeneracy accompanied the porcelain-skinned, sexless archetype of the middle-class Victorian lady. The idea, at least, of the fornicating Victorian bogeyman is supported by popular literature of the time, particularly that of Dickens, and by British legislation. The 1857 Divorce Act, according to Joan Smith, provided that a man could divorce his wife on the grounds of her adultery. A husband, however, who was 'a little profligate', in the words of one of the act's sponsors, could not be penalised by divorce.

There is certainly evidence that out of Victorian repression arose a great hunger for fantasy sex besides real prostitution. Sales of pornography boomed, albeit smut of a pitiful standard compared to John Cleland's. Victorian pornography, with one possible exception as we shall see later, was largely joyless and suffused with violent sadomasochism. None of it suggests that its buyers were sexually sophisticated or sensual by today's frames of reference.

Yet familiar as the archetype is of the hypocritical, waxed-moustached, middle-class Victorian seducer, it is far easier to pinpoint evidentially the anguished Ruskin type, suffering from the male version of the vapours because of his own sex's dissolute excesses. Thus we find John Addington Symonds, the gay (but vehemently anti-gay in public) poet, critic and historian, writing of his time at Harrow School: 'Talk in the dormitories and the studies was incredibly obscene. Here and there one could not avoid seeing acts of onanism, mutual masturbation, the sports of naked boys in bed together. There was no refinement, no sentiment, no passion; nothing but animal lust in these occurrences. They filled me with disgust and loathing.'

Or there is Gerard Manley Hopkins, so obsessed with sex as sinful that he recorded every instance of his own masturbation in his journal under the code 'O.H.' (Old Habits). He used the same abbreviation for nocturnal emissions, which he seemed to regard as even more heinous. Then we have Gladstone, famed for his missions to rescue prostitutes; although there is no account of his ever having sex with them, he acknowledged feeling inappropriate temptation. His diaries duly record his practice of self-flagellation. The shamelessly carnal poet Swinburne was another notorious flagellant but, as a bohemian, had no qualms about getting others to do his flagellation for him. He was, accordingly, a customer of a North London brothel renowned for its birching facilities.

Other progressive Victorian era literary characters were less apologetic about their sexuality. This is Walt Whitman, radical American journalist, formerly an innovative teacher who permitted his students to call him by his first name and devised learning games for them in arithmetic and spelling, writing in 1860 in his poem 'Children of Adam':

It is I, you women, I make my way, I am stern, acrid,
large, undissuadable, but I love you,
I do not hurt you any more than is necessary for you,
I pour the stuff to start sons and daughters fit for
these States,

I press with slow rude muscle,
I brace myself effectually, I listen to no entreaties,
I dare not withdraw till I deposit what has so long
accumulated within me.

Another Victorian male who bravely admitted his erotic interest and, by the by, changed the face of sexuality in the Western world was the retired British Army officer Richard Burton who, along with his fellow members of the Karma Shastra Society, such as Algernon Charles Swinburne, translated and published the *Kamasutra* in 1883.

Burton, a Devon-born Arabist, had been by turns an explorer, linguist, ethnologist, professional treasure hunter and Crimean War spy. While in India, he undertook with Forster Fitzgerald Arbuthnot, a colonial civil servant whom he had befriended, a study of brothels staffed by boys and eunuchs. Burton converted to Islam and underwent adult circumcision. Back in England he enjoyed, by all accounts, a free-ranging and very modern sexual relationship with his wife Isabel, and with other women and other men, including possibly the now-retired Arbuthnot whom he always referred to as 'Bunny'. Isabel was reportedly the envy of all who heard the rumours because Burton was reputed to excel in the art of cunnilingus. When he died, however, Isabel burned all his unpublished work and gave free rein to a prudish streak that must have been present all the time.

It was doctors and lawyers, practically all male, who formed the vanguard of Victorian sexual repression. Literary Victorians, and prolific 'ordinary' diary and letter writers, provide the surviving evidence that these pompous professionals' outpourings on sex were not necessarily taken very seriously.

Most Victorian doctors considered sexual desire in women to be pathological and warned that female sexual excitement and indulgence could damage their reproductive organs and urinary system. Even enlightened doctors who knew the 'secrets' of women's desire considered it inappropriate to let them speak about it themselves. They conducted research and diagnoses by asking husbands instead. Freud's breakthrough in the next century was primarily in letting women speak for themselves about their sex life, even if he did then re-cast much of their testimony in the likeness of his own Victorian prejudices.

Victorian physicians were generally quite clued up on anatomical fact. An 1836 gynaecological textbook reports that 'the lower part of the vagina and the clitoris are possessed of a high degree of sensibility'. It explains that while in some women these erogenous areas are 'the seat of venereal feelings from excitement, in others, such feelings are altogether absent'. Other medical authorities confirmed that the clitoris underwent erection, and even acknowledged female ejaculation. The way the socio-political prejudices of the age reasserted themselves, however, was in the near total consensus that women were quite capable of playing host to all this physiological turmoil without being disturbed by any feelings unbecoming to their sex. A Parisian doctor, Adam Raciborski, argued in 1844 that 75 per cent of women only 'endure' sex with their husbands.

In England Dr William Acton, FRCS, a prominent venereologist, set out in 1857 the most direct manifesto yet for the idealised sexlessness of the modern woman: 'Many of the best mothers, wives and managers of households, know little of or are careless about sexual indulgences. Love of home, children, and of domestic duties are the only passions they feel . . . A modest woman seldom desires any sexual gratification for herself. She submits to her husband's embraces, but principally to gratify him and, but for the desire of maternity, would far rather be relieved from his attentions . . . The married woman has no desire to be placed on the footing of a mistress.'

Curiously, it is Dr Acton who debunks the contemporary idea that prostitutes became so as a result of their excessive sensuality; that was too ridiculous even for him.

The United States Surgeon General, an ex-Army man called

William Hammond, was in accord with Acton's principal statements, averring himself of the belief that decent women felt not the slightest pleasure during intercourse. And Isaac Ray, an American gynaecologist, opined in 1866: 'In the sexual evolution, strange thoughts, extraordinary feelings, unseasonable appetites, criminal impulses, may haunt a mind at other times unviolent and pure.'

Roy Porter's *Medical History of Humanity* identifies a more ominous stage yet in the growth of prudery: the point when theory turned to practice and surgical 'cures' for women's sexuality 'grew widespread'. Surgery, specifically unnecessary hysterectomies, clitoridectomy and cauterisation of the clitoris, began to be sought by husbands concerned about their wives' sexual feelings. 'Abuse of gynaecological surgery to control women culminated in the work of Isaac Baker Brown (1812–73), a London surgeon who specialised in clitoridectomies on women whom he or their husbands judged oversexed, as evinced by masturbation or "nymphomania",' Porter recounts.

Other researchers have felt the need to point out, however, that Baker's procedures were not even remotely routine, and that he was drummed out of the London Obstetrical Society – albeit for operating without the correct consents and for self-promotion, not for performing the operations. An inspection of the records of the meeting at which he was expelled also reveals that a good part of Baker Brown's offence to medicine was to have implicitly insulted British womanhood by suggesting they were at all prone to masturbation! Brown is believed to have emigrated to the US, where it was easier to promote fringe medicine unregulated. There, for instance, a doctor called Battey had popularised an operation called 'normal ovariotomy', the removal of healthy ovaries in women diagnosed as hysterical (i.e. orgasmic) or neurotic.

Contraception received the condemnation of the medical profession from the 1860s, when the shocking truth of its spread began to become apparent. After Charles Goodyear's invention of vulcanised elastic rubber in 1843, a number of companies adapted the development for the manufacture of thick, re-usable condoms. Surprisingly, no one business patented the rubber condom, probably since the basic design, in a variety of animal materials, had been available for centuries.

In France condoms were, and still are, called *les capotes anglaises*, 'English raincoats'; in English-speaking countries the French letter. This is perfectly in keeping with the neighbourly convention that the British call anything they regard as faintly exotic 'French' (knickers, kissing, manicure, etc), while the French call anything they see as faintly risible (condoms, homosexuality, syphilis, boiled food, even strikes) 'English'.

In the nineteenth century, condoms as a commercial product were always more openly sold in the US than in Britain, where little evidence survives of their underground existence. But advertisements for 'Dr Power's French Preventatives', among other brands, appear in the *New York Times* and other newspapers by the time of the Civil War in 1861. Other contraceptive methods, especially douches and sponges, were widely used too. Women in the Northern States were frustrated during the war by a sponge shortage; all the fisheries were in Confederate territory, in Florida.

The American Social Hygiene Association fought vigorously against condoms as immoral and un-Christian. Their eccentric view was that anyone whose behaviour put them at risk of getting VD, deserved it. What would now be called 'natural' contraceptive methods were also touted. In 1866 Dr Russell Thacker Trall published a bestselling book in the US in which he claimed to have discovered in both Tonga and Iceland women with '. . . that flexibility and vigour of the whole muscular system that they can, by effort of will, prevent conception': Trall advised the women of America that 'sometimes coughing or sneezing will have the same effect'. He also recommended running, jumping and dancing as contraceptive methods.

The reaction from less adventurous doctors in the US and Britain was stony. The *Lancet* sneered in 1869: 'A woman on whom her husband practises what is euphemistically called "preventative copulation" is necessarily brought into the condition of mind of a prostitute. As regards the male, the practice, in its character and in its remote effects, is in no way distinguishable from masturbation.' In the States, Congress passed the Comstock Act, which banned 'obscene materials' – meaning condoms – from the mail.

In 1877, in London, Charles Bradlaugh and Annie Besant were tried at the Old Bailey for republishing Dr Charles Knowlton's forty year-old textbook on birth control, *The Fruits of Philosophy*. The prosecution argued, '. . . this is a dirty, filthy book, and the test of it is that no human being would allow that book to lie on his table; no decently educated English husband would allow even his wife to have it . . .' The jury ruled that a reissue of Knowlton's book was 'calculated to deprave public morals', but exonerated the defendants of the charge of having corrupt motives. On appeal, Besant and Bradlaugh won the right to publish.

Contraception still failed to receive official sanction in Britain, however. In 1879, C.H.F. Routh in *The Moral and Physical Evils Likely to Follow Practices Intended to Act as Checks to Population* argued: 'First, let me state that I look upon conjugal onanism as a great moral crime. Masturbation is mean and bad enough, and much to be reprehended, because it is fostered by a filthy spirit which can no longer control the sexual impulses. But here, at least, there is no partner in the sin, and no pure woman is degraded thereby. Conjugal onanism places both the man and the woman below the instincts of the brute creation.' As late as 1894, a Dr Alice Stockham was recommending that a woman should *never* copulate non-procreatively, *never* more than once a month, and *never* during menstruation or pregnancy. Ten years later, John W. Taylor, President of the British Gynaecological Society, was warning that 'mechanical shields' resulted in 'purulent vaginitis' and 'brain fag' – an archaic term for cerebropathy, a hypochondriacal condition in those who feel their brain has been worn out.

Rubber contraceptives and 'outercourse' (as *coitus interruptus* has latterly been called) were unacceptable to these moralists; cunnilingus and fellatio did not even show up on their radar. It was masturbation, nonetheless, that remained the ultimate taboo throughout the Victorian era and long beyond. In 1850, an editorial in the *New Orleans Medical & Surgical Journal* inveighs against self-abuse: 'Neither plague, nor war, nor smallpox, nor a crowd of similar evils, have resulted more disastrously for humanity than the habit of masturbation: it is the destroying element of civilized society.' An 1878 book, *Psychological Medicine*, by Bucknill and Tuke, warns: 'Onanism is a frequent accompaniment of insanity and sometimes causes it.' The previous year the Reverend Edward Lyttelton in *The Prevention of Immorality in Schools* states that: 'Solitary vice is dangerous and deplorable because it was learnt, not instinctual. Boys are innocent of masturbation until inspired to foul practices by other boys who spread corruption through the school.'

The influential German neurologist Richard von Krafft-Ebing's *Psychopathia Sexualis* appeared in 1886; it boldly linked masturbation to criminality. Women who exhibited 'excessive' sexual desire, for Krafft-Ebing, were nymphomaniacs, while, 'If [a woman] is normally developed mentally and well-bred, her sexual desire is small. If this were not so, the whole world would become a brothel and marriage and a family impossible.'

Masturbation was so offensive to Victorian sensibilities, in spite of the probability that most men still practised it, that extreme attempts were made to 'cure' it. An 1887 tract, *Spermatorrhoea* by J. L. Milton, advocated metal-spiked penis rings to prevent erections. It even leant on the developing field of electrical technology to prevent children from falling foul of the habit. Milton published plans for a penis switch that could activate an electric bell in the parents' bedroom if their son had an erection while asleep or awake. (No details were given as to what the parents should then do.) Even less fortunate boys would – or so we are told, and there is no evidence that anyone ever followed these nuggets of advice – have leeches applied to their genitals, a (supposedly) doctor-recommended method of subduing erections. Others still could allegedly find themselves wrapped tightly in sheets or chained to walls to prevent masturbation.

Dr Sturgis's *Treatment of Masturbation* (1900) encouraged an even more pointed disincentive, with what he termed a male chastity belt: 'The prepuce is drawn well-forward, the left forefinger inserted within it down to the root of the glans, and a nickel- plated safety pin introduced from the outside through the skin and mucous membrane, is passed horizontally for half an inch or so past the tip of the left finger and then brought out through the mucous membrane and skin so as to fasten from the outside. Another pin is similarly fixed on the opposite side of the prepuce. With the foreskin looped up, any attempt at erection causes painful dragging on the pins and masturbation is effectually prevented. In about a week some ulceration of the mucous membrane will allow greater movement and will cause less pain . . . but the patient is already convinced that masturbation is not necessary to his existence . . . If the penis erects while the wearer is asleep and he is awakened by the jab of metal on flesh, he should first remove the device. Next, soak the organ in cold water until it has subsided, and once again affix the apparatus to the genitals. The wearer can then return to his innocent sleep, assured that a moral as well as a material victory has been gained!'

In Britain, female masturbation was so unthinkable it was not considered worthy of attack. In America, though, a physician called Smith published in the *Pacific Medical Journal* a cut-out-and-keep guide for doctors on how to detect female masturbation by examination. Labia which were longer on one side than the other was, for Smith, an indicator of indulgence. But in keeping with the technological fervour of the age, he went on to develop an electronic test too. He would pass a mild current through the urethra in a way which purportedly helped him determine whether women were overly responsive to sexual stimuli, and hence masturbating.

A rare ray of sunshine filters through at this time from an unaccustomed source – the Catholic Church. One French Church official, the Vicar General, M. Craisson, endorsed women bringing themselves manually to orgasm if they did not achieve it during sex. In an 1870 book, *De Rebus Venereis ad Usum Confessariorum* (On Sexual Matters for the Guidance of Fathers Confessors), Craisson wrote: 'The fourth question is as to whether, if the husband should withdraw after ejaculation, before the wife has experienced orgasm, she may lawfully at once continue friction with her own hand in order to attain relief.' Most moral theologians, Craisson answered, permit this: 'In the same manner, it is lawful for the woman to prepare herself by genital stimulation for sexual union, in order that she may have orgasm more easily.' Perhaps it was in the light of advice like this leaking out of the Church that an English convert to Catholicism, the poet and essayist Coventry Patmore, felt free to masturbate. Patmore's method was, however, a little unusual. He perfected a form of masturbation without ejaculation, so he could enjoy the pleasures of arousal without, so to speak, a result; modern Tantric sex aficionados would later attempt the same feat, but in their case as a means of sexual circuit training.

The question that has to be addressed in view of the nineteenth-century flood tide across the Western world of sexual advice, practically all of it misleading or malevolent propaganda, is whether it was really meant to be followed or if it was just sensational, hack pseudo-pornography, informed by dubious personal psychological motives on the authors' behalf and designed to drum up book sales rather than practical guidance. Françoise Barret-Ducrocq, in her 1992 book *Love In The Time Of Victoria*, makes the point that what would now be called the chattering classes – the reformers, Christians and educators – showed at this time a suspicious level of interest for its own sake in other people's sex lives and bodies. 'They went into great detail about sexuality they deemed to be pathological. Their suggestions for such deviancy were harsh and they went into details such as what was the perfect number of coitions per month,' she observes.

Some of the more reliable instances of Victorian prurience, such as the unnecessarily detailed probing into the sex lives of single mothers by the governors of the Foundling Hospital, suggest that the Devil was truly in the detail for these (one imagines) monumentally priapic, but frustrated, Victorian men. They developed a talent for teasing the pornographic out of anything. In English law courts early in the century, it was necessary in order to obtain a separation to prove that 'criminal conversation' had taken place between one's wife and another man. 'Crim. con. cases were a steady source of prurient details, and the salaciousness of some of the judges who supervised them was notorious,' comments Gordon Rattray Taylor in his 1953 *Sex in History*.

Michel Foucault argues that in France, the restriction and the policing of sex at this time was in fact a sexual obsession, and like all obsessions it was overdone. His example does not chime very well with modern sensitivities, however, even though it is less than thirty years since Foucault wrote his *History of Sexuality*. He relates a story about what he describes as the absurd overreaction when a simple-minded farmhand 'begged a favour' of a little girl in 1867 in a village called Lapcourt. 'As luck would have it, the man was called Jouy, which sounds like the past participle of the verb *jouir*, which colloquially means "to come". They would play the familiar game called "curdled milk".

'So he was pointed out by the girls' parents to the mayor of the village, reported by the mayor to the gendarmes, led by the gendarmes to the judge, who indicted him and turned him over first to a doctor, then to two other experts who not only wrote their report but also had it published. What is the significant thing about this story? The pettiness of it all; the fact that this everyday occurrence in the life of village sexuality, these inconsequential bucolic pleasures, could become, from a certain time, the object not only of a collective intolerance but of a judicial action, a medical intervention, a careful clinical examination, and an entire theoretical elaboration.'

These, then, were the public faces of Victorian sex. But there is a strengthening trend today to revise our view of nineteenth-century sexuality, to argue that the received wisdom is overly influenced by writings that were a sham – that the real Victorian man and woman were sexually much more like us than we imagine.

To put things into perspective, the population of London was five times greater at the end of the nineteenth century than at the beginning. That, as many historians now point out, represents a lot of orgasms. Henry Mayhew reported that among costermongers in London, fewer than a tenth of couples living together were married, and 'legitimacy or illegitimacy of children was a matter of little concern'. A study of the middle class by a Cambridge University historian, Dr Simon Szreter, notes a simultaneous sharp decline in the birth rate in professional families at the same time, a decrease that cannot be accounted for merely by contraception. He concludes that abstention must have been the principle way in which the economically hard-pressed Victorian bourgeoisie kept their families to a manageable size. And yet, Victorian middle-class families were still enormous by our present-day standards. A significant amount of sex still had to be going on behind those heavy velvet drapes!

Or consider masturbation; in 1980, it was estimated that 97 per cent of males and 78 per cent of females practised some form of masturbation. Why, if we think of humans as being on some kind of slow-moving evolutionary escalator, should the statistics have been significantly different a mere hundred years earlier? Because of cultural pressure against masturbation? Maybe, but there was huge cultural pressure to dissuade people from prostitution, yet even by the Victorians' own account, if it is to be taken at face value, it was widespread.

The historical evidence discovered in diaries, medical accounts and even a handful of rudimentary Victorian sex surveys tends to counter the idea that middle-class Victorians were sexually repressed, and that dour public ideology invariably dictated the private pursuit of sexual fun. One such survey was conducted discreetly by a Baltimore women, Clelia Duel Mosher (1863–1940), who asked 45 married women about their sex life of whom 34 reported experiencing orgasm. One woman complained, however, of the inconsistency of her orgasms and commented that 'men have not been properly trained'.

What is most significant about Mosher's survey is that it remained unpublished until 1980. The same fate befell other such pioneering works on Victorian female sexuality. Katherine Bement Davis (1860–1935) did extensive research before the turn of the century, but only published her book *Factors in the Sex Life of Twenty-Two Hundred Women* when she was an elderly lady in 1929. She had been wise to hold back in the prevailing social climate as her work broke two taboos: her sample was split equally into married and unmarried women, and she questioned women up to 83 years of age. A third Victorian researcher, gynaecologist, Robert Latou Dickinson (1861–1950), surveyed 4,000 married and 1,200 single women from the 1890s onwards but, again, published only four decades later, when he had retired.

Peter Gay, in his 1984 *Education of the Senses, The Bourgeois Experience*, provides further ammunition for the argument that the Victorians' prudery was partly a façade. Gay discovered that married men and women were advised by their doctors to enjoy foreplay as well as intercourse. The Victorians, according to Gay and others taking his view, were also as ingenious at finding places to enjoy sex as they were at dreaming up mechanical contraptions to put a stop to it.

The advance in the provision of cheap public transport was a highly significant factor in bringing about a more equal distribution of orgasmic delight. Couples in London, for instance, who would otherwise have been confined to grabbing sexual pleasure in corners of such unsatisfactory love nests as workrooms, were now able to enjoy daytrips out to places such as Richmond Park, Wimbledon Common, Battersea Park and Epping Forest, where a quiet corner could be found for outdoor sex. The reasonably private rear area of the new hansom cabs were also the site for many trysts, while horizontal rendezvous could be had at hotels, coffee houses and pubs, where beds could be rented by the hour.

This sexual playground view of Victorian mores also points out that parts of the law were quite hard on out-and-out male seducers. They could be prosecuted for breaching a 'promise of marriage' and pursued with paternity orders. As a result many 'bounders' vanished a few months after their Sunday assignation.

Prominent Victorians, too, were sometimes outspoken in favour of sexual pleasure. John Stuart Mill was imprisoned for a few days for distributing leaflets advising couples on *coitus interruptus* and the contraceptive sponge. Mill regarded contraception as analogous to putting up an umbrella in a rain shower. A British Socialist doctor, Henry Havelock Ellis, who would later publish a revolutionary seven-volume *Studies in the Psychology of Sex* (1899–1918), was convinced that the presumed sexlessness of Victorian women was a falsehood.

Havelock Ellis, an English doctor, was the first authority to suggest openly that sex and reproduction did not need to be bedfellows. 'Reproduction,' he wrote, 'is so primitive and fundamental a function of vital organs that the mechanism by which it is assured is highly complex and not yet clearly understood. It is not necessarily connected to sex, nor is sex necessarily connected with reproduction.'

Ellis also argued that in marriage one should ascertain not just the sexual needs of the husband but of the wife, too. His open-mindedness was informed by the occasional sexual peculiarity. He was an 'undinist', a man capable of being brought to orgasm by the sight of a woman urinating while standing up. (Another prominent undinist was Rembrandt.) This obsession had been prompted, Havelock Ellis said, by an incident at London Zoo when he was twelve and his mother was caught short and obliged to lift her skirts to urinate. In adult life, this was Havelock Ellis's chosen method of arousal. His wife, Edith Lees, was bisexual, and when she died, he married one of her lovers. Unsurprisingly then, Ellis was always fascinated by case histories of other people with fetishes. A prostitute on the Strand in London once told him about a client whose complicated and expensive treat was to bring her to orgasm as she and another naked colleague wrung the neck of pigeons. Another client enjoyed licking a prostitute's boots.

Charles Kingsley, author of *The Water Babies*, social reformer and committed Christian, was another Victorian who believed in the value of sexual intimacy. He spent his four-year courtship with his fiancée Fanny Grenfell imagining how sex would be. Fanny referred to the 'delicious nightery' of her 'strange feelings' and her troublesome 'spasms' – for which the doctor recommended getting married sooner. Charles reminded Fanny in a letter before their marriage, however, of the remarkable amount that can be achieved in an un-chaperoned instant. Other Victorian women who speak of their sexual feelings are rare, though. Coventry Patmore's daughter, Emily, became a nun, but after dreams of religious fulfilment recounted awakening with 'a throb of ecstasy'.

Kim Murphy, an American author and writer of an essay called 'Frigid Victorian Women?', has noted how in their confidential diaries it was not uncommon for women to mention their inequality in society – and their sexual feelings. Murphy has discovered a letter from a woman called Laura Lyman to her husband Joseph in which she says: 'How I long to see you . . . I'll drain your coffers dry next Saturday I assure.'

'By the 1860s,' Murphy writes, 'diaries and letters reveal more explicit and intensely erotic discussions among middle-class couples. Mabel Loomis [the author and friend of Emily Dickinson] wrote in her diary after spending a passionate night with her fiancé, David Todd, "I woke up the next morning very happy though, and feeling not at all condemned." Murphy notes how Loomis, in the privacy of her diary, admitted that she was content in sharing sexual intimacy with her fiancé.'

Karen Lystra, Professor of American Studies at California State University, has studied love letters of the nineteenth century to discover what courting and married couples were actually doing in parlours and bedrooms. 'Instead of finding stereotypical couples who lacked in sexual intimacy,' explains Murphy, 'she discovered the middle class beheld sex with "almost a reverence for sexual expression as the ultimate symbol of love and personal sharing". And while most women apparently resisted coitus before marriage, the sensuality in their letters revealed quite a range of acceptable erotic activity while courting.'

Dr Elizabeth Blackwell, according to Murphy, the first woman to earn a medical degree in the United States, wrote about the 'immense power of sexual attraction felt by women', while Ida Craddock, another advocate of women enjoying sex, wrote detailed instructions for men on how they should 'arouse' their wives, 'and study carefully every movement with reference to its pleasure-producing effect upon her'. Craddock, a spiritualist and stenography teacher from Philadelphia, was the first woman known to have written a sex manual. *Right Marital Living* was remarkably graphic even by today's standards: 'Bear in mind that it is part of your wifely duty to perform pelvic movements during the embrace, riding your husband's organ gently, and at times, passionately, with various movements up and down, sideways and with a semi-rotary movement, resembling the movement of the thread of a screw upon a screw.' Warming to her theme, she further advised women to, 'go right through the orgasm, allowing the vagina to close upon the male organ'. Craddock was arrested in 1899 for sending obscene material through the mail.

Another iconoclastic view of the Victorians' sexuality was mooted by Steven Marcus in a groundbreaking 1966 study, *The Other Victorians*. He looked to the subversive underworld of Victorian licence and pornography as evidence of widespread reaction to prudery and control. One of his best examples of the illicit Victorian pursuit of recreational orgasm was an immense doorstop pornographic book called *My Secret Life*, of uncertain vintage or authorship but a bestseller for decades.

The author, 'Walter', is thought to have been a nobleman or at least a wealthy businessman with an insatiable appetite for women. *My Secret Life* amounts to 2,360 pages of hardcore porn. Most of the women Walter sleeps with are 'gay women' (prostitutes) of working-class or rural origins. One woman, in Marseilles, he reports as having two vaginas. Walter does not seem a great proponent of sexual equality, yet many readers have detected a thread of genuine love of women as well as sexual pleasure amidst the forests of verbiage; others, alternatively, have detected nothing more than a misogynist fantasy-monger with a rapist's mindset.

'Walter' professed a moral justification for paying above the odds for sex with virgins: 'Verily a gentleman had better fuck them for money, than a butcher boy for nothing. It is the fate of such girls to be fucked young, neither laws social or legal can prevent it. Given opportunities, who has them like the children of the poor? They will copulate.' And he seemed genuinely to have adored women's genitalia, and giving his rented paramours pleasure, arguing that to do so was 'natural' – a radical attitude for his time: 'What more harm in a man's licking a woman's clitoris to give her pleasure, or of she sucking his cock for the same purpose, both taking pleasure in giving each other pleasure?' he wrote.

There was an heretical body in medicine, too, that surreptitiously chipped away at the 'official' consensus. In the 1850s, Pierre Briquet, a French physician researching 'hysteria' in female patients, alighted on the idea that the syndrome was caused by sexual frustration, the cure for which was (as van Swieten was privately advising Royalty a century earlier) '*la titillation du clitoris*'.

In 1872, Dr Joseph R. Beck of Fort Wayne made what is credited as the first scientific observation of a female orgasm. He was inserting a pessary in a woman who had asked him to take particular care as she was prone to having an orgasm in such circumstances; the mind only boggles at how she delivered this information to him. Fascinated rather than fazed by such a prospect, however, Dr Beck accidentally-on-purpose carried out a little experiment, attempting deliberately to induce an orgasm in her (and, who knows, maybe in himself as well) by sweeping his 'right forefinger quickly three or four times across the space between the cervix and the pubic arch'. Almost immediately, Beck recorded, the orgasm occurred.

Beck was not the first doctor to happen on the phenomenon of women roused to orgasm in the doctor's surgery. Early in the century, the introduction of the speculum as a standard part of a physician's kit had caused something of an uproar. Tales circulated among medics of women begging them for a vaginal examination and climaxing virtually the moment the instrument was brandished.

A British doctor, Robert Carter, wrote in 1853: 'I have . . . seen young unmarried women, of the middle-class of society, reduced by the constant use of the speculum to the mental and moral condition of prostitutes, seeking to give themselves the same indulgence by the practise of solitary vice; and asking every medical practitioner . . . to institute an examination of the sexual organs.' Ironically, the speculum had been invented by an Alabama gynaecologist, James Marion Sims, who admitted, strangely for one in his specialism, that he loathed 'investigating the organs of the female pelvis'.

Sims had probed a good few celebrated pelvises in his time, too. Having emigrated to Europe, he included among his patients Napoleon III's Empress Eugénie, the Duchess of Hamilton and the Empress of Austria. There is no record of any of these women getting unduly excited by the sight of the speculum, but if they had, he doubtless would have followed the example of Otto Adler, a German gynaecologist who experienced the classic Victorian examining table orgasm, but even then refused to let it dissuade him from his pet theory that most women were subject to 'sexual anaesthesia'. Adler bafflingly included in his sexually anaesthetised category not only the patient who climaxed in his surgery, but ten women who admitted that they masturbated to orgasm or were subject to strong sexual desires that could not be consummated by means of intercourse.

There was nevertheless in the decades following Dr Beck's 'experiment', a positive cult of male doctors purposefully fiddling – by consent, unlike in Beck's case – with their sexually frustrated female patients. The previously pointless 'sexual paroxysm' now seemed to be a state that offered temporary relief of a variety of 'hysterical' symptoms; indeed, the hysterical symptoms produced by Mesmer, in his apparent experiments in Paris with orgasmic women, were clearly the exultant product of 'hysteria' rather than hysteria (as in extreme sexual frustration) itself.

The female orgasm was slowly fighting its way to the recognition and respect it began to acquire in the next century. It sounds more than a little dubious today, but in 1878 Desiré Magloire Bourneville, a male doctor at the Salpêtrière, a large hospital in Paris, published a three-volume work containing photos of women in the process of orgasm. Some of the women were seen tweaking their nipples, others arching their backs, other rocking backwards and forth and tossing their head.

His clinical notes read more like soft pornography than medical research: 'Then her body curves into an arc and holds this position for several seconds. One then observes some slight movements of the pelvis . . . she raises herself, lies flat again, utters cries of pleasure, laughs, makes several lubricious movements and sinks down on to the vulva and right hip.' During all of this action the vaginal fluids are carefully accounted for: '*La vulve est humide*', or, '*La secretion vaginale est très abondante*' were such notes.

Doctors such as Havelock Ellis and Elizabeth Blackwell

achieved their conceptual breakthroughs without resort to such spectacular methodology. Havelock Ellis was the true Mateo Columbo of female sexuality at the turn of the nineteenth century. He observed that men's and women's orgasms are strikingly alike, but – a daring suggestion – that women are naturally inclined, if given the chance, to desire and to have more of them than men. 'Every woman has her own system of manifest or latent erogenic zones, and it is the lover's part in courtship to discover these zones, and to develop them in order to achieve that tumescence which is naturally and properly the first stage in the sexual union,' he wrote.

Blackwell, in a 1902 collection, *Essays in Medical Sociology*, concurred, stating that 'the unbridled impulse of physical lust is as remarkable in the latter [women] as in the former [men]' She referred to women's orgasms as 'sexual spasms', and opined that it is not intercourse that women crave principally but, 'the profound attraction of one Nature to the other which marks passion, and delight in kisses and caresses – the love touch'. By the 'love touch' she meant foreplay 'by hand and by tongue', which was, she avowed, a woman's version of 'physical sexual expression'.

Some of the most remarkable progress, however, in the story of the female orgasm, as well as the most positive proof ever found that a healthy orgasmic culture existed underground in the Victorian era, took place almost surreptitiously in the late-nineteenth-century. It was chronicled in an extraordinary study published in 1999 by Dr Rachel P. Maines and entitled *The Technology of Orgasm*.

In the late 1970s, Maines was an academic specialising in the history of technology and working on a book about nineteenth-century needlework, with reference to early-American sewing machines. Her research took her to the Bakken Museum of Electricity in Life in Minneapolis, Minnesota, which kept a unique collection of eleven domestic appliances from the birth of electricity.

Curators there had for many years been puzzled by a

mysterious collection in their basement, which they were unable to catalogue because they could not work out what the devices were used for. They were of widely different design, but all those that still worked purred gently when you plugged them in. The oldest was produced by the Weiss Instrument Manufacturing Company, Chambers Street, Manhattan. (Weiss are still in business in Holtsville, NY, making instruments for the refrigeration and airconditioning markets. They do not mention their background in sex toys in company literature.)

Maines was pretty sure what the curious early Weiss and other instruments were. She had previously noticed in turn-of-the-nineteenth-century magazines such as *Needlecraft*, *Home Needlework Journal*, *Modern Women* and *Woman's Home Companion*, puzzling advertisements for a large number of electrical home appliances. They were advertised as performing all sorts of labour-saving chores like powering a fan, but always illustrated by drawings of women applying the machines to their necks or backs, and accompanied by puzzling copy describing the effects of the instruments with words like 'thrilling' and 'invigorating', along with phrases such 'all the penetrating pleasures of youth will throb in you again'. When marketed to men, they were recommended as gifts for women that would restore their wives' bright eyes and pink cheeks.

The inexplicable appliances in Minneapolis were nothing less than the world's earliest electrical vibrators. The modern vibrator, it turned out, had been invented in the 1880s by a British doctor with an interest in the humane treatment of the insane, Joseph Mortimer Granville. His machine went into production as the Weiss vibrator. Hysteria was, by Granville's time, one of the most commonly diagnosed problems in women, and from the mid-nineteenth century it fell to a small number of progressive (but highly paid) doctors manually to masturbate their female patients back to good health. Not every doctor offered the treatment; the majority, after all, had probably never seen an orgasm in their wives, let alone in their patients.

The vibration cure for 'hysteria' had been studied by Dr Jean-Martin Charcot, a physician at the Salpêtrière, at the same time as Bourneville was painstakingly working on his collection of photos of women having orgasms. Charcot and his team experimented doggedly with shaking machines, swings and train rides, massage and electrotherapy. He disliked administering the massage most, complaining bitterly at how time-consuming and tricky it was.

Even pre-electric power, in both Britain and the US, there had for several decades been wind-up clockwork vibrators made from wood and ivory, manual massagers that squirted oils, massagers with water jets, models with rollers, and padded tables with holes in them through which a sphere vibrated, controlled by a steam machine. Some of the less cumbersome very early vibrators were portable, allowing doctors to take them on house calls.

The Weiss electric vibrator was a boon initially to the specialist masturbation doctors, who could now do in five minutes what used to take up to an hour, but could charge their patients the same, typically two dollars. The resulting powered orgasm was also more intense. Once vibrators were offered as home appliances, most of the doctors were put out of that particular part of their business. They tried to warn women away from what one physician described as 'these mere trinkets which accomplish little more than titillations of the tissues'. But the truth was, lucrative though it may be, masturbating women was not a business they greatly cherished and most were glad to see the back of it.

The first electric device advertised directly to the public and noticed eighty years later by Dr Maines was the Vibratile, which appeared in an 1899 edition of *McClure's* magazine. The Vibratile was offered as a treatment for neuralgia, headache and wrinkles. More than fifty more electric vibrators were swiftly invented, the technology taking off exponentially with the introduction of alternating current power. They weighed anything between five and fifteen pounds, and cost from $5 to $20 for luxury models with brass fittings and a velvet-lined box. None came with instructions; it was assumed that if you ordered one, you knew what to do with it.

The home electric vibrator was developed at a fitting time, at the very end of the prudish but sometimes surprisingly frisky Victorian era. It has to be emphasised, however, that even if its use as a clitoral stimulator was covert, it was still not blatantly sexual. In her heart, the vibrator owner wanted her little device to do what it said in the ads – improve her health – even though robust health in a woman was in itself regarded as vulgar and a little too Socialist for refined Victorian ladies of the old school.

Yet even some of the health lobby pointedly rejected the use of the vibrator. The new Germanic-inspired fad of nudism, the health movement's most radical wing, was in full sway by the end of the century. Yet not a single sexual nuance was allowed to seep through the wholesome text of *Health & Efficiency*, the British nudist magazine started in 1899 and still published. Its first editor, Charles Thompson, regarded masturbators as 'moral imbeciles'.

It remains one of the Victorian era's most ironic postscripts that turn-of-the-century knitting magazines contained advertisements for several brands of intimate vibrators, while *H&E*, the masturbator's Bible for the next fifty years with its pages of nude photographs and gossamer-thin rationale as a journal promoting health, contained not a single acknowledgement that there was anything remotely sexual about people prancing about naked in the great outdoors and being photographed as they did so.

14

The Orgasm from Freud to Lady Chatterley

'The sole criterion of frigidity is the absence of the vaginal orgasm'

Sigmund Freud,
Three Essays on the Theory of Sexuality

Just as it did not start with her ascension to the throne, the constipated 'Victorian' attitude to all things sexual across the whole Western world, and the substantial tracts of the rest of the globe that were its colonial subjects, did not die with Queen Victoria. Almost every progressive step forward in the new twentieth century was accompanied by one step backwards in deference to the ways of the old era. There was frequently more the echo of Victorianism to be heard in the brave new Edwardian age than there was the shrill cry of modernity.

The puzzling old obsession with masturbation, for one thing, showed scant sign of being superseded by even a slightly more enlightened view. The founder of the Boy Scouts, Lord Baden Powell, then one of the most pervasive new influences on young men worldwide, wrote of masturbation in *Scouting for Boys* (1908): 'It is called in our schools "beastliness" and this is about the best name for it . . . should it become a habit, it quickly destroys both health and spirits; he becomes feeble in body and mind and often ends up in a lunatic asylum.'

He explained his theory further in his *Rottering to Success* (1922): '[Masturbation] cheats semen getting its full chance of making up the strong manly man you would otherwise be. You are throwing away the seed that has been handed down to you as a trust instead of keeping it and ripening it for bringing a son to you later on.'

Baden Powell's beliefs on the matter may be fairly predictable, but you would not expect the founder of the Boy Scouts and the leading lights of radical Women's suffrage to have much common ground. Yet both are clearly the wagging tail of Victorian values at the beginning of the twentieth century. Melanie Phillips, in her study of the suffragist movement, *The Ascent of Woman*, maintains that the essence of the 'Suffragettes', as the *Daily Mail* dubbed them at the time, was that men's rampant sexuality – as evidenced more than anything by prostitution – was ruining the world, both through their immorality and the spread of VD. Female sexuality, which was more restrained and controlled, was the antidote, and by women getting the vote, the whole world could attain new, spiritual heights.

The suffragists, Phillips argues, were hugely impressed by Charles Darwin, whose idea that mankind was on a trajectory from animalism to a superior, spiritual, state of being in which sexuality was almost nullified, struck home particularly. With their belief that women were intrinsically more spiritual and superior to men – many agreed with Mary Wollstonecraft from a century earlier, who held that women pursuing sexual pleasure were selfish – they would have horrified when mid-twentieth century feminists re-conceptualised sexual equality as meaning that they could be identical to men in the sexual area as well as elsewhere.

The Suffragettes had some bizarre theories, Phillips recounts. One, Elizabeth Wolstenholme Elmy, believed that even menstruation was the result of unrestrained male lust. Another, Frances Swiney, believed that men were only 'rudimentary females' and their sperm was 'a virulent poison'. She was particularly opposed to cross-racial and cross-class breeding, a theme taken up with relish by the female founders of the modern birth control movement.

Victorian suspicion of sexuality extended its tendrils well into the sexually liberated middle of the twentieth century. Laws against oral sex, even between married couples, remained theoretically in force, and in some US States have yet to be repealed. In a 1961 book, *Sexual Behaviour: Psycho-legal Aspects*, Frank Caprio and D. R. Brenner recount this sad twentieth-century case: 'A husband was performing cunnilingus on his wife in the privacy of their bedroom. One of three children in the family, unaware of the sexual activity of the parents, opened the door and observed what was going on. The child, frightened by what he had seen, ran to a neighbour with the story. The police were called and the husband arrested. He readily admitted the act and stated that he did not see anything wrong with it. He further said that the wife did not object to what he was doing and that, in fact, she encouraged him. Armed with this confession, a conviction was obtained and the man sentenced to prison for five years.'

Victorian sexual attitudes and their accompanying strand of hypocrisy extended to the upper reaches of American society. The passionately anti-homosexual FBI director, J. Edgar Hoover, said publicly in the 1950s, 'I regret to say that we of the FBI are powerless to act in cases of oral-genital intimacy, unless it has in some way obstructed interstate commerce.' Hoover might have passed for a conviction-driven moral campaigner and ultimate family man were he not at the same time as making such stern pronouncements having a long-term gay affair with his deputy, Clyde Tolson, and attending Washington parties in women's clothing; at one, according to a fellow reveller, Hoover wore a red dress with feather boa.

In a sense, even turn-of-the-century radicals like the artist Egon Schiele (1890–1918) were still dancing to the Victorian tune. Schiele painted self-portraits of himself masturbating as a protest against the conservatism of Austrian society. It would have been pointless had masturbation not been taboo. The same small but glowering Victorian storm cloud might still be said to have been hovering seventy years later over John Lennon and Yoko Ono when they staged their bed protest in Amsterdam – or even Richard Branson, who gave an entire business empire the name Virgin in the 1960s because it was still considered a slightly daring word to use amongst a public still influenced by Victorian moral values.

As in any age, in the mid-to late twentieth-century, radicalism was a minority view. More in tune than a Schiele or a Lennon with the establishment consensus during most of the century were the sentiments of Dr O.A. Wall. He wrote in 1932: 'A well-bred woman does not seek carnal gratification, and she is usually apathetic to sexual pleasures. Her love is physical or spiritual, rather than carnal, and her passiveness in regard to coition often amounts to disgust for it; lust is seldom an element in a woman's character, and she is the preserver of chastity and morality.'

But evidence that Victorian sexual repression was finally crumbling, or perhaps never existed for the less 'well-bred' at least, can be found in the charming reminiscences of Laurie Lee, describing his sexual debut before World War I in *Cider With Rosie* – or the franker boyhood recollections of Harry Daley, a policeman originally from Lowestoft, who had an affair with E. M. Forster and published his reminiscences in a 1986 autobiography, *This Small Cloud*:

> Another boy took out his large cock, the first I'd seen with hair round it, spat in his hand, and started to masturbate in the proper manner. After a minute or two he said he was tired and asked me to do it for him, which I did with pleasure. Thus began one of the happiest periods of my life; the real beginning of my happy life, the first awakening to knowledge of the pleasure and warmth in other people's bodies and affection; the realisation that physical contact consolidates and increases the pleasure and happiness to be got from mutual affection . . . It

was all open and uncomplicated . . . Whenever in our wanderings we came to a secret place, a wood, a shed or a deserted building, we would merrily wank away . . . Nowadays, for some reason or other, this traditional experience is thought to be undesirable . . . We continued happily and unworried for a long time, until the sort of people one finds in the fringes of church life, noticing the dark rings under our eyes, warned us that boys who played with themselves went mad and had to be locked away. This was a typical mean, dirty-minded trick, for they had been boys themselves and knew it was not true. In any case it didn't stop us. Henceforth we wanked and worried, whereas formerly we had experienced nothing but satisfaction and contentment.

Then we have Frank Harris (1855–1931), a poor boy from County Galway who went to America, came back to England, became a sub-editor on the *London Evening News*, an editor of literary journals and a notorious Edwardian rake. This passage is from his 1923 book, *My Life and Loves*:

The next moment I was with her in bed and on her; but she moved aside and away from me. 'No, let's talk,' she said . . . To my amazement she began:

'Have you read Zola's latest book Nana?'

'Yes,' I replied.

'Well,' she said, 'you know what the girl did to Nana?'

'Yes,' I replied with sinking heart. 'Well,' she went on, 'why not do that to me? I'm desperately afraid of getting a child, you would be too in my place, why not love each other without fear?'

A moment's thought told me that all roads lead to Rome and so I assented and soon I slipped down between her legs. 'Tell me please how to give you most pleasure,' I said and gently, I opened the lips of her sex and put my lips on it and my tongue against her clitoris. There was nothing repulsive in it; it was another and more sensitive mouth. Hardly had I kissed it twice when she slid lower down in the bed with a sigh whispering: 'That's it; that's heavenly!'

For an earthier confirmation still that orgasmic pleasure was alive and well, Victorianism notwithstanding, in the early-twentieth century, we have this First World War marching song, sung by British troops to the tune of 'Do Ye Ken John Peel':

When you wake up in the morning and you're feeling grand,
And you've such a funny feeling in your seminary gland,
And you haven't got a woman, what's the matter with your hand?
As you revel in the joys of copulation.

While the twentieth-century trend in openness about sexuality generally progressed in a forward direction, albeit principally in the Western world, which was taking the sexual lead at this time, in the case of the vibrator, time ran backwards. From being virtually out in the open in the late-Victorian and Edwardian periods, the vibrator quite abruptly dived for cover at the start of the 1920s. The reason for this is thought to be its starring role in the first 'blue' or 'stag' movies, which dictated that the distance between the safely medicalised practice of powered female masturbation and the reality that it was a wholly sexual habit could no longer be maintained.

The earliest known stag film made in 1915 was called *Free Ride* and was vibrator-free. The film concerned a man picking up two female hitchhikers and having sex with them – both vaginal and anal. Even bestiality made its screen debut before masturbation. In another film from the same period, three girls agree to have sex with a boy but on the stipulation that it is through a fence. They then substitute a goat for themselves. The boy, however, is unaware of the switch and exclaims, as patrons read on the caption card: 'That's the best girl I've ever had in my life.'

A generally positive, if oblique, spinoff of the Victorian era, meanwhile, was the rise to prominence of Sigmund Freud and psychoanalysis, which became world famous with the publication in 1905 of his *Three Essays on the Theory of Sexuality*. Freud was, like the Suffragettes in Britain, very much a product of his nineteenth-century past, but his techniques, unquestionably informed, and continue to inform, the greater part of today's psychology and psychosexual medicine.

The young Freud studied 'hysteria' under Charcot at the Salpêtrière. As a result of the slightly humiliating interrogations of women that he witnessed there, he invented the idea of the private, sacrosanct psychoanalytical couch, where infantile sexual traumas that caused dreadful symptoms in adults could, through conversation, be gently drawn like a poison from patients.

Freud encouraged women to become doctors, and further shocked the medical establishment by referring to the sexual organs in modern German, rather than Latin, as was the custom. Such iconoclasm drove doctors to form picket lines outside lecture halls in Vienna where Freud spoke.

All this sounds most encouraging, but the problem with Freud is that significant tracts of his work on sex simply fail to resonate with us today. He insisted on the importance of 'penis envy' in women; he devised the wholly discredited view that the clitoral orgasm is 'immature', that the developing girl must direct her clitoral eroticism and her penis envy into feelings of longing for a child – and that any woman who failed to graduate from clitoral to 'vaginal' orgasms – these during intercourse rather than masturbation – was 'frigid'.

He viewed masturbation as something one grew out of (he advised his sons to refrain), and that led to self-loathing and frigidity. He believed *coitus interruptus* had a harmful effect on the mind: 'It is a question of a physical accumulation of excitement – that is, an accumulation of physical tension. The accumulation is the result of discharge being prevented. Thus anxiety neurosis is a neurosis of damning up, like hysteria: hence the similarity,' he wrote.

Freud tends, therefore, to be more remembered for his methodology than his often over-egged puddings, such as this less than palatable pronouncement on a common cosmetic practice employed by some couples: 'Pressing out the contents of the blackhead,' he wrote in *The Unconscious* (1915), 'is clearly to him a substitution for masturbation. The cavity which then appears owing to his fault is the female genital.'

While Freud's talking cure, such nonsense notwithstanding, worked away in one way at restoring the orgasm as a force of Nature for both sexes to enjoy freely, a more practical approach was being taken by the prime movers of modern birth control in the first decades of the century, Marie Stopes (1880–1958) and Margaret Sanger (1879–1966).

Sanger (née Higgins), working initially in the tenements of New York, was the first of the two to agitate and campaign for birth control – a phrase she invented. Sanger was a working-class New York Irish midwife whose mother had eighteen children. Margaret, the sixth, became a Socialist, and, seeing the number of women dying of back-street abortions, started publicly declaring in a self-published newspaper the value of sex and orgasmic pleasure, maintaining that contraception, backed up by safe abortion, was the basis for both sexual and social happiness.

Her campaigning was partly the result of social conscience and partly of her knowledge of women's struggle to assert their right to sexual enjoyment. In a 1931 book, *My Fight for Birth Control*, she recounted the story of one of her patient's attempts to seek medical advice on contraception.

> 'Yes, yes, I know, Doctor,' said the patient with trembling voice, 'but,' and she hesitated as if it took all of her courage to say it, 'what can I do to prevent getting that way again?'
>
> 'Oh, ho!' laughed the doctor good-naturedly. 'You want your cake while you eat it too, do you? Well, it can't be done. I'll tell you the only sure thing to do. Tell Jake to sleep on the roof!'

Doctors Lena Levine and Abraham Stone, working in Sanger's birth-control clinics, became aware of the mixture of sexual discontent and ignorance in their clients and began to offer practical sex counselling using a model of the female genitals. Very few of their patients knew what or where the clitoris was. After Sanger's own rather more helpful advice on sexual satisfaction and birth-control appeared in her newspaper in 1915, she was charged with publishing an 'obscene and lewd article', and fled to Britain.

It was in London that she met Dr Marie Carmichael Stopes, daughter of a radical intellectual Edinburgh family. As a child, Marie had announced that she would spend the first twenty years of her life in science, the second twenty working on social projects, and the final twenty writing poetry – and she did precisely that. Stopes became Britain's youngest doctor of science in 1905. She took a double first in botany (specialising in fossilised plants) from University College, London, and engaged in a chaotic love life. After a string of unsuccessful love affairs, she married a Canadian geneticist, Reginald Ruggles Gates, in 1911. It was as disastrous a choice of husband as she could have made. Not only did Gates hold highly traditional views on women's role in society and the behaviour to be expected from them, and not only did he vehemently oppose his wife's membership of the feminist Women's Freedom League, he was also impotent. Marie had the marriage annulled in 1916 on grounds of non-consummation.

Margaret Sanger ignited Stopes's interest in contraception, and the firebrand young scientist decided to start a campaign for birth control in Britain. Knowing the relatively recent experiences of the Carliles (who were persecuted in the 1820s for their sexual and political radicalism) and Charles Bradlaugh and Annie Besant (prosecuted in 1877 for re-publishing an old birth control manual), Stopes pressed on regardless, and shortly after her divorce published a concise guide to contraception, *Wise Parenthood*.

The book predictably infuriated both the Church of England and the Catholic Church, but she escaped prosecution. During the First World War she had also started work on a book which, strictly speaking, she was less qualified to write. It was called *Married Love*, and argued, amidst a doughty early-feminist manifesto, that marriage should be an equal relationship between husband and wife. She declared: 'I believe it is my destiny, to tell [young married couples] how to make love successfully.'

Which aspect of *Married Love* – the sex or the politics – was the most offensive to wartime British sensibilities is hard to assess, but finding a publisher for the book was certainly difficult. Walter Blackie, of Blackie & Son, sent her manuscript promptly back with the message: 'The theme does not please me. I think there is far too much talking and writing about these things already . . . Don't you think you should wait publication until after the war? There will be few enough men for the girls to marry; and a book like this would frighten off the few.'

Some of Stopes's writing was on the florid side ('The apex of raptures sweeps into its tides the whole essence of the man and woman, vaporises their consciousness so that it fills the whole of cosmic space'), but it was the political content that Blackie objected to: 'Far too often,' one such passage read, 'marriage puts an end to women's intellectual life. Marriage can never reach its full stature until women possess as much intellectual freedom and freedom of opportunity within it as do their partners.'

It was Stopes's before-their-time beliefs on sexual pleasure that made *Married Love* such a remarkable book. 'By the majority of "nice" people woman is supposed to have no spontaneous sex impulses. By this I do not mean a sentimental "falling in love", but a physical, a physiological state of stimulation which arises spontaneously and quite apart from any particular man. It is in truth the creative impulse, and is an expression of a high power of vitality. So widespread in our country is the view that it is only depraved women who have such feelings (especially before marriage) that most women would rather die than own that they do at times feel a physical yearning indescribable, but as profound as hunger for food.' (Naomi Wolf has pointed out that almost seventy years later, the writer Sallie Tisdale, author of *Talk Dirty to Me: An Intimate Philosophy of Sex*, shocked modern America by remarking in the book that she sometimes felt a sexual desire as sharp as hunger.)

In March 1918, Stopes finally found a small publisher willing to take its chances with *Married Love*. It turned out to be an inspired gamble. The book was a global sensation, selling millions of copies by the mid-1920s. It was published in America but declared obscene and banned. Marie was the first feminist media star, revelling in every awkward turn of a flat-footed establishment's attempts to silence her. A Catholic doctor, Halliday Sutherland, called for her to be imprisoned via an article in the *Daily Express*. In fact, she was never even prosecuted, but two of her supporters, Guy and Rose Aldred, a prominent anarchist couple who published a pamphlet by Margaret Sanger, were found guilty of selling an obscene publication.

In 1921 Stopes founded the Society for Constructive Birth Control, with financial backing from her wealthy second husband, Humphrey Roe, a philanthropic Manchester manufacturer who was an enthusiast for contraception, having seen the sufferings of the female workforce from having too many children. Marie also opened the first of her birth-control clinics at 61 Marlborough Road, Holloway, North London, in a converted house between a sweet shop and a grocer's. It was designed by Marie to be homely and welcoming 'to mothers or fathers', but still only attracted a handful of women for many months; those who came were often afraid to give their name.

The letters Marie Stopes received, 40 per cent of which were from men, provide an eloquent grassroots statement of ordinary people's sex lives in her time. One, dated 1921, read: 'So many

Englishwomen look upon sexual intercourse as abhorrent and not as a natural fulfilment of true love. My wife considered all bodily desire to be nothing less than animal passion, and that true love between husband and wife should be purely mental and not physical . . . Like so many Englishwomen she considered that any show of affection was not in keeping with her dignity as a woman and that all lovemaking and caresses should come entirely from the man and that the woman should be the passive receiver of affection.' Another letter, from an elderly man, recounted how when he was a young husband in 1880 and his wife had an orgasm, he 'was frightened and thought it was some sort of fit'.

There was just one unfortunate and hugely embarrassing area where these pioneering sexual reformers let their Victorian petticoats show. A clue to this is to be found in the full name of the Society for Constructive Birth Control. A point mysteriously missing from the website of today's Marie Stopes International Global Partnership, and very likely unknown to the readers of the *Guardian* newspaper, who in 1999 voted Stopes their Woman of the Millennium, is that the radical organisation she founded was actually called the Society for Constructive Birth Control *and Racial Progress.*

Sadly, Marie Stopes's views on class would make her an intellectual leper, a consummate hate figure, in today's world, and her attitudes to racial issues would render her liable to immediate prosecution. In 1920, for instance, she wrote: 'Society allows the diseased, the racially negligent, the thriftless, the careless, the feeble-minded, the very lowest and worst members of the community to produce innumerable tens of thousands of stunted, warped, inferior infants . . . a large proportion of these are doomed from their very physical inheritance to be at best but partly self-supporting, and thus to drain the resources of those classes above them who have a sense of responsibility. The better classes, freed from the cost of institutions, hospitals, prisons and so on, principally filled by the inferior racial stock, would be able to afford to enlarge their own families.' Stealing a march from Scrooge, who at least conceded that the poor had the right of being fed, albeit in prison or the workhouse, Stopes advocated that 'the sterilisation of those totally unfit for parenthood [be] made an immediate possibility, indeed, made compulsory'.

Marie Stopes's inspiration, Margaret Sanger was no more sound on these difficult matters. Her 'mission statement', according to one biography, was: 'More children from the fit, less from the unfit'. She opined that birth control can be '. . . nothing more or less than the facilitation of the process of weeding out the unfit, or preventing the birth of defectives'. Sanger's medical views were equally creaky at times. We all acknowledge today that lack of orgasm is a great sadness and frustration for billions of women, but few go quite as far as Sanger who, in her 1915 pamphlet *Family Limitation* (ten million copies sold in thirteen languages) declared that failure to give a woman an orgasm would lead to the 'disease of her generative organs'.

Even D. H. Lawrence considered one of the early-twentieth century's standard bearers of liberated sex, carried the burden of the nineteenth-century's core anti-sex attitudes. Of masturbation, he wrote in 1929: 'Instead of being a comparatively pure and harmless vice, masturbation is certainly the most dangerous sexual vice that a society can be afflicted with, in the long run . . . in masturbation there is nothing but loss. There is no reciprocity. There is merely the spending away of a certain force, and no return. The body remains, in a sense, a corpse, after the act of self-abuse. There is no change, only deadening.'

Lawrence was also, much more importantly, caught in the mantrap that most characterises the first seventy years or so of twenteenth-century progressive thinking on sex. This is the idealistic, but wrong-headed and naïve, notion that the only form of orgasm that 'counts' is when man and woman climax simultaneously, and do so exclusively by the mechanism of penetrative sexual intercourse.

Lawrence, like so many other sexual liberators, had his heart in the right place. In *Lady Chatterley's Lover*, written in the mid-1920s, one character, a parson, says to Clifford Chatterley, 'My good man, you don't suppose for one moment that women have animal passions like ours?' It was clear that Lawrence did not accord in the least with this view. Here is a typical account from the book of a glorious, if slightly implausible, mutual orgasm:

> . . . she felt the soft bud of him within her stirring, and strange rhythms flushing up into her with a strange rhythmic growing motion, swelling and swelling till it filled all her cleaving consciousness, and then began again the unspeakable motion that was not really motion, but pure deepening whirlpools of sensation swirling deeper and deeper through all her tissue and consciousness, till she was one perfect concentric fluid of feeling, and she lay there crying in unconscious inarticulate cries . . . He sat down again on the brushwood and took Connie's hand in silence. She turned and looked at him. 'We came off together that time,' he said. She did not answer. 'It's good when it's like that. Most folks live their lives through and they never know it,' he said, speaking rather dreamily. 'Don't people often come off together?' she asked with naïve curiosity. 'A good many of them never. You can see by the raw look of them.'

It is hard, with the best will in the world, to escape the conclusion that enlightened authors penning such accounts, as well as progressive doctors and sexologists endlessly promoting simultaneous orgasm late into the century, had either never had sex as they describe it – or, in the case of men like Lawrence, were mistaken, or even gulled by a well-meaning conspiracy among women by which they would attempt to please their men (or at least appear modern) through the skilful faking of orgasm to look something like Lawrence's fanciful description.

Marie Stopes idealised mutual orgasm as 'the co-ordinated

function'. Dr Eustace Chesser in *Love without Fear* (1939) stated, 'Both parties should, in coitus, concentrate their full attention on one thing: the attainment of simultaneous orgasm.' But the greatest and most passionate twentieth-century advocate of the simultaneous orgasm fantasy was the most outspoken, radical, influential and successful (though long forgotten) sex manual writer of the period, a Dutch gynaecologist called Theodore Hendrik van de Velde.

Van de Velde's extraordinarily explicit book was called *Ideal Marriage: Its Physiology and Technique* and first appeared in Holland in 1926, in the days, or so we are assured by a lot of elderly people today, when most people knew as much about orgasms as they did Microsoft Windows. The real extent to which sexually active people in the early twentieth-century were ignorant of the potentiality for both sexes to enjoy sexual pleasure is impossible to gauge; there were no surveys on the matter, no magazines urging women to demand the Big O every time.

At best, assessing this important matter is a judgement call. The only empirical evidence is that sex was not discussed in public, that it did not figure in the mass media, that literature which dealt with sex was marginalised and banned – and that activists like Marie Stopes gained the strong impression that both women and men were, for the most part, sexually ignorant. It is credible – and some literature, diaries and letters support this – that there was a body more substantial than we smugly acknowledge today of 'underground' knowledge and experience of sex, especially among the middle class and members of the intelligentsia who had secreted away copies of the *Kamasutra* and *Married Love*. But the dour decades that led Philip Larkin to conclude, however facetiously, that 'sexual intercourse began in 1963' cannot be said to have seen a democratisation of orgasmic pleasure to compare with that which unfolded in the mid- to late-twentieth century.

Unlike with *Married Love* or *Lady Chatterley*, there was no public fanfare or attempted prosecution when *Ideal*

Marriage slipped out in the UK in 1928. The ostensibly low profile of the Dutch book was accounted for by its being an import published by Heinemann's medical books division, which made it nominally an obscure textbook for doctors. It had barely any illustrations, and was therefore unlikely to appeal to or fall into the hands of schoolboys. And as a further safety measure, in case the book, with its deceptively bland title, might yet attract the wrong sort of reader, it was emphasised inside that the author's observations and advice applied solely to married people. For good measure, the author came close in his introduction to apologising for the book's very existence. Like the early researchers on female sexuality mentioned previously, he found it prudent to publish his book when he was old. In a notably downbeat personal introduction, he explained from his retirement home in Switzerland that he was only able to write it because he was close to the end of his life.

News of van de Velde's book nevertheless travelled quickly. It was reprinted forty-three times in English alone, as late as 1960 – when it was still the only manual of its kind, with the comparable (but in many ways more inhibited) *The Joy of Sex* still some years away. In 1928, *Ideal Marriage* was the only modern book to date by an authoritative male to endorse cunnilingus and fellatio. Van de Velde advocated monogamy, but believed the way to make fidelity work was for husbands to learn to satisfy their wives in bed.

Van de Velde left nothing to the imagination. Neither *Lady Chatterley* nor, for that matter *Fanny Hill*, written in 1749 (neither of which could be read uncensored by the general public before the 1960s), had much to say, for instance, on oral sex – there was one fleeting allusion in Cleland's work and none at all in *Lady Chatterley*.

Here, then, is Dr van de Velde in the midst of a scholarly medical discourse on vaginal lubrication; the language may be dated in parts, yet it is far too fruity for a modern tabloid newspaper: '. . . the most simple and obvious substitute for the inadequate lubricant is the natural moisture of the salivary glands . . . and during a very protracted local or genital manipulation, this form of substitute must be applied to the vulva not once, but repeatedly. And this may best, most appropriately and most expeditiously be done without the intermediary offices of the fingers, but through what I prefer to term the genital kiss, by gentle and soothing caresses with lips and tongue . . . Lack of local secretion ceases to be a drawback, and even becomes an advantage . . . The acuteness of the pleasure it excites and the variety of tactile sensation it provides, will ensure that the previous deficiency is made good – i.e. that sexual excitement and desire reach such a point that – either by these means alone or aided by other endearments – distillation [orgasm] takes place, heralding psychic and bodily readiness for a sexual communion successful and satisfactory to both partners.'

Van de Velde's was an evangelistic, and also a remarkably humane work. If a wife, the doctor wrote, 'chooses not to give access to the husband's caressing hand', and consequently that, 'there is not the necessary excitement and desire on her part to cause swelling of the labia, dilation of the vulva and erection of the clitoris, then, as these manifestations are both normal and desirable before coitus, it is both stupid and grossly selfish of the husband to attempt it if they are absent'.

Yet even in this astonishing early textbook on orgasm, for *Ideal Marriage* was no less, the most fundamental premise was quite hopelessly incorrect. In van der Velde's book, not only was the importance of simultaneous orgasm paramount, but some women (supposedly) quoted claimed they could *only have* an orgasm once they had felt their partner's seminal fluid released. One elucidated in what was a startling explicit manner for the era: 'Then, I feel the liquid torrent of the ejaculate, which gives a perfectly distinct sensation, as gloriously soothing and refreshing at the same time.'

This is odd in the extreme. While a very few women in other studies have occasionally confirmed, or believed, that they can just about make out, as if it were a distant radio signal, their partner's ejaculation, the idea of a woman's orgasm being dependent on the sensation is so unusual as to prompt the modern reader to wonder whether Dr van de Velde made it up – that it was his own fantasy.

A lot of the work for even mid-twentieth-century sexual pioneers consisted of a simple naming of parts. Helena Wright, a British gynaecologist, published *The Sex Factor in Marriage* in 1930, and then revised it in 1947 to take up the cause of the clitoris with even greater precision. The earlier instructions, she had realised, were simply not encouraging enough for women whose socialisation had forbidden them to touch themselves at all. 'Arrange a good light and take a mirror,' Dr Wright instructed: 'The hood can be gently drawn backward by the finger tips and inside will be seen a small, smooth, rounded body . . . which glistens in a good light.' Then touch it, she advised. ('Any small, smooth object will do.') She promised that 'the instant the clitoris is touched, a peculiar and characteristic sensation is experienced which is different in essence from touches on the labia or anywhere else.'

Even with such specific anatomical information becoming available, however, the old propagandists for synchronised mutual orgasm were still not quite dissuaded. In the same way as Freud's view persisted among Freudians for fifty years, as recently as the 1970s, old-school doctors and newspapers articles on sex were still reassuring men that a woman who failed to achieve orgasm with him was suffering from a physical or a psychological problem. It was not until the emergence of Dr Alfred C. Kinsey in the late-1940s and early-1950s that the reality was finally laid bare that penetrative sex rarely, if ever, produces female orgasm – and that simultaneous orgasm is a myth.

15

The Orgasm Comes of Age: From Kinsey to the Swinging Sixties

'There cannot be many who would hesitate before admitting that the present age is, in the sexual sense, a period of freedom'
Burgo Partridge, *The History of Orgies*, 1958

Alfred Kinsey's name is practically synonymous with the post-Second World War liberation of sex in the West. His achievement at the Institute for Sex Research at Indiana University, the painstaking delineation and mapping of a generation's and a culture's hitherto unknown private sexual beliefs, experiences, predilections and practices, was a landmark in social science. Yet Kinsey was by training a zoologist who had taught in his core discipline and biology at Harvard and, by the age of thirty-five, was the world's foremost authority on the gall wasp.

His first book, *Sexual Behavior in the Human Male* appeared in 1948 in a welter of publicity. It ran to eight hundred pages of tinder-dry statistical and scientific material and commentary, with not a line of unseemly passion or a tendentious remark that strayed beyond commentary on the evidence. Yet 'the Kinsey report', as it was called by a sex sensation-hungry media, sold half a million hardback copies in the US,

even at the price of $6.50 – the same as ten or fifteen paperback novels. A single copy reputedly found its way to the Soviet Union, where there was a room in the Kremlin in which authorised personnel could study the decadence of the West. Even books *about* the Kinsey report sold in hundreds of thousands. The subject patently lit a fuse with publics far away from America; even reading at an ocean's distance about the behind-the-drapes goings-on in American bedrooms seemed to titillate. Kinsey's follow-up, *Sexual Behavior in the Human Female*, came out five years later. It did not sell as well, but only because huge excerpts were published by newspapers and magazines whose editors knew from the experience of 1948 that the very word 'Kinsey' in a headline could guarantee extra sales.

Kinsey appeared to approach his research entirely free from moral bias. Like Desmond Morris in the next decade, he regarded *Homo sapiens* as just another species which, in the case of the male, seemed to display as a primary behaviour the pursuit of orgasm. In a decade of evidence-gathering for their two books, Kinsey and his team interviewed (unlike Shere Hite in the 1970s, who sent out questionnaires) over 12,000 men and women on 200 separate areas of their sexual history. The interviews could take several hours, after which the numbers were processed by a punch-card-reading computer.

The conclusions of *Sexual Behavior in the Human Male* would not even make the cover story of a women's magazine today. Up to 70 per cent of the population, Kinsey found, used the missionary position exclusively. Yet overall people were having more sex and of more adventurous varieties than was thought. It was revealed '. . . that there is no part of the human body which is not sufficiently sensitive to effect erotic arousal and even orgasm for at least some individual in the populace'. The average man, it emerged, reached the peak of virility at 16 or 17. Males between 16 and 20 typically sustained an erection for 42.88 minutes. The time declined immediately after that. Men who had sex early continued having it later in life.

Over a third of the men questioned had 'some homosexual experience between the beginning of adolescence and old age', yet only 'about 6.3 per cent of the total number of orgasms is derived from homosexual contacts'. Educated men were more likely to perform cunnilingus than uneducated. Working-class men had extramarital affairs early in marriage when they were healthy and virile, whereas white-collar men had them later, when they had fat stomachs but fat wallets to match. And, the figure most enjoyed by successive generations of schoolboys and students, some 17 per cent of boys raised on farms in the US in the 1930s and 1940s admitted to having experimented sexually with livestock.

The second study proved more of an aphrodisiac for the 1950s man. Kinsey's comments, never less than academically rigorous, were nonetheless perceptibly bolder. One of the most surprising findings of *Sexual Behavior in the Human Female* was that of 5,940 American women questioned, 62 per cent admitted to masturbating: 'We have recognised very few cases, if indeed there have been any outside of a few psychotics, in which either physical or mental damage had resulted from masturbatory activity,' Kinsey commented.

His other remarks were equally incendiary in their dry way: 'There were wives and husbands in the older generation who did not even know that orgasm was possible for a female; or if they knew that it was possible, they did not comprehend that it could be desirable.' Then there was: 'All orgasms appear to be physiologically similar quantities, whether they are derived from masturbatory, heterosexual, homosexual, or other sorts of activity. For most females and males, there appear to be basic physiologic needs which are satisfied by sexual orgasms, whatever the source.'

Kinsey's second study was a perfect example of fieldwork demonstrating how, whatever the posturings of medical professionals, moralists and churchmen, ordinary people frequently draw their own conclusions as to what is and is not pleasurable and/or advisable to do with their own bodies. 'It cannot be emphasised too often,' Kinsey concluded at one point from the data crunched from his team's interviews, 'that orgasm cannot be taken as the sole criterion for determining the degree of satisfaction which a female may derive from sexual activity. Considerable pleasure may be found in sexual arousal which does not proceed to the point of orgasm, and in the social aspects of a sexual relationship. Whether or not she herself reaches orgasm, many a female finds satisfaction in knowing that her husband or other sexual partner has enjoyed the contact, and in realising that she has contributed to the male's pleasure.'

The effect of Kinsey was seismic; comparisons were made with Newton and Darwin. He was seen as the architect of the sexual revolution. His statistics encouraged people of both sexes to realise that practically whatever they enjoyed was OK, at least in the sense that a lot of other folks were doing the same. A new liberation theology, as it practically was, was unleashed among the educated classes. The British critic Kenneth Tynan's review of John Osborne's *Look Back In Anger* singled out Jimmy Porter's 'casual promiscuity' as one of the ways he was typical of Britain's youth post-war. Among the less educated, too, there was a sea change in sexual morality; a sharp rise in the illegitimacy rate was seen, especially amongst teenagers.

The group most profoundly affected for the better by Kinsey's findings was the thirty-plus woman, who discovered what she'd privately suspected but dared not admit was in fact true – far from being a de-sexed wife and mother destined to gossip over the garden fence for the rest of her with a mouth full of clothes pegs, she was, in fact, an elegant, sensuous sexual creature in her prime. The Mrs Robinsons of the world had reason to thank Kinsey for generations to come.

As seems to be par for the course with the sexual pioneers of the last century, however, all was not quite normal in the private life in Bloomington, Indiana, of Dr Alfred and Mrs Clara Kinsey. The publicity version of the bow-tied Kinsey was of a regular family guy, a Republican voter, partial, according to Clara, a former student of Kinsey's, to a helping of 'persimmon pudding, highly spiced and topped with whipped cream', when he got home from the lab.

The Kinseys, however, according to a highly controversial 1997 biography of Alfred by James Howard Jones, an historian at the University of Houston, were quietly involved in a long-term *ménage à trois* with one of the guys from the lab. Kinsey, says Jones, was bisexual but with a preference for men. He had once planned to marry another woman but was unsure at the time how to consummate the relationship. He had a taste dating from adolescence for masochistic practices including insertion of objects into his urethra. He was also subject to depression and once, in such a state, circumcised himself with a penknife, in the bath and without anaesthetic. And he was also a voyeur, once paying a black interviewee a dollar for a peep at her clitoris, which she had claimed measured two inches.

None of this, as Jones is keen to emphasise, is to devalue Kinsey's extraordinary legacy to the second half of the twentieth century. Other, flashier sexual pioneers of the period have left us with nothing but dross sillier even than that of the Victorian quacks. One such is the (still) much-trumpeted Wilhelm Reich, a Freudian analyst born in Austria in 1897, whose life's mission was to try to prove that 'libido', the expression Freud coined for sexual desire, was a substance, sexual energy. The orgasm, he argued, was more than a sugar coating to lure humans to procreate. Its real function was to release sexual tension that literally built up in the atmosphere.

According to Reich, only a very few individuals (himself naturally included) had the requisite 'orgiastic potency' to dispel the dangerous clouds of sexual energy in the air by liberating rival clouds of 'orgone'. Society's anti-sex attitudes, monogamy, conservative sexual morality and pre-marital chastity were therefore endangering the planet by filling the air with undispelled sexual energy. What was needed, then, was more orgasms being fired off into the biosphere by orgiastic individuals such as him. The good orgone sexual energy could then be harnessed to cure a number of medical conditions: the cold, cancer, frigidity and impotence.

At his base in rural Maine, where he moved in 1939, Reich built special zinc-lined accumulators where orgone could be 'stored'. Patients would sit pointing a collector at their genitals as they experienced orgasm. As these experiments were proceeding, the cultists came to believe that menacing clouds of sexual tension were forming above Reich's institute. Trees were reported to have blackened. Reich decided that he had started inadvertently to produce a sort of orgasmic antimatter that he called DOR – Deadly Orgone Radiation. More beneficial OR – Orgone Radiation – was urgently needed to cancel out the DOR. Reich duly built what he called his 'Cloud Buster', a set of skywards-facing aluminium pipes on a revolving platform from which he fired orgone from the accumulators into the atmosphere.

The amusing thing is that this silliness was taking place simultaneously with Kinsey's dogged research, and was taken just as seriously by some. Reich's book *The Oranur Experiment*, covering his work from 1947–51, was published in 1951. Many of his other books are still available. Orgone accumulators can still be bought today from Internet vendors and are reputedly in use in Germany, Austria, Mexico and Brazil. Reich died of a heart attack in jail in 1957 while serving a sentence for contempt of court. His institute is now the Wilhelm Reich Museum, or 'Orgonon'. Holiday cottages can be rented where he worked, and his tomb viewed.

There was a growth of unrestrained, often pretentious, thinking on the orgasm in the post-war liberation period, too. Here is Simone de Beauvoir in *The Second Sex* (1949): 'We have seen that the act of love requires of woman profound self-abandonment. She bathes in a passive languor; with closed eyes anonymous, lost, she feels as if borne by waves, swept away in a storm, shrouded in darkness: darkness of the flesh, of the womb, of the grave. Annihilated, she becomes one with the Whole, her ego is abolished. But when the man moves from her, she finds herself back on earth, on a bed, in the light; she again has a name, a face.'

Thankfully, a strand of more grounded progressive thought on sex than Reich's or de Beauvoir's was coming to light as well. Concerning masturbation, for instance, the radical psychiatrist Thomas Szasz remarked astutely in 1946, 'Masturbation: the primary sexual activity of mankind. In the nineteenth century it was a disease: in the twentieth century, it's a cure.' And in 1955, Dr Abram Kardiner, an internationally respected psychoanalyst at Columbia University, commented how times had changed to the extent that, '. . . Today, there are parents who are alarmed when they discover that their adolescent boy is *not* masturbating. Several parents who have consulted me about such adolescents are quite concerned that the young person's sexual development is not proceeding normally.'

The ground was also prepared by the Kinsey reports for the most important advance towards increasing what Lionel Tiger's 'gross national pleasure', or what Hugh Hefner described as the change in the status of sex 'from procreation to recreation'. This was the invention of the contraceptive pill, which was approved by the US Food and Drug Administration in May 1960. This open-Sesame for the most unrestrained multilateral orgasmic pleasure potential in human history came so hot on the heels of the first mass-market contraception that it is likely future historians will fail to distinguish between the initial motivation for birth control, public health and social responsibility, and the subsequent liberation of private sexual gratification on an unprecedented scale.

Sexual pleasure, as we have seen, was not entirely missing from the agenda of the likes of Marie Stopes, and although she declared herself against women putting in their vagina anything they would not care to put in their mouth (she was thinking of chemical pessaries at the time), she would have been a great advocate of oral contraception. But before 1960 no such near-ideal solution existed. At the turn of the nineteenth-century, the most common form of contraception was the dually unsatisfactory *coitus interruptus*. An anonymous woman of 1920, quoted in Elizabeth Roberts's 1986 book, *A Woman's Place*, gave an idea of just how primitive contraception was within the majority of marriages: 'We were on the bus and Harold knew the conductor and he asked Harold if we were married. He said, "Don't forget, always get off the bus at South Shore, don't go all the way to Blackpool." That was how they kept the family down. It was just that the men had to be careful.'

Between 1910–30 in Britain use of mechanical and chemical methods (the new latex sheath, the Dutch cap and pessaries) rose from 9 per cent of middle-class couples to 40 per cent, and from 1 per cent to 28 for the working class. Even though as late as 1938 the Birth Control Advisory Bureau in London needed to remind clients, 'Many people imagine that birth control is practised only by those who don't like children. This is not so', by 1935, two hundred types of contraceptives were available in the Western world. These were used in addition to the rhythm method, which the Pope approved for Catholics in 1930. The Anglican Church had meanwhile approved birth control if further pregnancy was judged to be potentially detrimental to the mother's health.

The other, even less satisfactory, contraceptive method that was growing apace was abortion – either of the effective but dangerous back-street kind, or the safe but ineffectual patent remedy variety. A 1940 American study of the subject in Britain revealed that a hundred newspaper advertisements a week were appearing for 'abortions thinly disguised'. One such, in the *Newcastle Evening News*, purported to offer 'a secret Remedy for the Prevention of Large Families. Guaranteed Infallible'.

With women in an ever more sexual age attempting a rear-guard action against their own fertility by dint of a mixture of enforced criminality and quackery, the need for up-to-date advice on birth control was never greater. The Family Planning

Association was founded in 1950 by, among others, Lady Helen Brook, later founder of the Brook Advisory Centres. The FPA's prim name was part of its strategy for getting help to the maximum number of women without attracting unhelpful interference from moralist busybodies. Theoretically, then, the FPA dealt only with married women, but in practice saw 'premaritals' – girls who had a wedding date fixed, or at least, as Helen Brook used to urge them, were prepared to give the doctor a date – any would do. Other contraception clinics at the time demanded a wedding-dress receipt before issuing contraception. (Helen Brook, a wealthy banker's wife from an arty, bohemian family, was a contemporary of Stopes and Sanger and rumoured to hold similar racist and eugenicist views. She had always been a passionate believer in women having equal sexual rights to men. 'When I was a girl,' she once said, 'being brought up as a pure young virgin and hearing about men sowing their wild oats and having experience before marriage, I thought, but what is the difference?')

It was Margaret Sanger, in her eighties, who was the godmother of the Pill. There had been various futuristic dreams in her early days as a contraception campaigner to regulate women's endocrine systems artificially. It had been discovered in the 1930s that hormones could prevent ovulation in rabbits, but it was considered unethical to conduct such experiments on humans. In the 1940s, research in organic chemistry led to cheap methods of synthesising hormones. Accordingly, in a 1945 essay, Fuller Albright, an endocrinologist at Harvard and Massachusetts General Hospital, proposed a concept he called 'birth control by hormone therapy'. His idea was later dubbed 'Albright's Prophecy'.

In 1950 Sanger met Dr Gregory Pincus, a reproductive biologist and endocrinologist from Clark University in Worcester, Mass. Through wealthy libertarian contacts, Sanger raised $150,000 to start Pincus's research into a universal contraceptive. Pincus was already interested in studying the mode of action of female steriod hormones and had done tests on animals to see if the new synthetic steroids could suppress ovulation. He joined forces with John Rock, a clinical Professor of Gynaecology at Harvard Medical School – and also, surprisingly, a strong but dissident Catholic who, in a double irony, had previously dedicated himself to searching for a solution to infertility. Rock later left Harvard to found the Rock Reproductive Clinic, a free drop-in centre for women in the city of Boston where disseminating birth control advice was still technically illegal.

Pincus and Rock's first clinical trial took place in 1954 on a hundred volunteer women. The results were excellent: not one of the women ovulated, there were few side effects, and when the women came off the experimental drug, they swiftly regained fertility and had children. For a larger trial, Rock and Pincus went to Puerto Rico and Haiti, where again the results were positive.

As the new decade began, the Pincus/Rock pill came on to the market under the brand name Enovid. It was welcomed as a miracle, proved 99 per cent effective and caused fewer side effects than those of a normal pregnancy. It soon emerged tragically, though, that the inventors had overestimated the necessary dosage by a factor of ten. Of the 2.3 million early adopters of Enovid, 11 died and over a hundred developed blood clots. With dosages adjusted and various 'mini-pills' on the market, by 1974 fifty million women worldwide were on the Pill.

Part of the history of progress is that every step forward has unexpected reverse spinoffs. Although the Pope did not directly allude to this when he condemned the Pill in 1968, there was an unexpected cultural change wrought by the drug. The Pill was seen as a lifeline by women already struggling under the burden of large families and desperate not to have more babies; it also enabled single women to explore their sexuality free from the fear of pregnancy. But whereas previously men had the responsibility of 'not going all the way to Blackpool', now it was entirely up to women to organise contraception. They did not even have to go through a routine in bed to remind their men of the precautions they were taking; it was done privately in the bathroom at a different time from intercourse.

Paradoxically, then, in a variety of Western social milieux, from the student campus to the urban working class, it became harder, not easier, for young women to choose not to have sex. Instead of empowering women and easing their path to unalloyed sexual pleasure, the Pill had the effect of liberating men. So while feminist campaigners for sexual equality like the American Clare Booth Luce could correctly declare: 'Modern woman is at last free as a man is free to dispose of her own body, to earn her living, to pursue the improvement of her mind, to try a successful career', men were free simply to assume that most women were on the Pill. As the Australian critic Clive James said jokily: 'Of course I'm in favour of the Pill: it puts more crumpet on the market.'

The sixties were swinging, in London more than anywhere in the world. The city's moribund Soho was springing up as a sex *quartier*, and Kenneth Tynan was saying 'fuck' on prime-time TV. (To be precise: 'I doubt if there are any rational people to whom the word fuck would be particularly diabolical, revolting or totally forbidden.') Even the Church was loath to be left behind as the bandwagon departed: the Archbishop of Canterbury, Dr Coggan, was interviewed on the BBC's *Meeting Point* programme in January 1962 by a hip young pop star called Adam Faith: 'I'm one of those people who think sex is a thoroughly good thing, implanted by God,' said Dr Coggan. 'I'm not one of those who belong to the generation who thought it was a sort of smutty thing that you could only talk about hush-hush.' The phenomena, in short, were in train that would impel the religious thinker Malcolm Muggeridge towards his immortal conclusion in 1966 that, 'The orgasm has replaced the Cross as the focus of longing and the image of fulfilment.'

The Sixties images of the genie of orgasm emerging from its bottle and the birth of 'the permissive society' – Ann Summers sex shops, the musical *Hair*, Serge Gainsbourg and Jane Birkin singing '*Je t'aime . . . moi non plus*' and so on – are so familiar to everyone today that they barely need repeating. All branches of the mass media pushed and pulled at boundaries in a concerted attempt to dislodge and demolish them for ever.

The literary colossus, Penguin Books, became an unlikely agent of sexual revolution when it tried to bring out the first unexpurgated edition of *Lady Chatterley's Lover* and found itself in court. The Bishop of Woolwich, in evidence, declared: 'What I do think is clear is that what Lawrence is trying to do is portray the sex relationship as something essentially sacred . . . as in a real sense an act of Holy Communion.' The prosecution counsel, Mervyn Griffith-Jones, managed to sing in three sentences the swansong of the entire Victorian age, even though he was only fifty-four at the time of the trial, having been born in 1909. 'Ask yourselves the question: would you approve of your young sons, young daughters – because girls can read as well as boys – reading this book? Is it a book that you would have lying around the house? Is it a book you would wish your wife or servants to read?' Penguin was found not guilty of obscenity in November 1960. The public besieged the bookshops, buying two million copies of the Penguin edition in a year, thus vastly outselling the Bible. A flood of sexually explicit books such as *Kamasutra, The Carpetbaggers*, the first unexpurgated *Fanny Hill, Last Exit to Brooklyn* and *Portnoy's Complaint* subsequently came out one after the other.

There were inconsistencies and injustices in the Swinging Sixties. Although *TIME* magazine observed in March 1963, 'On the island where the subject has long been taboo in polite society, sex has exploded into the national consciousness and national headlines', and the *Daily Herald* asked at the same time, 'Are We Going Sex Crazy?', the 1960s in Britain (and no less elsewhere) were sexually speaking still very much a 'boys' party'. Men could every night enjoy the unrestrained orgasmic bliss of unprotected sex, an enjoyment their fathers might only have experienced once or twice in their lives when they were actively trying to conceive. Another *TIME* reporter, John Crosby, in a 1965 piece on Swinging London, became quite excited as he typed, enthusing about 'young English girls who take sex as if it is candy and it's delicious'. Soft-porn magazines also introduced the previously unseen sight of female pubic hair to innocent men who might never before have seen the stuff in photos or in real life. *Penthouse* magazine was introduced in London in 1965, but its mould-breaking crotch shots did not reach US newsstands until 1968. It was November 1974 before *Hustler* first showed what it termed a 'pink shot', with the interior of a model's vagina exposed as near fully as possible.

Not everything about the Swinging Sixties was what we would recognise today as sexually progressive. The Freudian fixation with the vaginal orgasm, for example, was still widely accepted orthodoxy. Stephen J. Gould wrote: 'This dogma of transfer from clitoral to vaginal orgasm became a shibboleth of pop culture during the heady days of pervasive Freudianism. It shaped the expectations (and therefore the frustration and often misery) of millions of educated and "enlightened" women told by a brigade of psychoanalysts and by hundreds of articles in magazines and "marriage manuals" that they must make this biologically impossible transition as a definition of maturity.'

Everyone was caught up in the vaginal orgasm fad. Here is Doris Lessing in her 1962 novel, *The Golden Notebook*:

> Paul began to rely on manipulating her externally, on giving Ella clitoral orgasms. Very exciting. Yet there was always a part of her that resented it. Because she felt that the fact he wanted to, was an expression of his instinctive desire not to commit himself to her . . . A vaginal orgasm is emotion and nothing else, felt as emotion and expressed in sensations that are indistinguishable from emotion. The vaginal orgasm is a dissolving in a vague, dark generalised sensation like being swirled in a warm whirlpool. There are several different sorts of clitoral

orgasms, and they are more powerful (that is a male word) than the vaginal orgasm. There can be a thousand thrills, sensations, etc, but there is only one real female orgasm and that is when a man, from the whole of his need and desire, takes a woman and wants all her response. But when she told him she had never experienced what she insisted on calling 'a real orgasm' to anything like the same depth before him, he involuntarily frowned, and remarked: 'Do you know that there are eminent physiologists who say women have no physical basis for vaginal orgasm?'

'Then they don't know much, do they?'

A stubborn belief also persisted in the possibility of routine simultaneous orgasm despite what must have been the routine disappointment of real life, if only from the frequency with which people liberated by the Pill were now having sex. L.J. Ludovici, a radical former RAF Squadron Leader who wrote passionately in the sixties on women's right to sexual equality, gave a good summary in a 1965 book, *The Final Inequality*, of what might be called the enlightened, but non-hippy, view of the orgasm's progress to the date when he was writing.

'While great changes are now taking place, it still remains true that our cultural values expect from women a passive role, especially in sexual intercourse. Husbands tend to regard their wives as the instruments of their pleasure and demand coitus of them as a "right", but few would concede that their wives have any "right" to demand coitus of them. Investigations show that the participating wife, as distinct from the passive wife, actually offends many husbands who are conditioned socially to regard a wife's sexuality as "indecent", or "wanton", or even as positive evidence of past "sinfulness" which she may be prone to repeat in the future. In extreme cases husbands diagnose the sexuality of their wives as a malady bordering on nymphomania. Many wives on the other hand still admit that they would rather die than show sign of sexual desire.'

Yet within a few paragraphs, Ludovici is back in the wishful-thinking 1930s: 'It is Woman herself who is able to effect simultaneous orgasm by postponing acceptance of her partner, however eager, until she is sure that she is herself verging on climax which she will reach by the time the man begins to ejaculate. Thus, Woman is the determinant of properly-fulfilled sexual intercourse and not Man as is too widely believed. If a man falls into the rhythms and tempo set by her, and if she times her acceptance of him correctly, they are unlikely to be disappointed. They will achieve true mutual sexuality and equality in the act of love which will set the tone for equality in other departments of their lives, for it is the subtle balance of sexual realities between them which lays the foundations of all their harmonies.'

Thus women were still not generally treated to an equal automatic upgrade of their sexual experience – only of their safety from pregnancy. The feminist movement worked hard at getting women a better deal from the new, freer atmosphere of sexual expression. Commentators like Sally Cline had a valid point when they made such frequently derided observations such as hers that orgasms are a 'form of manipulated emotional labour which women worked at in order to reflect and maintain men's values'. It was quite true to say that women's sexual satisfaction had not kept pace with the sexual revolution.

William Masters and Virginia Johnson were allies of the new women in this regard. The central tenet of their work was that women should enjoy sex. Their 1966 classic *Human Sexual Response* was a powerful plea for understanding of the role of the clitoris. Masters and Johnson cemented the notion for good that a woman's ability to have an orgasm was every bit as developed as a man's, and reinforced for a generation born after the Kinsey hoo-hah that clitoral orgasms could, should and would be much more likely to be gained by manual stimulation and masturbation than penetrative sex. The clitoris finally became the undisputed gold standard of female orgasm. Men were incessantly advised in magazines and newspapers that the way to sexual harmony was to 'find the man in the boat'.

We have omitted thus far in this account of the orgasm's progress in the first two-thirds of the twentyith-century the role of the entertainment industry in promoting and popularising the democratisation – for such it was – of orgasmic pleasure. Simulated orgasm on film and stage was actually quite rare, not surprisingly given the power of the censors. Mae West served eight days in jail for obscenity in 1926 after the New York City authorities closed down a play, *Sex*, that she had written and starred in for 375 performances. Valentino's last film, *Son of the Sheik*, in the same year, contained a scene showing him seducing Vilma Banky in which his face looked vaguely orgasmic even after the censors had their way with it. A 1933 Czech film, *Ecstasy*, starring one Hedy Kiesler (later known as Hedy Lamarr), had its nude scenes savaged by the censors, but they allowed a close-up of her face in orgasm to remain; since she was a woman, there is every chance they did not realise what was supposedly being portrayed. *And God Created Woman*, featuring Brigitte Bardot as an eighteen-year-old nymphomaniac and directed by her husband, Roger Vadim, was an orgasmic on-screen tonic for 1957. Vadim said of his work: 'For the Americans, it was the first declaration in a film that love for pleasure is not a sin. After *Et Dieu*, they accepted the idea that love could be filmed erotically without being pornographic.'

Rock and roll, with all those phallic guitars and crashing chords following endless drum solos, was the forum in which orgasmic messages were more regularly pounded out to a highly receptive young public. As Burgo Partridge observed in the age of Bill Haley, when he was writing: 'Now a new form of worship has appeared and quite recently, in the shape of a sequence of new varieties of music forms – skiffle, rock 'n' roll, etc which, particularly in the case of the latter, involve a rhythm and movement which individually, but more particularly in combination, produce an impression suggestive in the extreme. The "deities" of this new cult appear to be exclusively male, the outlet being provided for members of the opposite sex. The surplus energies of the males seem to be exuded in different forms, sometimes in activity punishable under the criminal law.'

Even the relatively wholesome offerings of the Beatles were orgasmically imbued for those with an ear to hear it. The nightly hysteria of screaming female fans peaked interestingly as the Fab Four (and plenty of other groups) shifted into a frantic falsetto climax at the end of many songs. 'You don't have to be a genius,' said a consultant at a London hospital to Christopher Booker, author of *The Neophiliacs*, a seminal book on the pleasure-seeking fifties and sixties in Britain, 'to see parallels between sexual excitement and the mounting crescendo of a stimulating number like "Twist and Shout".'

The pop idols' personal lives, understandably, were the epitome, as well as the best-documented examples, of the primacy of orgasm in the post-war world. This is not strictly a rock and roll reminiscence, but it is still very illuminating of how the cult of sex had exploded in fifty years or so. The Beatles as teenagers, Sir Paul McCartney has recounted, had a friend called Nigel Whalley who was in the original band out of which the Beatles grew, the Quarrymen. Nigel's father was a Chief Superintendent of Police, which meant that his son was often alone in the house at night. 'We used to have wanking sessions when we were young at Nigel Whalley's house in Woolton,' Sir Paul says. 'We'd stay overnight and we'd all sit in armchairs and we'd put all the lights out and, being teenage pubescent boys, we'd all wank. What we used to do, someone would say, "Brigitte Bardot . . . Oooh!" That would keep everyone on par, then somebody, probably John, would say, "Winston Churchill" . . . and it would completely ruin everyone's concentration.'

Another friend of the band, Pete Shotton, also mentions the communal masturbation in his book, *John Lennon In My Life*:

'... John and I got into the habit of tossing off in the bushes on the way home from school. We also enlisted our entire gang in a few mutual masturbation sessions, giving us all the opportunity to compare sizes and shapes. Lest any reader get the wrong impression, our fantasies, at least, were strictly heterosexual.' Shotton went on to describe again Lennon's Winston Churchill joke, 'as we furiously pommeled our hard-ons', and how it 'rather deflated the proceedings'. Lennon, it might be noted, later exploited this group masturbation as the basis for a sketch, 'Four In Hand', in Kenneth Tynan's *Oh! Calcutta!* Tynan, however, substituted the Lone Ranger for Winston Churchill, as well as not quite getting round to taking up Lennon's suggestion that the actors should masturbate for real on stage.

There was a further subtle twist during the early- to mid-twentieth century in the Western attitude to sexual pleasure, which although it was in fact occurring from the 1920s is purposely considered here as an afterthought. It concerned something which had so far failed to trouble even the sexually aware in previous centuries – the future of sex.

An interest in 'the future' as a subject worthy of informed scientific speculation and study had been developing across the world from the seventeenth-century onwards, but had never previously touched on sexual relations. As we have seen, sexual freedom and enjoyment was regarded from the time of the first Christians to the Age of Reason as a regressive force, a reversion to the ways of ancient, uncivilised peoples, or even of animals. In the twentieth century came a curious switch, however. As, in the backlash against Victorianism, it became slowly more acceptable for educated, middle-class people to admit to an enjoyment of sex, the notion began to take root that the future might see an increase in sexual fulfilment rather than its withering away. For the first time ever, universal sexual pleasure became an ideal of the future, rather than the past. Advocating sexual freedom was no longer a back-to-Nature argument, but an example of enlightened futurism.

Like most prognostications, predictions of how sex would

be in the future were never less than entertaining, especially as with increased longevity those 'futures' became knowable and we could in one lifetime measure past projections against contemporary reality.

The most common prediction in literature and film, spurred by liberationist and socialist philosophies, was that the orgasm would become a bodily function about which people would be totally uninhibited – a scenario more reminiscent of the Ancient Greeks than of futuristic predictions of human beings living in Lurex space suits. Sex, it was predicted, would in the future become more of a social, communal event than a personal one.

Some predictions were not an extrapolation or exaggeration of current sexual behaviours, but simply an inversion of them. A good example of this is seen in *Brave New World*, Aldous Huxley's dystopian 1931 novel, which describes a society organised exclusively around the pursuit of pleasure, in which twentieth-century sexual mores were pretty much turned on their head. Promiscuity in this world is promoted as being socially constructive, whilst meaningful sex that encourages emotional attachment is disparaged.

Movies in Huxley's new world were called 'the feelies', usually with a high sexual content which could be enjoyed sensually by the audience: 'The plot of the film was extremely simple. A few minutes after the first Oohs and Aahs (a duet having been sung and a little love made on that famous bearskin, every hair of which, could be separately and distinctly felt . . .' *Brave New World* citizens attended 'orgy-porgies', where, amongst other things, they would have sex. When one of the characters, Bernard, does not want to have sex with the gorgeous Lenina on their first date, he is deemed extremely odd. Sex, after all, was to be uninhibited and guilt-free, aided and abetted by the mood-altering drug Soma.

Huxley's book was more satire than an attempt at accurate prediction, but while it is true that some of the phenomena he sketched were aspired to and achieved in the free love sixties, it is always worth remembering in that decade that whatever was happening in enclaves such as Carnaby Street, Haight Ashbury and Greenwich Village, far more numerous than the highly publicised swingers were the equally motivated campaigners for Victorian values, such as the sexually fixated Mary Whitehouse and her National Viewers' and Listeners' Association in the UK, and the right-wing creationist conservatives who, in the 1980s, coalesced into the Moral Majority in the US.

Brave New World, however, took the inversion of Huxley's contemporary values one step further than woolly declarations of free love for all. He delineated a world in which the great obscenities were love, marriage and parenthood, where pregnancy was the worst humiliation imaginable and, consequently, where reproduction took place not in the womb, but on a conveyor belt. Even sex between children was encouraged.

George Orwell's 1949 novel *1984*, a pessimistic satire of the even nearer future, also contained a vision of how sexual mores might come to affect children. Orwell imagined things turning out a little differently from Huxley, however:

> The aim of the Party was not merely to prevent men and women from forming loyalties which it might not be able to control. Its real, undeclared purpose was to remove all pleasure from the sexual act. Not love so much as eroticism was the enemy, inside marriage as well as outside it. All marriages between Party members had to be approved by a committee appointed for the purpose, and – though the principle was never clearly stated – permission was always refused if the couple concerned gave the impression of being physically attracted to one another. The only recognised purpose of marriage was to beget children for the service of the Party. Sexual intercourse was to be looked on as a slightly disgusting minor operation, like having an enema. This again was never put into plain words, but in an indirect way it was rubbed into every Party member from childhood onwards. There were even organisations such as the Junior Anti-Sex League, which advocated complete

> celibacy for both sexes. All children were to be begotten by artificial insemination (artsem, it was called in Newspeak) and brought up in public institutions.

Orwell's Junior Anti-Sex league may seem a future fantasy too far, especially when one considers that he dreamed it up almost at the moment the Kinsey report was launched on a sex-hungry world. Yet in 2003, we have the extraordinary spectre of a high school and campus-organised national abstinence movement in America called True Love Waits, devotees of which sign the following statement: 'Believing that true love waits, I make a commitment to God, myself, my family, those I date, and my future mate to be sexually pure until the day I enter marriage.' The cult's official website was announcing at the time of writing that True Love Waits will be going global in 2004 and suggests to youngsters worldwide that they '. . . have the chance to be part of an international stand for purity'.

Orwell's *1984* posits a nation where sex for personal pleasure is a crime – investing Huxley's idea this time – and the only sex which is tolerated is intercourse which produces 'new material' for the Party. The orgasm, too, has become a target, representing as it does the most personal of private passions:

> In our world there will be no emotions except fear, rage, triumph, and self-abasement. Everything else we shall destroy, everything. Already we are breaking down the habits of thought which have survived from before the Revolution. We have cut the links between child and parent, and between man and man, and between man and woman. No one dares trust a wife or a child or a friend any longer. But in the future there will be no wives and no friends. Children will be taken from their mothers at birth, as one takes eggs from a hen. The sex instinct will be eradicated. Procreation will be an annual formality like the renewal of a ration card. We shall abolish the orgasm. Our neurologists are at work upon it now.

Cinematic representations of sex in the future often suggest that science and technology will find us the ultimate orgasmic satisfaction. Take the fun fantasy of *Barbarella*, Roger Vadim's 1968 film in which a Space Age nymphet heroine played by Jane Fonda has sex by taking a pill – but is also capable of making love in the old fashioned way, as well as having sex with her elbow. She even manages to overcome the 'Orgasmatron' or 'Excessive Machine', which is supposed to pleasure people to death. Barbarella, however, proves too much of a woman for any machine and it catches fire.

The Orgasmatron had a guest comeback role in Woody Allen's 1973 film *Sleeper*, which was predicated on Allen re-materialising in 2173. Here, the Orgasmatron was a home appliance that provided instant pleasure, bringing to mind the multi-tasking vibrator-cake mixer combination marketed in turn-of-the-twentieth-century women's magazines.

What, it may validly be asked, was happening to the orgasm's status outside the Western world during the early part of the twentieth-century? For the greater part, with so much of the world under colonial sway, the unsteady Western path from Victorian hangover to cautious advance was similarly followed. In the communist world, as it expanded from the USSR to take in Eastern Europe and China, a rigid prudery steamrollered any remaining tradition of sexuality which had previously survived the Western influence. Yet in the kind of colonial outposts that would have been regarded in the West until the 1960s as 'uncivilised' or 'primitive', a certain unfettered delight in orgasmic pleasure had been quietly continuing.

The sexiest area in the world was unquestionably French Polynesia, as Westerners *en masse* would learn in the early 1960s musical *South Pacific*, even if anthropologists were already infesting the region before Rogers and Hammerstein had given it much thought. The sexual Nirvana that Yale University psychologists Clellan Ford and Frank Beach discovered in the Pacific was eyebrow-raising stuff, even in the era of Kinsey. Among the Pukapukans and Marquesans, they reported, discussions about sex with children were so open and frank that all children were aware of orgasm, and the role of the penis (*ure*) and the clitoris (*tira*) in sexual arousal. Delayed ejaculation was a valued expertise because of the way it facilitated female pleasure. Multiple orgasms were also sought by both partners. The Marquesans were particularly keen on cunnilingus and fellatio, while the Pukapukans, unusually for a traditional society, had no preference between sex during the day or at night; each was equally popular.

A few years after Clellan and Beach, another American anthropologist active in French Polynesia, Robert I. Levy, found sex in the Pirae area of Tahiti to be centred on the female orgasm. A man would be humiliated if he failed to bring his partner to orgasm. The women boasted about what they believed was their unique capability to contract and relax the vaginal muscles during coitus – an ability also known in the Hawaiian islands as amo' amo – the 'wink-wink' of the vulva, that could 'make the thighs rejoice'.

16

*A Little Coitus Never Hoitus**: *from Fear of Flying to Sex and the City*

**(Dorothy Parker)*

'Sexual intercourse began
In nineteen sixty-three
(Which was rather late for me) –
Between the end of the Chatterley ban
And the Beatles' first LP'

Philip Larkin, '*Annus Mirabilis*'

'The deli scene' in Rob Reiner's Nora Ephron-scripted 1989 film *When Harry Met Sally*, in which Meg Ryan spectacularly fakes an orgasm for the benefit of an embarrassed Billy Crystal, was very funny. It is also iconic in the history of the orgasm.

As the journalist Ian Penman explained, in a contemporary article in the London *Independent* on the genre of cinematic portrayal of the orgasm: 'The man is incredulous but the woman proves it to him – right in mid-bite, in the middle of this crowded deli – girls fake it all the time. It's something learnt, known by heart (cruel expression), second nature, like ironing out skirts and dented egos. It comes naturally (even crueller expression). Faked orgasm is convenient, like microwave food – the same result (same satisfied partner) with half the time and bother. The laughter provoked by the scene is as much one of release as recognition.'

The fact that audiences laughed at the scene – that it was the main selling point of the film, indeed – makes it more important for our purposes here than bolder instances of celluloid sex such as those in *Last Tango in Paris, 91/2 Weeks* and so on. For the late-twentieth-century was the era when women in significant numbers finally confirmed the nightmares of the ancient patriarchs that females were not less but *more* sexual than men – and dangerously demanding with it. For the first time in history, outside of isolated pockets of female sanctuary in settings like Ancient Athens, women in large areas of the world were free to express their sexuality fully.

Now they wanted clitoral orgasms, and lots of them. This new mood of acquisitiveness was reinforced by regular, highly publicised interventions by sexologists such as Carol Travis and Carole Wade, who declared in 1984 that 'during masturbation, especially with an electric vibrator, some women can have as many as fifty consecutive orgasms'. No wonder that the orgasm, according to the American sociologist Daniel Bell, had 'overtaken Mammon [i.e. the false god of riches and greed] as the basic passion of American life'.

The new sense of orgasm as a woman's right was not just a Western phenomenon, either; nor in other parts of the world was it an exclusively middle-class, aspirational phenomenon. In the Indian magazine *The Week*, Mumbai sex specialist Dr Prakash Kothari was quoted in 1998 as saying: 'Women are increasingly demanding to know why they too are not reaching a climax. It is not just the convent-bred, pizza-gorging types who are curious about climax. A woman from the slums came to me and said her husband finished very fast and she does not get *nasha*.'

India provides one of the most interesting arenas for the introduction of more Western-style patterns of sexual behaviour –

even if such a concept is inherently ironic, since Indian culture has long forgotten more about rampant, unrestrained sex than the West has ever known.

Twentieth-century sex in India prior to the Western-influenced 'sexual revolution' was regarded predominantly as a shameful affair, summed up by Sudhir Kakar, the pre-eminent Indian psychoanalyst and *Kamasutra* translator as, 'No sex in marriage, we're Indian.' Kakar explains how the ludicrous prohibitions on when one could have sex in Medieval Christian Europe live on in Hindu tradition, and are still followed by millions in India. According to this tradition, a husband may only approach his wife during her *ritu* (season), a period of sixteen days of the month, but not on six of these. Of the remaining ten, only five are truly acceptable because sons can only be conceived on odd days – or rather nights, since daytime sex under these codes is completely beyond the pale. Then again, moonless nights and full moons are also off limits for sex, as are festival days for gods and ancestors. Kakar has conducted surveys among 'untouchable' women in Delhi as well as with religious Hindus of higher castes. The lowest caste women only have sex clothed, under sufferance and in fear of beating.

If this leads anybody to wonder how India still suffers from an over-population problem, Dr Promilla Kapur, a research psychologist and sociologist at New Delhi's India International Center, explains that sexual habits are markedly less inhibited in rural villages, where tribal groups, such as the Muria people practise near totally free sex and adults talk loudly and openly about sexual matters in front of children.

Additionally, as Kakar explains, lust finds a way even in prohibitive cultures. 'Despite these pervasive negative images of the conflict between the sexes in marriage, and the negative view of women and sexuality,' he writes, 'it must be pointed out that Indian sexual relations are not devoid of regular pauses in the conflict between man and woman. Tenderness, whether this be an affair with the soul of a *Mukesh* song, that is much quieter than a plunge into the depths of erotic passion known in Western culture, or sexual ecstasy of a husband and wife who have found their way through the forest of sexual taboos, does exist in India'.

Middle-class India is now adopting more Western (or, strictly speaking, Eastern) ways. There is widespread discussion of sex, especially in relation to getting the best orgasms, in academic journals, in Indian digital media and in the press and broadcasting. Satellite TV in particular has exposed children to sexually related material from an early age. Among children in urban areas, sexual play and exploration have, probably as a result of this, increasingly become a feature of growing up, even though parents are often unaware of it. Teenagers are also increasingly open about kissing and holding hands in public places. A recent sex study has accordingly shown that 30 per cent of respondents experience premarital sex, while 41 percent of men have sex before they are twenty. Unmarried women, too, are gaining sexual experience with male friends and work acquaintances; 43 per cent of women believe casual sex with someone you have no plans to marry is acceptable.

China, with its even sexier distant past than India, is struggling to overcome the puritanism of its immediate past, and succeeding to some extent. Suiming Pan, a sex researcher and Associate Professor of the Department of Sociology at the China Renmin University in Beijing, conducted seven social surveys of sex in the 1980s and 1990s. Investigators found it very difficult directly to elicit information on orgasms, but most couples reported they experienced 'sexual pleasure' – *kuaigan* – frequently. Of 1,279 men and women in 41 cities, Suiming Pan found men reach orgasm 7.2 times out of every 10 attempts, as against 4.1 times for women. In another 20,000-respondent survey by Professor Dalin Liu of the Shanghai Sex Sociology Research Center, a third of urban women and a quarter of rural women claim to experience *kuaigan* 'very often', while 58.2 and 76.8 per cent respectively enjoyed it 'sometimes'.

Parallel research in Hong Kong in 1996, however, showed women's knowledge of their own sexuality was particularly poor. A third of women interviewed did not know where the clitoris was located. It was noted, though, that the better educated people of both sexes were, the more they knew about the difference between female and male orgasm.

Thailand, regarded as something of a sexual paradise by visiting Westerners (and, sadly, by paedophiles, too), has in reality, surprisingly unsophisticated sexual habits. Sex is barely discussed, and not seriously when it is; a newlywed couple will routinely be teased and asked if they 'had fun' on their wedding night and how many times they 'did it'. But such banter distracts from a quite phallocentric society, living under the pervasive myth that men's sexual desires are boundless and unchangeable. As Sukanya Hantrakul, a noted Bangkok writer and social critic, says: 'Culturally, Thai society flatters men for their promiscuity . . . Women's magazines always advise women to tolerate the situation and accommodate themselves to it.'

In Indonesia, with its similar dichotomy between traditional attitudes (Islamic in its case) and modern, tourism-borne liberality, an analogous tension exists between old and new ways. Young women are increasingly eager to have sex with whomsoever they like without having to love the person, which upsets their parents' generation. But at the same time, there is a perception that sex is a secretive activity, in which women are like maids, subservient in everything, sex included. Some men make a point of having regular homosexual contacts because of a folk belief that they have supernatural powers which diminish during sex with women.

Indonesia does, however, have what would be regarded in the West as an unusually healthy acceptance of masturbation, which is almost universal among teenagers as a tension release. One study by Professor Wimpie Pangkahila, a reproductive health expert at Udayana University on Bali, found that 81 per cent of male adolescents and 18 per cent of females admitted masturbating. Some parents of young children reported happily watching their children masturbate to orgasm.

The dogged attempt by Indonesian young women to do their damnedest to strip emotion out of the sex equation, to take control and demand their due orgasmic pleasure, represents a significant change that characterised sex across the world in the very late-twentieth century – as evidenced by Frasier Crane's chastisement of the promiscuous Roz in one *Frasier* episode: 'Didn't your mother tell you that sex leads to things like dating?'

In 1973 Isadora Wing, heroine of Erica Jong's landmark tale of female sexual liberation, *Fear of Flying*, decides that she has come to 'that inevitable year when fucking [your husband] turned as bland as Velveeta cheese', and searches for the responsibility-and-guilt-free sex she calls 'the zipless fuck'. Also published in 1973 was *My Secret Garden*, a compilation of female sexual fantasies collected by Nancy Friday. In an NBC radio discussion at the time of its launch, Dr Theodore Rubin said to Friday: 'Your book *My Secret Garden* reduces women to men's sexual level.' Her response: 'Aren't women entitled to a little lust too?'

A woman as well as a man in this sexualised (and very slightly Huxley-esque) New World of Orgasm could be 'serious about her relationship with a partner', as Lionel Tiger explains in his history of pleasure, 'and yet have no intention of conceiving, or even having more than a semi-durable relationship based on playful pleasure-seeking. Neither is it inconceivable to us that a one-night stand might be the arena for thoughtful, considerate, generous sexual dialogue – an extraordinary advance, that, on animal behaviour, and very probably on early human conduct too.'

The New Woman – a cliché, yet justified in the late-twentieth century – developed a sexual acquisitiveness and outspokenness that would have astonished even Chaucer's Wife of Bath. Here is Germaine Greer, a respected Warwick University academic as well as a professional controversialist, in hearty voice in her 1970 book *The Female Eunuch*. 'At all events a clitoral orgasm with a full cunt is nicer than a clitoral orgasm with an empty one, as far as I can tell at least.'

While *Cosmopolitan* et al made commercial hay by turning the pursuit of the 'Big O' into a glossy monthly serial – which still continues – feminism, some of the quite prim Mary Wollstonecraft-ish hue, also got to work on the orgasm. Heedless of the mocking of Camille Paglia, who opined that, 'Leaving sex to the feminists is like letting your dog vacation at the taxidermists', feminists asserted that the primary cause of women's oppression lay in men's sexual dominance. The most vehement, such as Valerie Solonas in her 1970 *SCUM Manifesto* (SCUM was the 'Society for Cutting Up Men') went further, arguing not merely that women should withdraw from sex with men, but that heterosexual women were as 'dangerous' as men.

It was only a matter of time before male-style competitiveness about the female orgasm began to enter the picture. With the spotlight glaring on the orgasm there was an almost tangible media and peer pressure at large to be having them more frequently, or at least talking about them if you were not enjoying them. This new, almost consumerist, spirit can be found in the most diverse societies, often overriding deeply held taboos. In hygiene-obsessed Japan, *The Weekly Post*, the country's most widely read magazine, recently published a survey of 2,000 readers. Of the male respondents 51 per cent said they practise oral sex, and 8 per cent replied that they practise anal sex. Additionally, only 8 per cent of the women in the sample said they never experience orgasm.

Neither are just any orgasms good enough in much of the world today; orgasms have to be earth-shattering, mind-boggling, if a women is not to feel guilty of short-changing herself, and the rest of womanhood by extension. 'The discipline imposed is the discipline of the orgasm, not just any orgasm, but the perfect orgasm, regular, spontaneous, potent and reliable,' Germaine Greer has commented.

A glamorous young feminist sex-researcher came splashily on to the scene in 1976 to dampen down the flames of this cult of unrealistic expectation. The former Shirley Gregory from Missouri, a one-time model who appeared topless (albeit on a single occasion) in *Playboy*, had pursued an academic career between photo shoots, achieved two degrees and a PhD, and rebranded herself every bit as brilliantly as Max Factor would a new cosmetic. The name of the new product was 'Shere Hite'.

The prodigiously intelligent Hite then spent six years quietly sending out a 100-question sex survey to 3,000 American women and then collated the results into a modern, sexily written and presented version of Kinsey's second report. *The Hite Report: A Nationwide Study of Female Sexuality* was regarded as being revelatory and sensational, with its evidence that nearly 12 per cent of American women never had orgasms of any shape or size, and that only 30 per cent of those who did orgasm could do so without extra clitoral stimulation.

It was welcome re-confirmation at a timely moment that women's sexual satisfaction had failed to keep pace with the sexual revolution. But, as Naomi Wolf and others pointed out, it represented nothing remotely new, other than extremely clever marketing and a more attractive, TV-ready protagonist than was Alfred Kinsey. True pioneers of the same century, from Stopes to van de Velde to Helena Wright, Wolf reminded us, had said the same, even if Hite's book was 'somewhat less explicit than theirs'.

But with that irresistible combination of sexy looks, name and brain, Shere Hite was widely greeted as a pioneer rather than an inspired repackager, and even credited by some as having 'discovered' the clitoris. Criticism of her methods and credentials was discounted as so much sexist scattershot, pelted at her because she happened to be beautiful; Hite's supporters cleverly argued that her detractors simply objected to her conclusions as 'anti-male' – when, ironically, Hite may have *over* estimated the extent to which American men were managing to satisfy women. Her conclusion that even 30 per cent of women were able to orgasm without extra stimulation of the clitoris seems, after all, to err heavily in men's favour.

The Hite Report was nonetheless nominated in London's *Times* at the turn of the millennium as one of the hundred key books of the twentieth century. Even so, Anthony Clare, Clinical Professor of Psychiatry at Trinity College, Dublin, must surely have regretted such tabolid hyperbole as his statement in the *Daily Telegraph* that, 'Western society owes a debt to Shere Hite that will be well nigh impossible to repay.' The Reuters reviewer who commented that 'Hite's work has left an indelible mark on Western civilisation' was nearer to a fair assessment.

In concert with Shere Hite's singularly unremarkable but disproportionately influential deductions, the sexual practice that was the greatest beneficiary of the post-sixties sex boom was female masturbation, which became positively respectable. Vibrators first went properly public (as opposed to being disguised as vacuum cleaner attachments, orange juicers and the like) in 1973, when Betty Dodson, PhD, author of a new book called *Sex For One*, led a female masturbation workshop at a sexuality conference in New York. Earnest young women across the States began attending masturbation classes. A woman who took part in one of these in Berkeley, California, in 1973, remembers the group members being taught to inspect and manipulate their vaginas and bring themselves to orgasm. One woman admitted that her husband would have found it easier to cope with her being unfaithful than knowing she had learned to provide her own orgasms.

The following year, the sex therapist and author Helen Singer Kaplan was moved to write: 'The vibrator provides the strongest, most intense stimulation known. Indeed, it has been said that the electric vibrator represents the only significant advance in sexual technique since the days of Pompeii.' In 1983, *Playboy* commissioned a survey of 1,207 women that showed masturbation was 'the most reliably orgasmic sexual practice'. But female masturbation's real moment in the sun arguably came in 2001, when a Dutch mobile phone dealership, Tring, came up with the offer of a free vibrator with every vibrating Nokia 3330 phone when connected to the KPN network. Nokia and KPN soon persuaded Tring to drop the offer, Nokia calling it 'disgusting'.

Women, then, for the first time openly celebrated the wonders of the clitoris, flaunted their own superior orgasms, and (although probably *not* for the first time) openly laughed at men's failings. Larkin's words at the head of this chapter express a measure of pathos at being left behind by the sexual revolution, and this sense of climb-down by men to accompany the new female triumphalism is equally symptomatic of the spirit of the post-sixties age.

As Ian Penman, writing in 1989, pointed out: 'Male orgasm as represented in the movies is nearly always a cause for hilarity in a way that the female response just isn't. An endless hydraulic drollery, the slapstick of male orgasm consists not in faking but delaying it – the oscillation between arousal and deferral summed up by Madeline Kahn's chanteuse in [the Mel Brooks film] *Blazing Saddles*: "They're always coming and going and going and coming . . . and always too soon".'

None of the manifestations of the permissive society improved men's image particularly in women's eyes. ('When did God make men?' ran a feminist joke of the period. 'When She realized vibrators couldn't dance.') It was more as consumers of pornography and the sex industry than as 'sex gourmets' of the kind Dr Alex Comfort idealised, that male sexuality developed in the seventies and eighties. So we see the Pussycat chain of 'adult' cinemas in Britain thoughtfully removing every second seat so men can masturbate in privacy, and VCR sales simultaneously booming principally for the viewing of porn in the home. Masturbation, similarly, was the embarrassing and unacknowledged motor force behind the exponential growth of the World Wide Web as somewhere that the likes of the American porn star Annie Sprinkle could offer a global audience 'room cervix' – a photographic tour of the interior of her vagina.

'Pornography is the only place where "real" orgasm is found on screen,' explained Penman in his *Independent* piece on orgasm in the cinema. 'Porn itself is a strange hybrid of the real and faked. Its narratives are patently absurd; the real frisson comes from the hard fact that the bodies actually did what you see them doing. As a record of something that "really happened" it's a perverted, hijacked form of documentary. In hardcore porn nothing is left out, or, in one sense, left in: what, in the trade, is inelegantly known as the "Come Shot" is the leading member's contractual obligation – proof that the End really was reached. The women participants might pant away for the duration – faking it or not – but it is down to the male outburst to validate the act as fact.'

'Come shots' are still a little way off for the developing world's biggest cinema audience, in India, but Mumbai's Bollywood film industry, which entertains an audience of fifteen million paying customers per day, is currently turning its attentions to superseding the rose-petal-tinted, romantic but chaste view of sex it has peddled for decades. A new, raunchier Bollywood is leading the charge towards a franker sexuality for all Indians, not just the Mumbai elite. Sexually explicit Hindi films released or due to be released at the time of writing include *Oops*, an Indian version of *The Full Monty*, set among the world of Mumbai toy boys and male strippers, *Jism* – 'Body' – which replaces the traditional sari-clad, sexually subservient Bollywood heroine with a bold, sexually aggressive female career-girl lead, and *Khwahish* – 'Desire' – which features seventeen kissing scenes.

Film sex, of course, relies on effectively conjuring up sexual feeling by visual means. Donald Symons, Professor of Anthropology at the University of California and author of the renowned *The Evolution of Human Sexuality*, was impressed by a research finding which demonstrates the extent to which men can be sexually aroused by quite different stimuli from women.

'The profoundly different natures of men and women are dramatically illustrated by Bryant and Palmer's 1975 study of masseuses in four "massage parlours",' Symons writes. 'The primary service these women offer their male clients is masturbation, but in the process the clients are allowed to massage or fondle the naked masseuses. The purpose of this is to arouse the clients sexually as quickly as possible and hence to generate maximum business: the masseuses' motto is "get 'em in, get 'em up, get 'em off, and get 'em out". Although masseuses regularly look at, masturbate, and are masturbated by naked men, and although most of the women expressed a positive attitude toward their clients, only one masseuse reported that she herself experienced sexual arousal during her work, and this apparently occurred as a result of being massaged rather than massaging. To overcome her arousal, she would stand up, look at her client, and lose interest.

'The ability to engage in these activities without being sexually aroused represents a uniquely female adaptation,' observes Symons. 'It represents this ability not to become sexually aroused simply by the stimuli of male bodies *per se*. Few heterosexual men could massage and masturbate naked women without being themselves sexually aroused.'

Another unconventional form of one-way sexual pleasure distinctly more popular with men than with women is sex with animals. Kinsey's discovery about American males' sexual experimentation with farm animals is, as we have seen, the stuff of legend, but it may be that simple sex with sheep and goats represents only the more conservative end of the scale. We heard earlier about the practice of avisodomy, which involves a man putting his erect penis into a hen's anus and then breaking its neck. The poor creature's death spasms are said by aficionados to feel uncommonly pleasant to its torturer. The Marquis de Sade claimed that turkeys were used for the same purpose in Parisian brothels, and, according to the encyclopaedia, famous practitioners of the same practice have included Tippoo Sahib, an eighteenth-century Sultan of Mysore, also famous for introducing the British to the use of rockets as military armaments.

There is even, within the modern annals of the still weirder, evidence of men attempting to have sex with animals on a supposed basis of mutual orgasmic pleasure. The question of whether animals can enjoy sex at all is moot; zoologists have ample evidence that male animals enjoy ejaculation for its own sake, or at least find it pleasantly tranquillising. Male elephants stimulate their penis with their trunk. Male porcupines have been observed walking on three legs, with one forepaw on their penis, and male dolphins have been known to hold their erect penis in the jet of a water intake in their pool, and also to attempt homosexual intercourse. As for female animals enjoying sex, we can only be sure of the famously sexy bonobo ape, whose womenfolk give every impression of enjoying prolonged bouts of sex for the pure pleasure of it; other primates have been observed to give an orgasm-*like* response, but only after prolonged laboratory stimulation.

This leaves the interesting question of female dolphins. According to the anecdotal accounts of some students of cetaceans, female dolphins have been observed to manoeuvre themselves into a position in which they *appear* to be masturbating to orgasm. And this seems to have been enough to encourage the mermaid fantasies of some disturbed male individuals.

One such (literal) animal lover, an anonymous veterinary student when he was writing in 1996, maintains on a still extant website, *www.dolphinsex.org*, that he has loving relations on a continuing basis with female dolphins. 'I have been extremely lucky on two occasions with wild dolphins, and my current mate is a dolphin who lives in the harbour of my resident city,' he reported, going on to detail how he brings female (and male, too) cetaceans to orgasm – and they, him.

The sexual marketplace seems almost too crowded in these

busy times for Victorian attitudes to sex even to get a look in any more; yet right up to the present day, a chill wind still blows from time to time over the orgasmically literate modern West. Legislatures are still distinctly nervous of public displays of orgasm. In April 1998, Alabama passed an addition to the state's obscenity law making it 'unlawful to produce, distribute or otherwise sell sexual devices that are marketed primarily for the stimulation of human genital organs'. The law maintained that such sex toys are obscene and appeal to a 'prurient interest'. There is, State officials argued, 'no fundamental right to purchase a product to use in pursuit of having an orgasm'.

Masturbation in other forms, too, has continued even into the twenty-first-century to be a source of enormous cultural and moral awkwardness. Almost at the very moment that President Clinton was being fellated by Monica Lewinsky at the expense of the taxpayer (the lights were on and someone was paying his salary, after all), Clinton was obliged to fire the US Surgeon General Joycelyn Elders, after months of controversy over her remarks at an AIDS conference that masturbation 'is part of human sexuality and it's a part of something that perhaps should be taught'.

Elders maintained that she meant that children should be taught about masturbation in sex education courses – something a highly respected academic anthropologist at Rutgers University, Lionel Tiger, had been saying for years. Tiger has written amusingly, yet with an edge of common sense, about his own first experience of masturbation as a boy in 1940s Montreal.

> The truth has to be told that I was a sexual success that night. I mastered myself during an act of theft from a gloomy culture that embargoed pleasure. The earnest student became an autoerotic autodidact. And the question of who has a right to pleasure and how much and why, and what communities think about all this and under whose control, has vexed and interested me

ever since. Why are high-school students induced to enjoy Elizabethan drama but in virtually no school system offered helpful hints on enjoying physical pleasure by themselves, at no economic cost, without fear of unwanted pregnancy, without interfering with someone else's schedule of homework?

The future of sex continued to be perceived in technological terms as Huxley and Orwell had suggested. The Orgasmatron remained a frustratingly distant prospect, but, mindful of Desmond Morris's sixties observation that even in a technological age we remain essentially the same animals – 'The space ape still carries a picture of his wife and children with him in his wallet as he speeds towards the moon,' Morris wrote – attempts to help humankind restyle the orgasm for the space age continued apace.

Most futurists interested in space travel gave over some time during the permissive society's heyday to having great thoughts about sex in space, according to a 1983 article by the scientist Robert A. Freitas, Jr in the journal *Sexology Today*. Isaac Asimov, Freitas reported, wrote an article entitled 'Sex in a Spaceship' in 1973, and Arthur C. Clarke once commented, 'Weightlessness will bring new forms of erotica. About time, too.' In 1992, Elaine Lerner, a Sunday school teacher in Easton, Massachusetts, patented a system of straps and loops that would allow one partner to exercise control of the movements of the hips of the other partner during astronautic lovemaking. Raymond Noonan of the Sex Institute in New York argued that in-flight intercourse using Lerner's system would help relieve astronauts of in-flight stress. And the science writer G. Harry Stine recounted a rumour of one couple's attempts at sex in a NASA weightlessness test; they found they needed a third party close at hand, Stine said, '. . . to push at the right time and in the right place'. At the time of writing, thirteen years later, Lerner was still awaiting an official response from NASA to her invention.

Pierre Kohler, a respected French scientific writer, reported in a book published in 2000, *The Final Mission: Mir, The Human Adventure*, however, that both American and Russian astronauts have had sex in space as part of research into how human beings could survive for years on lengthy interplanetary missions. Kohler cited a confidential NASA report on a project codenamed STS-XX, which was supposedly carried out on a space shuttle mission in 1996. The project's objective was to explore the sexual positions possible in a weightless atmosphere. Twenty positions were tested by two astronauts, the results allegedly videotaped but considered too sensitive to release. The conclusion: that only four positions were achievable without the kind of 'mechanical assistance' suggested by Elaine Lerner. The other sixteen positions required as sex aids a special elastic belt and an inflatable tunnel, like an open-ended sleeping bag. 'One of the principal findings,' said Kohler, 'was that the classic so-called missionary position, which is so easy on Earth when gravity pushes one downwards, is simply not possible.'

With the success in 1998 of Viagra, the first oral medication designed to overcome impotence (it works by enhancing the effects of nitric oxide, a chemical that relaxes muscles in the penis during sexual stimulation, allowing increased blood flow to erectile tissue in the penis), a stream of attempts – some reportedly successful – to prescribe the drug or its variants for women were made. (Viagra for women is not universally regarded as an orgasmic cure-all, however. Dr Rosie King, a leading sex therapist and author who treats couples at the Australian Centre for Sexual Health at St Luke's Hospital in Sydney, has been involved in the international clinical trials of the drug for women, and has her doubts. 'Women's sexuality is a lot more complicated than popping a pill,' Dr King says. 'No amount of medication can make up for an unhappy relationship, poor sexual technique or a tired, exhausted, stressed-out woman. . . . There is no pill yet that will create sexual arousal. But there is one sure aphrodisiac – love.')

A current frontrunner, Dr. King's caution aside, as the female

Viagra equivalent is the provision of testosterone supplements for women lacking sexual desire or experiencing weak orgasms. In a study reported in the *New England Journal of Medicine*, of 75 women aged 31–56 and suffering from lack of libido, those given extra testosterone were two to three times more likely to have sexual thoughts and actions.

Nonetheless, supposed orgasm pills and creams continue to appear, all predicated on the masturbatory idea that modern women, in the main, prefer sex as a soloist than with a man. Cybersex, too, represents nothing less – or more – than a huge and eloquent global statement of the primacy of masturbation at this stage in human development. Jane Brody, writing in the *New York Times* in 2000, quoted a psychologist who styled Internet sex 'the crack cocaine of sexual compulsivity'.

If watching sexual images from the Internet is as compulsive as crack, then we can be confident that 'virtual sex', with the assistance of the panoply of orgasm-inducing body suits and the like that have been promised for the past twenty years, would be the most addictive leisure pursuit ever invented. There is no prospect yet of the kind of virtual sex beloved of science-fiction writers, but one of the world's most ardent propagandists for robotics, Kevin Warwick, Professor of Cybernetics at Reading University, has not only had an experimental computer chip surgically inserted in his own arm, but persuaded his wife to have such an implant too. By streaming the information from the two chips over the Internet, the couple have achieved a measure of interaction a little reminiscent of telepathy. Professor Warwick believes that there will in the future be sexual possibilities within such man-machine robotics. If we could reliably monitor by electronics exactly what a sexual partner is feeling, there could be some extremely interesting ramifications for sex.

Rather closer to the 'traditional' notion of the Orgasmatron, in 2001 Dr Nicolae Adrian Gheorghiu, from the town of Voluntari, near Bucharest, Romania, claimed he had invented a device that can give a woman sixteen orgasms in a minute.

He said the machine had been tested on women whose only complaint was that the thrill was too strong. 'It's more effective for a woman than having thirty men', Gheorghiu said.

In the same year, the *New Scientist* magazine carried a report on an electronic implant which would enable women to orgasm whenever they wanted. The device, patented in America, was aimed at those who found it difficult to achieve orgasm naturally, although, as with Viagra for men, it could also be used for those wanting to further boost an already healthy sex drive.

Stuart Meloy, a surgeon at Piedmont Anaesthesia and Pain Consultants in Winston-Salem, North Carolina, came upon the idea by accident. During a routine operation to give pain relief to a patient with a bad back, he was implanting into her spine a device with electrodes designed to emit an electrical pulse that normally interferes with pain signals. However, in this case, due to an electrode making contact temporarily with the wrong nerve, the device caused her, to use Meloy's coy description, to start 'exclaiming emphatically', before adding, 'You're going to have to teach my husband to do that.' Meloy expounded on how the device, a titanium-cased generator with more than forty electrodes and almost the size of a pack of cigarettes, could be suitably modified, inserted in the buttocks and controlled by a hand-held remote. Meloy conceded that his implant might need also to be programmed to limit its use.

There was a predictably ribald response to news of Dr Meloy's invention. The novelist Jeanette Winterson, in an article for the *Guardian*, predicted that electronic sex was just 'another way of faking it'. And she went on to consider some of the possible technical snags. 'Electrical pulses are sensitive to their environment. At the moment of ecstasy, will you set the car alarm off? Will the car alarm set you off? I don't want to be routinely doing the beep in Waitrose carpark, only to find myself writhing over the windscreen like a photo shoot for Nude Readers' Wives.

'I have not read whether the device comes with a timer,' Winterson added. 'Could you programme it like those burglar light switches that go on and off at random? That would at least leave the element of surprise that comes with real sex. It might be a turn-on, wondering where you'll be, and with whom, when the first tingle starts.'

Germaine Greer was even less bowled over by the concept: 'Giving yourself a few jolts from a titanium implant is about as bad as sex can get,' she commented.

Men around the world were less dismissive of the prospects of electronic orgasm implants for women, sensing that the technology would relieve them of an onerous duty. One US website largely aimed at men prefixed a report on the Gheorghiu Orgasmatron with the comment, 'No pushing now, get in line. Have your check books ready . . .' Of course, the enthusiasm *may* simply have been that the prospect of women becoming as effortlessly orgasmic as men appealed to men's sense of fair play.

Electronics and sex in equal measure continue to fascinate the stranger religious cults: the egregious Quebec-based Raelians, for instance, believe that one day, all our sexual needs will be attended to by microscopic, nanotechnology robots. It is interesting that in Japan, normally first in line for any robotic technology, there is no discernible market or research interest in Orgasmatron-type advances; sex dolls are the limit of technological invasion of the sexual sphere there.

Understandably less bullish than either Stuart Meloy or the Raelians about the prospects for Orgasmatron machines was a German electrician, Manfred Lubitz, who spent years perfecting a device for men. It incorporated a vibrating mat, massage pads and electrodes wired directly to the penis. Mr Lubitz was found dead in 2003 at his retirement apartment in Malaga, Spain. A police source quoted by the *Sun* in London said: 'He was watching a film called *Hot Vixen Nuns*. Unfortunately there seems to have been a power surge. And the flat was damp.'

On the less technological side, an interesting development in the institutionalisation of masturbation – especially surprising in light of the contemporary momentum for a pharmaceutical approach to improving women's orgasmic pleasure – was the low-key launch in the UK in 2003 of a new mass-market sex aid for women with the commercial name Vielle.

First made public in a several-page article in the *Daily Mirror*, the Vielle's marketing slogan was 'Discover yourself'. It was a non-vibratory masturbation aid for women. Made of PVC, the new device is in the form, the *Mirror* explained, of a rubbery finger puppet designed to enhance the effect of a mere digit. Most remarkably, the Vielle, it was reported, was expected to be stocked by high street stores such as the Boots chain.

The Vielle was invented by Liz Paul, a middle-class mother of three living in the sedate country town of Ilkley, West Yorkshire. Its development – noteworthy again considering that the Vielle would have been unmentionable and even illegal just a few years earlier – was partly funded by a loan from Yorkshire Enterprise, a venture capital group partly owned by five local councils. Most astonishing of all, a few days after the *Mirror* made public the invention, Mrs Paul was presented with the British Female Inventor of the Year (Health Section) award at a lunch at the Café Royal in London. (The overall winner however, it may be noted, was a new, lightweight fork for mucking out stables.)

The Vielle is worn on the index finger of a woman or her partner and has a circle of eight nodules protruding from the pad of the fingertip. These are located to ensure that they stimulate the sides and top of the clitoris. 'A finger alone would normally reach only the top,' explained the *Mirror*. 'There is also a dip in the middle of the nodules where, if needed, a lubricant can be applied.'

A trial on the Vielle by Alan Riley, Professor of Sexual Medicine at the Postgraduate School of Medicine and Health of the University of Central Lancashire in Preston, showed that orgasm rates were improved from 82.8 per cent to 95.3 per cent, and the time taken to achieve orgasm was reduced from, typically, 13.57 minutes to 5.05. According to the trial results on the company's website (*www.vielle.info*), 68.75 per cent of subjects reported that using the Vielle made orgasm attainment easier, while 31.25 per cent felt it made no difference to their ease of orgasm. (These figures look slightly less impressive when it is noted that the trial consisted of just 16 women, meaning that the device worked for 11 and did not for 5.)

The marketing approach for the Vielle was, interestingly again, hybrid; it was offered both as a traditional sex aid and to women suffering from various forms of female sexual dysfunction (FSD), from Orgasmic Disorder – difficulty in having an orgasm even when a women is stimulated and aroused – to Sexual Arousal Disorder – in which sex is not pleasurable due to a lack of vaginal lubrication – to Hypoactive Sexual Desire Disorder – a simple lack of libido.

The Vielle may 'work', but the problem, from the perspective of the *British Medical Journal*, as outlined in an article appearing in 2002, is whether FSD itself is 'real' or, as the journal argued, 'a corporate-sponsored creation' invented by drugs companies to make money. A typical medical products company claim, it was said, is that 43 per cent of women suffer from an FSD.

The commentator Christina Odone, writing in the *Observer*, supported the *BMJ*'s view and examined one such claim: '. . . there is something distinctly suspect about the figures they've come up with,' Odone wrote. 'This figure is based on a sample of just 1,500 women, who were asked whether they had experienced any of the symptoms of FSD for two months or more. If they answered yes to just one symptom they were categorised as "dysfunctional". This isn't just dodgy science: it's part of a dangerous tendency to "medicalise" the female condition. In recent years scientists have outlined a host of new "illnesses" from GAD (Generalised Anxiety Disorder) to FAD (Freefloating Anxiety Disorder). Far more women than men are diagnosed with these conditions . . . The truth is, feeling stressed, like losing interest in sex, is just part of life. Turning these ordinary emotions into "conditions" will only make women feel even more inadequate than they already do.'

New FSDs continue nonetheless to be identified, and not necessarily by 'Big Pharma', as the medical version of Dwight Eisenhower's 'Military Industrial Complex' tends to be known today, but by disinterested medical academics. Professor Riley in Preston, for example, in his *Daily Mirror* interview, said: 'A relatively new cause of orgasmic dysfunction affecting women in their early-twenties is a fear of letting go. Often they say they have seen women having orgasms in blue movies and find it disturbing. This is a new attitude which certainly didn't crop up when we started studying orgasms in the late 1960s. There are now far more sexually explicit films.'

Another novel, if arguably gimmicky, approach to FSD – or simply to improving existing orgasms – is being pioneered by cosmetic surgeons. It involves 'plumping up' the G-spot (or the area where it is expected to reside) with injections of hyaluronic acid, a chemical widely used in anti-aging procedures. The results of the ten-minute 'G-Delight' operation, as it is known in the US, last for several months. Some women report becoming so sensitised that they start having orgasms simply by walking. Amy Anderson, a professional woman in her thirties who wrote about her experience of the procedure for the London *Evening Standard*, reported: 'I had sex with my boyfriend 48 hours later, and experienced the most intense orgasms I've ever had . . . My orgasm was longer, deeper and stronger than I have experienced before. I found I was much more sensitive during sex, and it was much easier to reach orgasm through intercourse. I found that I was also more sexually aroused before sex began – it was what I imagine taking Viagra must be like for men . . . it made me feel more aroused while I was going about my everyday business, such as driving to work.'

A psychosexual counsellor from RELATE appended to

Anderson's *Evening Standard* article the advice that the £850 G-Delight operation was still to be regarded as controversial; equally likely to succeed to some degree in improving orgasm, she wrote, were pelvic-floor exercises, vibrators, a good run, a ride on a roller-coaster or a cup of coffee (all of which kick start the central nervous system with a shot of adrenaline) – or even long baths, scented oils and a bit of massage.

It was not that male sexual dysfunctions, principally erectile problems, were being ignored in some sort of pharmaceutical conspiracy to blame women for their own lack of fulfilling orgasms. Business in the last part of the twentieth-century was working hard at inventing solutions to men's sexual problems, too, if not fabricating the problems themselves. For men who did not benefit from Viagra, there was Penile Injection Therapy – the injection by fine-gauge needle directly into the side of the penis of a combination of drugs prescribed by a urologist. The drugs relax muscles and increase blood flow to create an erection. The treatment produces in 80 per cent of men a firm erection within ten to fifteen minutes which can last up to an hour.

PIT and Viagra aside (as well as the *eight* so-called 'me-too' Viagra-like drugs for erectile dysfunction in development by Big Pharma at the time of writing), commercial prescriptions for male sexual dysfunctions of various sorts tended strongly towards gadgetry. They ranged by the start of the twenty-first-century from the Vacuum Constriction Device (a non-surgical, external implement that induces erection by applying negative pressure to fill the penis with blood and then trapping it with a rubber ring at the base of the penis), to the Implanted Penile Prosthesis (a simple, semi-rigid device that produces a permanent erection), to the Inflatable Penile Prosthesis (a more complex system with inflatable cylindrical balloons in the penis that can be pumped up or deflated on demand), to microvascular surgery (correction of erectile dysfunction by correcting abnormal blood flow to and from the penis).

It is clearly as a response to the embarrassment felt by men about their sexual dysfunction that the boom in 'Tantric' sex techniques occurred from the 1980s onwards. As we have discussed earlier, the premise of Tantrism, revived in its many forms from Ancient Indian and Chinese roots, is that male ejaculation can and should be consciously withheld to prolong erection and give women a better penetrative sex experience. The *coitus obstructus* technique, as it was known before 'Tantra' became the buzz phrase, was seen in various other parts of the world before its orientalist-inspired revival. It was practised by Turks, Armenians, the islanders of the Marquesas in the Pacific, and the North American Cherokees. Masters and Johnson, in the late 1970s, also argued the benefits of men training themselves to orgasm – preferably several times – without ejaculation.

The charge against such practices in the modern era is that Tantra is really an under-the-duvet power politics play, the long-lost cousin of Viagra and therefore, rather too often, amounts to not a great deal more than male egoism and competitiveness thinly disguised as sensitivity to women. Like taking Viagra, men are inclined to practise Tantric techniques less to combat impotence than premature ejaculation, and from the desire to keep an erection at all costs even after orgasm – based on the very male conception that what women want is to be incessantly humped by a penis like a piston in a cylinder. One construction that can be placed on Tantric sex makes it appear as just another way for men to exercise their sexual selfishness; a lot of the modern Tantric literature emphasises not so much the added enjoyment it might provide women, but the enhanced experience it affords men – the idea, in other words, of travelling in hope rather than suffering the disappointment of arriving.

Men certainly imbue Tantric sex practices with some ambitious hopes. It is claimed that adept Tantric lovers can bring one another to orgasm simply by staring intently across a room. One current Japanese website, publicising what its author Houzan Suzuki calls Zen Tantrism, argues that even the supposedly sublime delights of the ejaculation-free, wait-for-it orgasm are not the final aim of Tantrism: 'Through these experiences of orgasm at multiple levels, the ego and sense of individuality will disappear in the end, and there will be no more distinction of male or female. There will be perfect state of "Union". However . . . the ultimate goal of sex is the vanishing away of the Union-Existence itself. Therefore, the ultimate sex goes beyond the transcendence of gender and the union of man and woman. It aims at the same thing as Zen: the transcendence of the existence, in other words, "Nothingness". "The death of ego".'

There is, it has to be said, no definitive 'truth' yet regarding any aspect of Tantric sex, even the basic principle that orgasm and ejaculation in men are separate events. Many men in a scan of websites advocating Tantrism report that their orgasms are not as intense as the exploding sensation of orgasm with ejaculation – more like a quiet, 'held-in' sneeze rather than a full-blooded *ker-chow*.

One authority, the British psychosexual counsellor and writer Julia Cole, points out that the Holy Grail of delayed ejaculation is otherwise well-known as a sexual dysfunction, in which, as she explains, 'He simply never quite gets there despite thrusting inside the woman for a very long time, during which the poor woman can be getting sore and uncomfortable.

'I think that it is true that the moment of orgasm and the moment of ejaculation, biologically in men are separate,' Cole argues, 'but the distance is so infinitesimal most men wouldn't be able to tell the difference; it can be something like half a second, a tiny time lapse. But some men do believe that through holding on and holding on and holding off the orgasm for as long as possible, they can achieve orgasm without ejaculation, and I think for some men that is what they are actually experiencing. Or they might have a release of a small amount of ejaculate and then be able to hold back. That *could* be nice for their partner, because they would be able to go on and stimulate her more, but I am not fully convinced, reading the material that I have, that these men are having a full orgasm without ejaculation, because as far as a man is concerned, the prostate is such a big part of the sexual response that not to have any release of liquid at all at orgasm is really difficult and I wouldn't think that most people could do it. Anyway, the myth about the man being erect for hours or whatever, and that this must be what women want, is not really the case. It is the use of hands and lips that is much more important.'

Tanya Corrin examined the Tantric sex fad in an article for the New York *Observer* in 2003, and drew a similar conclusion – that the sex can become a little burdensome for the woman but was often seen as a great achievement (and, to be fair, a rewarding one) for men. 'It was break-up sex with a former girlfriend of mine,' one male Tantric enthusiast, a thirty-five-year-old university professor in New York, told Corrin. 'I knew we were saying good-bye, but subconsciously I didn't want to say good-bye. I wanted it to last. And I lasted for a mighty long time. I think we made love for one and a half hours. And then I had this warmth going up my spine. I was thinking, "Hey, wait – something else is happening here! I think it's similar to what women have".' But another interviewee, a twenty-nine-year-old yoga teacher and graphic designer, had failed to get his girlfriend at all interested in his Tantric practices. 'It becomes a drag,' he confessed to Corrin. 'She'll have an orgasm, and then she'll be like, "Are you done yet? What the hell is wrong? What's wrong? Why didn't you come? Could you please stop with all this?"'

'I think what men are talking about with this is that it makes them feel more competent. Men want to feel competent, in control and powerful,' Dr Frederick Woolverton, a clinical psychologist and director of the Village Institute for Psychotherapy, told Corrin. 'Women are far more enthusiastic about intimacy than men are. This has just been demonstrated to me so many times. But despite everything, men yearn for intimacy. The problem is, when they get it, they don't know what to do with it. Intimacy, while desired, becomes threatening, and men sort of have to find their way out of that conflict.'

Chiara Simonelli, Associate Professor in Clinical Sexology at the Sapienza University in Rome, has a different, but still sceptical, slant on Tantric sex: 'Unsatisfied people look for easy solutions to difficult problems,' said Professor Simonelli. 'It is very difficult to use a Tantric vision as a mere technical manual. Many Tantric sexologists are very simple-minded people who use Tantra as many have already tried to use other Eastern philosophies like Buddhism and Daoism.

'There is a common problem in our culture of people escaping from pleasure and being more interested in power and performance. A lot of male ejaculations seem to me to be realised with little pleasure, sometimes none at all, even in the absence of specific male sexual dysfunctions. Many young men prefer to be engaged in other activities – meeting friends, playing with computer games, watching TV, drinking or dancing – and sex is not in pole position for them: they have good erections and ejaculations, but not what could be called a good orgasm, because they don't realise that quality of contact is more important than the number of ejaculations, their size and so on.

'So the problem for me with Tantric sex is that in our culture the body seems to be a tool for performance rather than a way to exchange pleasure and love with another person. The extreme lack of time we suffer in Western countries leads us to improve and organise every human activity, and too many of us try to find a place in their agenda for sex simply because it seems to be important for their health.'

Rabbi Shmuley Boteach, the American writer on sex, is more positive than others about Tantric practices. 'I have spoken to many Tantric practitioners and they can't get enough. It's not just about semen retention, but rather the erotic thrill of sex having no culmination, the delight of living in passion rather than building up to climax. There are actually people who believe in passion and welcome it rather than feeling it to be a burden. And women love a man who is continually excited about them. An all-out night of sex shows a women that he is so excited about her that he can't calm down.'

Perhaps the very last word on Tantric sex should come from Sting, the pop singer who in the 1990s did the most to popularise it when he made it known that, thanks to Tantric practises, he and his wife were able to enjoy eight-hour sex sessions. During a drunken night out in 2003 with fellow singer Bob Geldof, according to Geldof, Sting finally confessed: 'I think I mentioned I could make love for eight hours. What I didn't say was that this included four hours of begging and then dinner and a movie.'

The conviction amongst modern Tantrists that they have 'discovered' some absolute, eternal truth about sex, nonetheless, is very typical of a widespread modern belief that evolution has somehow finished its course, that the development and maturing of the human orgasm is no longer a work in progress but a done deal, a process that has reached its terminal velocity.

Such 'arrogance of the present', as this author terms it, or 'the snobbery of chronology' as C.S. Lewis described the syndrome, is endemic to every generation, and, in practically every case, wrong. We have seen how the Christian suspicion of sex and revulsion at orgasmic pleasure was the intellectual modernism of its day. Victorian prudery and shrinking from sexual pleasure too, hypocritical and sometimes insane though it seems to us, was the 'political correctness' of its day. Freud, at the turn of the century, seemed to be quite deluded and working from his own distinctly 'male chauvinist' agenda – yet was a revolutionary in the cause of orgasmic enjoyment. The fixation on simultaneous orgasm in the Freud-influenced first half of the twentieth-century, similarly, seems almost embarrassing now in its naïvety, yet was an enormously progressive social and political 'cause' too in its time.

In the same way as the Victorian world view on almost every subject from empire to engineering is continually up for revision, if we can be sure of one thing, it is that our twentieth and early-twenty-first century smugness about sex will be the subject of academic theorising and counter-theorising for many decades yet. One pervasive contemporary idea about sex that could, for example, be due a rethink is the perception that we have sex as much as we say we do.

In a survey in 2001 of 18,000 people across 27 countries by the condom manufacturer Durex, respondents claimed they were having sex typically twice or thereabouts a week. Annual 'scores' ranged from 132 times a year in the US, to 122 times in Russia, 121 in France, 109 in the UK, 98 in Australia, 86 in New Zealand, down to 37 times in Japan. But these figures are, naturally, self-reported and hence unreliable for any number of reasons. Just one of these is that modern people tend to have 'binge sex', doing it every day or more for a short while, then going weeks or months without.

A reaction to the assumption that we are all having sex all the time was already underway at the time of writing. Indeed, the competitive conversation in which couples, citing work, children and tiredness, try to out-brag one another over the time since they last had sex – longest since wins – had almost become a standard feature of the modern thirty- and forty-something dinner party. The jokey acronym DINS – Double Income No Sex – was increasingly being touted across the Western world in 2003, and only semi-ironically.

This candid twenty-first-century admission that we are usually too tired for adventurous, multi-orgasmic sex – or any sex at all – is remarkably uniform across cultures. Research in 2001 for *Top Santé* magazine in Britain showed one in five women said they were too tired or busy for sex. The figure was one in three in a survey the following year for the Chartered Institute of Personnel and Development. Only 16 per cent of women in the *Top Santé* figures said their sex life was 'fantastic', the highest proportion – 32 per cent –

characterising it as 'OK'. The National Sleep Foundation in the US polled 1,004 adults in the same year to discover 52 per cent have less sex than they did five years ago, 38 per cent have sex less than once a week – and 12 per cent of married couples sleep separately.

Lack of sex and a resultant slump in the birth rate is seen as a national crisis in Singapore, one of the world's wealthiest countries. Women are having an average of just 1.4 children, against the 2.1 demographers say is necessary for a population to replace itself. The government there has tried a variety of schemes to boost sexual activity, from tax breaks for married couples to a speed-dating service sponsored by an official government matchmaking agency, the Social Development Unit.

But Professor Victor Goh, from the Department of Obstetrics and Gynaecology at the National University of Singapore, in a 2002 study of 133 men and 460 women aged 30–70 found Singaporeans aged 30–40 still have sex only some six times a month, and from 41–55 four times. 'At the end of the day, when all their other responsibilities have been fulfilled, Singaporeans just feel too tried to perform,' reported Professor Goh. Yet his research showed that most Singaporeans were happy with this. Only 25 per cent of men and 10 per cent of women under 40 said they wanted more sex.

The problem even pervades societies thought of as being more sensual than most. The top sexual problem reported in Hawaii in 2002, according to an article in the *Honolulu Advertiser*, was women confessing to therapists that they are too tired for sex. In Italy, the polling institution IPSA has found women allowing less than an hour for lovemaking every fifteen days and 40 per cent of wives unhappy in their marriages. The spectre of the Platonic marriage is as common in India and the Indian diaspora, according to *Desi Match Maker*, a web magazine on marriage for South Asians living in the US. Professor Aroona Broota, a clinical psychologist at Delhi University, comments on the site that long-term lack of sex can become a self-fulfilling prophecy: 'Infrequency leads to a fear of performance. People forget a marriage is about partnership.'

In Australia expectations of sexual delight have a habit of disappointing people. Dr Rosie King, speaking to the *Sydney Morning Herald* in 2001, talked about what she saw as a new myth that everybody must want and enjoy sex, twenty-four hours a day, seven days a week, and the perception that sex is an Olympic sport where everybody should go for gold on every occasion. 'People can get uptight about their sense of entitlement to sex,' commented Dr King. 'It's a bit like salaries. Everyone thinks everyone else is getting more than they are.'

Despite such evidence that there is no great evolutionary call for better and more rampant sex, it is conceivable that, five million years down the evolutionary road, both human genders could be evolving a mechanically more efficient orgasmic response, with women better adapted to receive orgasmic pleasure and males developing, after generations of cultural pressure, the ability to slow down their hair-trigger ejaculatory mechanism. There is an argument that the species would benefit from bodies better built for sexual pleasure; an equal, speedy sexual response, whereby the majority of humans could copulate face-to-face and both sexes orgasm swiftly, reliably and simultaneously, would arguably be as beneficial a development as the entire world population speaking a common language.

What, on the other hand, do we *really* have by way of demonstration that our practice and appreciation of orgasm have improved significantly – or that the sex experts of today are not disseminating as much nonsense as they were fifty, a hundred, or a thousand years ago? And if they are, what unimagined and untold damage could we be doing to ourselves by believing we finally know 'everything we ever wanted to know about sex but were afraid to ask'?

Just as it is almost certain that, however well you believe

you are bringing up your children, you are still probably doing your bit to keep the next generation of psychiatrists in business, the chances are we have it wrong about sex too and will ultimately suffer for it. We always have. The psychiatrist Thomas Szasz bitterly described in 1973 how the misconceptions of his generation had affected them: 'The modern erotic ideal: man and woman in loving sexual embrace experiencing simultaneous orgasm through genital intercourse. This is a psychiatric-sexual myth useful for fostering feelings of sexual inadequacy and personal inferiority. It is also a rich source of psychiatric "patients".'

The poet and novelist Al Alvarez made a similar observation in a 1982 book, *Life After Marriage: Love in an Age of Divorce*. 'At the centre of that religion of marriage was a cult every bit as hallowed as that of the Virgin: the cult of the orgasm, mutual and simultaneous. It descended to the young people of my generation from both Lawrence and Freud as the "Inner Mystery", something they all aspired to, a sign of grace. Because of it I had impossible expectations of my marriage, my sex life, myself. I was an absolutist of the orgasm before I had had enough experience to ensure even sexual competence.'

While we are fond of congratulating ourselves that the sexual sphere has been revolutionised in the past few decades, there are billions of women around the world who have yet to have the opportunity to enjoy their fair share of orgasm. Even in the sophisticated Western cultures, large numbers of us are still 'hung up', unfulfilled and embarrassed sexually. Young women continue to be marginalised for perceived 'sleeping around' whereas young men are not, sexual ignorance is still extraordinarily rife and sexual diseases by and large on the increase.

Sexual enjoyment continues to be significantly skewed towards men's needs, and women demonstrably conspire in this. A question in QueenDom's 1999 survey amply illustrates this. In answer to the question, 'What do you do if your partner orgasms and you haven't yet?' 52 per cent of men say that they continue with sexual activity, while just 25 of women keep going; 26 per cent of women said sex usually ends when their partner climaxes, but only 7 per cent of men. When the question was reversed to, 'What do you do if you have reached orgasm, but your partner hasn't?' 46 per cent of men admitted they stopped, but only 33 per cent of women. Taking all shades of response into account, QueenDom concluded that men are eight times more likely to say that they stop because sex ends with their orgasm, while women are twice as likely to give up and just let their partner come. Men were also three times more likely than women physically to walk out of the bedroom – statistical evidence, this, of the egregious post-orgasmic 'pork and walk' syndrome.

There has also arisen as a function of widespread sexual knowledge quite an extensive cult of orgasm-faking among women who are sexually educated and aware, but cannot always find as much enthusiasm for sex as men. In 1985 Ann Landers's newspaper column asked female readers what they felt about having sex. Over 100,000 women responded with 72 per cent saying they would rather be doing something else. The point was taken up by a writer who, in a 1995 edition of *Cosmopolitan*, suggested that faking orgasms was a matter of speed and politeness. 'When you have got to get up for work the next morning, who has two spare hours to make him feel better about not making you feel great?' she asked. In the QueenDom poll, 70 per cent of women and 25 per cent of men admitted to faking orgasm at least once.

A number of American campus questionnaires have revealed widespread faking by women at least some of the time. Male respondents, however, were frequently under the impression that no woman of theirs had ever faked orgasms. The conclusion of one such survey was that, 'Clearly, the refined performances which women are giving are extremely convincing.' One does not need to be a rigorous feminist to see orgasm-faking as bad both for sexual equality and for the internal dynamics of a relationship. The renowned sex researchers Dr Jennifer

Berman and her sister Laura, of the Female Sexual Medicine Center at UCLA, see women 'owning their sexual pleasure' as 'the last frontier of the women's movement'. Their most trenchant and succinct advice to women faking orgasm accordingly is: 'Don't'.

This seems to be judicious advice. Of the women among QueenDom's 15,000 respondents, 73 per cent claimed they can tell if their partner fakes it, but only 61 per cent of men notice women faking. However, since it is manifestly easier for a woman to fake orgasm right down to the vaginal contractions, it is not surprising that only 23 per cent of women confirmed that their partner can tell the difference between a real orgasm and a fake. The predominant reason cited for faking orgasm in the poll was selflessness and making a partner happy. But the figures also showed eloquently that faking orgasm is likely to make both partners feel worse about themselves and each other. And when asked directly how they would feel about a partner faking, 95 per cent of women and 92 per cent of men said that they would not welcome it at all.

It could be argued that the culture of hedonism, sexual equality, sexual licence – whichever you choose to style it – has spread to good effect from the spoiled, gluttonous West to the rest of the world. Sex advice and a focussed seeking to improve orgasms is a boom industry in countries like India, Russia, China and Indonesia, where there is an ever-swelling surfeit of agony aunts, sexual studies centres, sex manuals, pornography and shops selling sex aids. The problems that agony aunts in the developing world encounter, however, can make Western sex advice columns seem a little trivial by comparison. Although in fashionable Mumbai or Moscow society there is plenty of angst about the quality and quantity of orgasms, in China, the country's first agony aunt, Xinran Xue, was more likely to find among the two hundred anguished letters a day that flooded into her show, *Words on the Night Breeze*, harrowing accounts by women of being continually raped by their father or a party official. So moved was Xue by the letters that she compiled them into a book, *The Good Women of China*. It was first published in the West in 2002, and was due to appear in mainland China in 2004.

Globally, though, the trend towards wanting to live one's sex life to the full is inexorable. There is clearly both the potential and the impetus for humanity to become a great deal more orgasm-literate and, ultimately, to regard orgasm as a fundamental human entitlement on a par with legal and political rights.

One rarely reported modern instance of orgasm coming to be regarded as a right was seen in Switzerland in 2003, where disabled people in Zurich were offered professional sexual services as part of a trial project. 'There is a very big demand for this,' commented Angela Fürer, local director of a social welfare organisation, Pro Infirmis. 'We have been hearing about the problem for years, both from disabled people and from those working with the disabled.' Her organisation was recruiting ten 'touchers' to offer sexual and emotional relief to Zurich's disabled. Full sex and oral sex was not going to be included in the pilot scheme, but Fürer said registered prostitutes might be included in the service at a later date. 'For now, we will just be offering massage, body contact, stroking, holding and bringing people to orgasm, if that is what they wish.

'It can be very difficult for some disabled people to take off their clothes and show a body which is deformed and I feel you can only expect people to put that kind of trust in you and offer their vulnerability when you yourself are willing to be vulnerable. On the whole, though, the response we've had has been extremely positive with many disabled people calling us to say how happy they are. These are people who don't just want to spend their lives breathing and eating and being cleaned up. They have souls and feelings like everybody else and sexuality is a part of their lives, just as it is with any other human being.' It should be noted that Pro Infirmis's brave advocacy of the sexual rights of the disabled was suspended four months after it was announced, having provoked fury and threats of funding withdrawal from the good burghers of Zurich.

Another positive benefit of the current era's sexual liberation has been a perceptible trend towards a limited 'feminisation' of men. This is more than a merely stylistic fashion, with celebrities such as the footballer David Beckham openly adopting more feminine styles; nor is it a function of more inclusive attitudes towards gay men. The more hidden signs of feminisation are to be found in traditionally heterosexual men. These indications range from increased interest in personal grooming, to acceptance of a greater role in childrearing to a greater willingness to discuss personal feelings and the intricacies of loving relationships. While men are still by far the more likely sex to pursue better and different orgasmic experiences in opportunistic extra-marital sexual relationships, that situation too is shifting seismically.

In the macho society of Argentina, it has been noticed by researchers recently, oral sex and anal intercourse have lost their status, especially among younger male Argentines, and come to be thought of as activities to be indulged in with prostitutes. Mutual masturbation with a female partner is increasingly viewed as a means of sharing pleasure and orgasm without penetration.

Some fascinating data unearthed by Dr Shirley P. Glass, a clinical psychologist from Baltimore, in a 2003 book on infidelity, provided evidence that men are looking for more from such affairs. 'The old "sex first" definition of men's affairs is changing,' said Dr Glass. 'In this new crisis of infidelity, more men are now following what has traditionally been a female pattern, that of emotional bonding first and sex later . . . an increased number of unfaithful husbands have deep emotional connections to their affair partner.'

'In my clinical sample,' she wrote, '83 per cent of involved women and 61 per cent of involved men characterised their extramarital relationship as more emotional than sexual.'

The trend was almost as marked in a poll Dr Glass conducted in 1980 at Baltimore Washington International Airport. Here, she handed out 1,000 questionnaires to travellers. An amazing 300 mailed them back. Of the airport sample, 71 per cent of involved women and 44 per cent of involved men described their extramarital affair as 'more emotional than sexual'. Dr Glass went on to characterise such relationships as more dangerous to marriages than old-style affairs between, typically, older men with time, money and opportunity to spare and younger, sometimes vulnerable, single women curious to experiment. 'The most threatening kind of infidelity combined a deep emotional attachment with sexual intercourse,' she commented.

Hand in hand with such feminisation of men came what critics such as Shmuley Boteach have described as a masculinisation of women. Shirley Glass noted in her book that, 'In a sexual addiction recovery program, 16 per cent of all of the sex addicts were women', but Boteach has publicly taken issue with the sexually rapacious ethic exemplified by the women characters in *Sex and the City*. He believes that 'women can be everything men are, and more', but argues that in ancient times women were lauded for their power and sexual prowess, and that the fight for equality has lost them their souls. Equality for women, Boteach says, was a step back for women. 'Real greatness,' he says, 'is where you don't have to prove yourself constantly.'

It is interesting, incidentally, that in order to make the *Sex and the City* women plausible as sexual huntresses with the pursuit of orgasm their ultimate goal, the show's creators had to insulate them from any intrusive sense of reality. They live in the rarefied upper echelons of Manhattan, never worry about money, never seem to do much by way of work and have little to do but eat and drink endlessly (but naturally, never put on an ounce in weight). A shard of the reality normal women face, one can easily conclude, would be like a piece of broken glass under their delicate feet. Yet, as evidenced by sentiments such as the *Cosmopolitan* writer who fakes orgasms to save time, or Ann Landers's 72 per cent of a huge sample of women who would really rather be doing something other than having sex, *Sex and the City* was never anything more than make believe.

Better supported by real-life evidence, however, is a profound shift in older people's attitudes towards sex. If you use the media as your sole indicator, it would appear that something quite drastic happened to all the longhaired lover boys and mini-skirted chicks of the 1960s when they hit sixty; they seem by most accounts to have lost their prodigious former libido. The orgasm must surely, a flick through any newspaper or magazine will show, be exclusively the privilege of young, firm-fleshed people.

Yet an ICM poll of 1,000 married couples published by the *Reader's Digest* in 2003 showed that sexual satisfaction was more important during mid-life than in early marriage, and was 'surprisingly important' during a couple's oldest years. The survey discovered that those aged 24–34 were the *least* concerned with sexual satisfaction, while those of both genders between 35–44 placed a higher value on having good and frequent sex – with wives rating it more highly than husbands. There was a dip in interest in sex (but not in love) for the 45–65 age group – but every aspect of a sexual relationship was still more important to men over 65 than it was to the youngest husbands.

Shere Hite made the point too in *The New Hite Report* (2000) that older women are more likely than younger ones to enjoy more multiple orgasms. 'Confusion between reproductive activity and sexual pleasure is playing havoc with our lives,' Hite wrote. 'It is true that the capacity to reproduce ends at menopause, and that vaginal lubrication can decrease, but women's sexual arousal or orgasm capacity actually increases.' Hite added that Hormone Replacement Therapy, which boosts oestrogen levels to pre-menopausal strength, can overcome the dryness question. Hite was writing before some of the negative aspects of HRT were widely known.

Such is the arrogance of youth through history, especially in the West where age is less venerated than in the East, that the idea of orgasm being of the remotest interest to older people can arouse feelings of palpable disgust in young people. The novelist Fay Weldon has said that she believes at young people don't like to contemplate sex at sixty years-plus because they are squeamish about 'crêpey flesh'. 'I think that is why I cried when I was thirty, because I didn't understand that attraction had not all that much to do with youth and firmness. It is only when you are at childbearing age that physical appearance is important. Once you are beyond procreation, you relate more as human beings. The male/female pairing off goes on, but it is more spiritual – less to do with lust and more to do with love, which is not a bad thing.'

Only today is the orgasm beginning to be seen as the province of older Western people also. But this is not such a new phenomenon in Southern Europe or South America. A Brazilian psychologist and gerontologist, Lucia Helena de Freitas, has studied the sexuality of a group of retired people who involved themselves in cultural activities at a social club. She made the discovery, alarming to some maybe, that 73.8 per cent of them still had sex, with 35.7 per cent claiming they made love two or three times a week. Almost all de Freitas's interviewees (90.5 per cent) felt sex was necessary to them; 95.2 percent believed that sexual desire does not end with age; 40 per cent said it *increases* with age; and a third averred that sexual *pleasure* increases with age. Almost 30 per cent said they were able to reach orgasm quickly, although 40.5 percent said they needed more time these days. Just 13.5 per cent of the women at the social club said they experienced any change in their sex life as a result of menopause; some said they now reached orgasm more quickly. Just 4.8 percent said they suffered impotency. De Freitas concluded that, in Brazil at least, the frequency of sex typically decreases with age – but that its quality does not.

In Britain the Pennell Initiative for Women's Health was set up in 1997 as a charitable trust to champion the cause of the health and, specifically sexual, needs of women over the age of 45. Chaired by the former editor of the BBC radio show *Woman's Hour,* Sandra Chalmers, the organisation took its name from a species of clematis, the Vyvyan Pennell, which blooms with a double flower in the summer and then again with a single flower each autumn.

Dame Rennie Fritchie, a Civil Service Commissioner, pro-Chancellor of Southampton University and Pennell's President, spoke publicly in 2001 at the launch of a study of the sex lives of older women about the misconception, that 'old women don't have sex': 'When I flick through my television channels late at night, I come across all these bodies and limbs and thrashings and gruntings. I'm fifty-nine and it makes me think what a limited view the younger generation has of sexuality. This report takes us beyond these thrashings, and highlights the fact that sexuality in older age is a whole chapter in itself.'

The study revealed that the main reason the misunderstanding had arisen was that research had always previously been restricted to the number of sexual encounters a woman reported in a given period, with the 'sexual encounter' taken to mean an act of penetrative sex. But, as Julia Cole, Pennell's Development Director, explained: 'A researcher asking direct questions might come away with the impression that older women have little sex, whereas the woman may merely be having less penetrative sex and instead enjoying plenty of other forms of sexual or sensual activity which are just as important to her.

'I have found that women in mid and later life have great sexual relationships. They may make love less often than when they were younger, but more frequently than people would imagine, and many go on enjoying sex into old age.' Older women, according to Julia Cole, are more aware of what they want, and better able to voice those desires, have an acceptance of their bodies, fewer inhibitions and no fear of getting pregnant – most of what is needed, that is to say, for a satisfactory sex life.

But, as Dame Rennie explained, sexuality for older women can encompass far more than sex. 'One of the things I love about the Pennell Report is that it lists having your hair combed as a sensual act,' she commented. 'If you live on your own, as many older women do, you may desperately miss being touched in a loving or caring way, and having someone comb your hair might be all it takes to make you feel good. If something makes you feel more womanly and good then it is a part of your sexuality. Sexuality in older women is not about grannies in hotpants. It is about feeling good about yourself. All of yourself.'

17

Epilogue: How Was It for Us?

'We're all in this together – by ourselves'
Comedienne Lily Tomlin

One of the few certainties about life is that there are no certainties. However, if there is one thing we can be reasonably sure about, it is that in many areas of previous uncertainty, especially concerning medical matters, the current century will be an age of closure.

We see this proved in our understanding of the physiology of the human heart, on which the book is now probably very near to being closed. Despite the ever-looming risk of being hugely embarrassed by some unimaginable scientific advance, we seem now to know practically everything there is to know about keeping this glorified pump working to the best of its design capability.

The book will most likely, in the next few decades, be similarly closed on the scourges of cancer, dementia, AIDS, many mental illnesses and infectious diseases. New problems naturally will arise; in their non-conscious way viruses, pestilences and plagues are problem-solving organisms, for whom we are the problem.

Will our understanding and enjoyment of sex also reach its terminal velocity in the coming century?

Will the greatest number of people find themselves enjoying the best possible orgasms for the optimum possible time?

The answer to this has to be, very possibly, yes. A broad scan of the sexual scene worldwide today suggests rather cogently that while the journey towards widespread optimal sexual enjoyment has a long way to go, humanity across a wide variety of cultures and socio-economic classes is very much on the right track. The right track towards what, though?

Towards getting the pursuit of orgasm into perspective, whereby it occupies neither too little, nor too much of our time. Towards ensuring that the desire for orgasm is equally felt and discharged by both sexes. Towards ensuring that ancient, pervasive myths about sex and orgasm are debunked for ever.

This may seem an over-optimistic prediction. But just look at ten instances of where we stood regarding the orgasm less than a hundred years ago.

- Most people, doctors and educated women included, believed there was no such thing as the female orgasm, and that women who enjoyed sex were mentally ill, morally degenerate or both.

- The only point of sex, for most people, was that the male should ejaculate as quickly as possible with a view to getting the whole sordid business over swiftly – and, ideally, impregnating the female.

- The harmless, largely beneficial practice of masturbation was regarded as sinful, psychologically corrosive and medically dangerous. Doctors were so worried about its effects that they happily invented evidence and syndromes to convince people to stop doing it.

- Most women were unaware they owned a clitoris, or where, if they knew of its existence, it was located.

- Contraception was regarded as a social, moral and medical evil. (Today, even many Catholics, for whom it is a sin, practise it nevertheless.)

- Progressives who believed the orgasm was not only natural but a human right for both sexes were, nevertheless, highly dubious over whether the working class should be let in on their secret.

- Progressives were also in thrall to a completely false belief that the only orgasm that was worth having was a simultaneous climax attained by penetrative sex alone.

- Educated people who believed in the female orgasm were convinced by Freud that the clitoral orgasm was 'immature' and undesirable and that the only valid, adult sexual response was the 'vaginal' orgasm.

- There was no sex education of any kind for children or teenagers, apart from misinformation.

- Women who were raped were regarded as sinful *themselves* and legally and socially marginalised.

Let us by contrast briefly scan the sexual scene as it is across the world in the early-twenty-first century, with special regard to orgasm, its status and pursuit. Not all these developments, as can be seen, are necessarily 'good news'; the important point is that they are *in* the news.

- Orgasm is out in the open, but it has retreated from its status as the absolute *sine qua non* of relationships. The Canadian website QueenDom's survey of 15,000 Western women and men reveal that 35 per cent of women and 29 per cent of men say they do not necessarily need to orgasm to enjoy sex.

- Alleged new forms of orgasm, from the 'whole body orgasm' to a new 'heart orgasm' – another borrowing from Tantric sex – are widely discussed in women's and men's magazines.

- Sexual foreplay is a matter of everyday media debate. One recent global sex survey by Durex asked what constitutes ideal foreplay. The top answer for men was oral sex. For women, the top choices were touching, feeling and kissing – and a romantic dinner *à deux*.

- The National Health and Social Life Survey, a study conducted by researchers at the University of Chicago and reported in the *Journal of the American Medical Association*, finds that 43 per cent of women complained of dissatisfaction in the bedroom, 10 per cent more than the men who participated in the study. They declare lack of libido 'an epidemic public health concern'.

- Vibrators are now widely acceptable and available in a huge variety of designs. The hard, round head of the device has metamorphosed into such variations as phallus-shaped dolphins and other animals. Electronics made possible tiny but powerful machines such as the Cybervibe, a 10-speed pseudo-phallus with a remote control. And plastics can now be supplemented by silicon, which retains body heat, and a life-like material called Cyberskin. Vibrators based on the vagina are also available. 'To have a phallus shape for the clitoris is silly,' says an American former art student, porn star and producer of feminist porn films, Candida Royalle, who is now big in the vibrator business. Her models include the Petite, which is about four inches long and looks like a peach-coloured, slightly curved mobile phone; the Superbe, a larger chartreuse model; and the Magnifique, a seven-and-a-half-inch version. In San

Francisco, a feminist sex shop, Good Vibrations, stocks almost two hundred different designs of vibrator. One of its fastest moving lines, the Magic Wand, is made by Hitachi.

- Where women used to be accused of 'frigidity', it is now men who feel the pressure to 'perform'. When Kinsey reported in the middle of the twentieth century, men thought one to two minutes was a reasonable time to elapse before ejaculating. Today, men asked at the Australian Centre for Sexual Health how long they think intercourse should last reply 'between ten and fifteen minutes'.

- In Germany, research shows that the proportion of women experiencing orgasm increased sharply in the 1970s. By the age of 27, 99 per cent of women there have experienced orgasm. The rate of orgasm during sex has also increased. There is, reports Humboldt University in Berlin, 'a growing aversion to orgasm achieved with all manner of tricks, and used as a measure of male or female performance, celebrated as a victory in joint conflict, and feared as a stress-obsessed prestige event. Instead, the individual quality of a steady relationship is sought, linked with closeness, trust, warmth, carefree pleasure, and unpredictable, uncalculating, uncalculated affection within the total erotic form. Cuddling is back in fashion; compulsive or cheap commonplace sex is out.'

- Women in Australia, counselled for the past twenty years by sex expert and author Dr Rosie King, used to ask simply how they could achieve orgasm of any sort. 'Now, they want G-spot orgasms, female ejaculation, multiple orgasms, preferably all of them at once, over and over,' says Dr King.

- The most common problem seen today at the same centre in Sydney is 'desire discrepancy', where one partner seeks more sex than the other. Says Dr King: 'Often when men say they want more sex, they are really saying they want more love and affection. It's easier to ask for sex than emotional comfort. The reverse tends to be true for women. While men need sex to become intimate, women tend to need intimacy to desire sex.'

- Sex is in crisis in fundamentalist Islamic communities. In Iran, for example, lay teachers and mullahs present sexual behaviour of all kinds as polluting, rendering the participant spiritually and physically unclean and obstructing spiritual readiness. While the polluting effects on body and spirit of excretion can be washed away in the bathroom, sexual contact or orgasm requires a ritualised bathing with spiritually cleansing words.

What, then, might be said about the orgasm's future in this era of closure?

- Implants and artificial, pharmacological methods, especially Viagra for women, will not work. Orgasmatron-like devices will always be a joke.

- The worldwide battle over sex between men and women will continue. It may be inherent in our design that there be a continual tension between the sexes.

- There will be, however, a levelling of the psychological battleground; women and men will increasingly understand what one another needs and require from sex. The media will continue to have a huge part to play in propagandising this process.

- The conquering, or, perhaps, mere disappearance of AIDS will trigger the biggest explosion in sex since the Pill in the 1960s.

- Some form of male pharmacological contraception will exacerbate this – but at the expense of sexual health, which will deteriorate even if AIDS disappears.

- Virtuality – cybersex and various remote methods of promoting orgasm – will become the masturbation aid of the masses. Orgasm will thus become an ever more lonely pursuit. The middle class, however, will be very superior about having 'natural', non-aided sex – with a live partner.

- The orgasm and sexual pleasure will continue to be the number one subject in everyone's life, to dominate the media, the arts, and most people's every waking moment.

- The elderly will stake out their right to join in the fun.

- Women in the Third World and fundamentalist regimes will begin, very slowly, to make progress towards getting their share of pleasure from sex, too. The Internet and satellite TV will play a major role in this, just as they have in the spread of democratic ideals, which are more common now than at any time in history.

- A dampening down of male aggression will accompany this sexualisation of women in developing countries. Much crime in the Western world is caused by sexually inadequate and unfulfilled men, and much of the raw violence in fundamentalist societies, too, must stem from male sexual frustration; in keeping women down, men in such societies have deprived themselves of the joy of shared orgasmic pleasure. When the blinkers come off, the world will become a calmer, better place. Orgasm, which has caused such trouble for so long, may yet be a powerful force in mankind's mental wellbeing.

Early in this full and, I hope, quite illuminating history of the orgasm, I put forward the proposition that the compound $C_{19}H_{28}O_2$ – better known as testosterone, the primary generator of sexual desire for both males and females – has been the single most influential chemical in human history.

For some, this may seem a hyperbolic claim. Other compounds, such as H_2O, or the complex of hydrocarbons which comprise gasoline, or a range of explosives, or even the simple element *Au* – gold – have had not exactly has an unprofound effect on everything from personal psychology to global geopolitics.

Yet I would urge readers at this late stage to reflect again, in the light of the proceding pages, on the extraordinary, unique power the desire for orgasm has exerted over the course of history.

From the Roman general Antony, who abandoned his career to pursue sexual pleasure with Cleopatra; to Edward and Mrs Simpson; to almost the entire Kennedy family; to Oscar Wilde and Lord Alfred Douglas; to Mata Hari (the Dutch Nazi spy who secured sensitive military information by seducing senior Allied officers); to John Profumo (the British defence minister who was found to be sharing a prostitute, Christine Keeler, with a naval attache at the Soviet Embassy in London, and ultimately brought down a government); to Prince Charles (who baffled most of mankind by abandoning marriage to one of the most desired women in the world in favour of a sexual liaison with a far older woman, who to most of the world looked exactly like a horse); in each case, the desire for sexual pleasure has significantly altered outcomes. There are many hundred similar examples that could be advanced.

Consider, too, the tidal pull that the urge for sexual delight has brought to bear on billions of ordinary people's lives. It is one of those easily overlooked but incontrovertible facts that every single one of us who has ever lived on Earth – some 110 thousand million souls, it is currently estimated – is the result of at least one person having wanted and achieved an orgasm. Even test tube babies owe their existence to a man masturbating quietly in a clinic booth somewhere.

Imagine for a moment, furthermore, that each baby born is in reality the result of on average – what? – a hundred, a thousand, sexual acts and we begin to appreciate that, if sexual energy could somehow build up in clouds, as that ludicrous charlatan, Wilhelm Reich, firmly believed, we could be shrouded in a permanent fog of the stuff. We would live and breathe orgasmic longing, orgasmic tension and orgasmic release; we would be surrounded by a sort of orgasmic ectoplasm, all day every day of our lives.

A civilisation living and breathing orgasmic longing, orgasmic tension and orgasmic release? Preposterous. Yet look around the world and what do we see? Funnily enough, it is billions of people entranced, obsessed, fixated on sex, spending large portions of their lives being dominated by the thought of having orgasms, of not having orgasms, of being desperate to have orgasms like the ones they read about all the time, of being medically or psychologically unable to have orgasms – or being forbidden by religious codes to have orgasms.

Even though Reich's 'Orgone', as he termed the physical matter of sexual energy, was patently a fantasy prompted by heaven knows what personal psychiatric crisis, there is a sense in which the old fraud may have had a point.

Bibliography and Webography

Paul Ableman, *The Mouth and Oral Sex*, Running Man, 1969; Sphere, 1972 (as *The Mouth*)

Federico Andahazi, *The Anatomist*, trans. Alberto Manguel, Doubleday, 1998

Amy Anderson, 'My G-spot Secret', *London Evening Standard*, 6 May 2003

Stephen Bailey (ed.), *Sex*, Cassell & Co., 1995

Françoise Barret-Ducrocq, *Love in the Time of Victoria: Sexuality and Desire Among Working-Class Men and Women in Nineteenth-Century London*, trans. John Howe, Penguin Books, 1992

Fanny Beaupré and Roger-Henri Guerrand, *Le confident des dames. Le bidet du XVIIe au XXe siècle: histoire d'une intimité*, La Découverte, 1997

Stephen Beckerman and Paul Valentine, *Cultures of Multiple Fathers: The Theory and Practice of Partible Paternity in Lowland South America*, University Press of Florida, 2002

Antony Beevor, *Berlin: The Downfall 1945*, Viking, 2002

Elizabeth Benedict, *The Joy of Writing Sex: A Guide for Fiction Writers*, Souvenir Press, 2002

Jennifer Berman, M.D., and Laura Berman, Ph.D., *For Women Only*, Virago, 2002

Simone Bertiére, *Marie-Antoinette l'insoumise*, Editions de Fallois, 2002

Geneviève Bianquis, *Amours en Allemagne à L'Epoque Romantique*, Librairie Hachette, 1961

Tim Birkhead, *Promiscuity: An Evolutionary History of Sperm Competition and Sexual Conflict*, Faber and Faber, 2000

Dr Sue Blundell, *Women in Ancient Greece*, British Museum Press, 1999

J.G. Bohlen, 'State of the Science of Sexual Physiology Research'. In C.M. Davis, ed., *Challenge in Sexual Science*, Philadelphia, 1983

Christopher Booker, *The Neophiliacs: The Revolution in English Life in the Fifties and Sixties*, Collins, 1969

Shmuley Boteach, *Kosher Sex: A Recipe for Passion and Intimacy*, Duckworth, 1998

Jacob Bronowski, *The Ascent of Man*, Little, Brown, 1973

Geraldine Brooks, *Nine Parts of Desire: The Hidden World of Islamic Women*, Penguin, 1996

Sir Richard Burton and F.F. Arbuthnot, trans., *The Illustrated Kama Sutra: Ananga-Ranga. Perfumed Garden: The Classic Eastern Love Texts*, Octopus Publishing, 1987

David M. Buss, *The Evolution of Desire: Strategies of Human Mating* (revised edition), Basic Books, 2003

David M. Carr, *The Erotic Word: Sexuality, Spirituality, and the Bible*, Oxford University Press USA, 2003

Brian D. Carroll, '"I indulged my desire too freely": Sexuality, Spirituality, and the Sin of Self-Pollution in the Diary of Joseph Moody, 1720–1724', *William and Mary Quarterly*, Vol. 60, No. 1, January 2003

Jolan Chang, *The Tao of Love and Sex: The Ancient Chinese Way to Ecstasy*, Wildwood House, 1976

Mantak Chia and Douglas Abrams Arava, *The Multi-Orgasmic Man: Sexual Secrets Every Man Should Know*, HarperCollins, 1996

Dr Eustace Chesser, *Woman and Love*, Jarrolds, 1962

John Cleland, *Fanny Hill: Memoirs of a Woman of Pleasure*, Wordsworth Editions, 2000

Julia Cole, *Stay Together For Ever*, Vermilion, 2003; *www.juliacole.org* and *www.emotionalbliss.co.uk*

Harry Daley, *This Small Cloud*, Weidenfeld & Nicolson, 1987
Emma Dickens, *Immaculate Contraception: The Extraordinary Story of Birth Control – From the First Fumblings to the Present Day*, Robson Books, 2000
E. Elkan, 'Evolution of Female Orgastic Ability – A Biological Survey', *International Journal of Sexology*, 1948
Verrier Elwin, *The Kingdom of the Young*, Oxford University Press, 1968
Roy Eskapa, *Bizarre Sex*, Quartet Books, 1987
Nicholas Farrell, *Mussolini: A New Life*, Weidenfeld & Nicolson, 2003
Helen Fisher, *Anatomy of Love: A Natural History of Mating, Marriage and Why We Stray*, Ballantine, 1992
Robert Flacelière, *L'Amour en Grèce*, Librairie Hachette, 1960
Michel Foucault, *The History of Sexuality, An Introduction*, Random House, 1978
Jonathan Franzen, *The Corrections*, Fourth Estate, 2001
Antonia Fraser, *Marie Antoinette*, Weidenfeld & Nicolson, 2002
Robert A. Freitas Jr., 'Sex in Space', *Sexology Today*, No. 48, April 1983, *www.rfreitas.com/Astro/SexxxInSpace.htm*
Peter Gay, *Education of the Senses, The Bourgeois Experience*, Oxford University Press, 1984
Elizabeth Gips, *The Scrapbook of a Haight Ashbury Pilgrim: Spirit, Sacraments, & Sex in 1967–68*, Changes Publishing, 1995
Shirley P. Glass, PhD, with Jean Coppock Staeheli, *Not 'Just Friends': Protect Your Relationship from Infidelity and Heal the Trauma of Betrayal*, The Free Press, 2003
Stephen Jay Gould, *Bully for Brontosaurus: Reflections in Natural History*, Vintage, 2001
Madeline Gray, *The Normal Woman*, Scribner, 1967
Shirley Green, *The Curious History of Contraception*, Ebury Press, 1971
Lesley A. Hall, *The Other in the Mirror: Sex, Victorians and Historians*, Wellcome Institute, 1998, *http://homepages.primex.co.uk/~lesleyah/sexvict.htm*
Frank Harris, *My Life and Loves*, Avalon, 2000

Gilbert H. Herdt, *Guardians of the Flutes, Idioms of Masculinity, A Study of Ritual Homosexual Behavior*, McGraw-Hill, 1980

Tom Hickman, *The Sexual Century: How Private Passion Became a Public Obsession*, Carlton Books, 1999

Humboldt University, Berlin, *Magnus Hirschfeld Archive for Sexology, www2.huberlin.de/sexology/*

Shere Hite, *The Hite Report*, Macmillan, 1976; *The New Hite Report*, Hamlyn, 2000

George Jacobs and William Stadiem, *Mr S: The Last Word on Frank Sinatra*, Sidgwick and Jackson, 2003

Wendell Stacy Johnson, *Sex and Marriage in Victorian Poetry*, Cornell University Press, 1975

James Howard Jones, *Alfred C. Kinsey: A Public/Private Life*, W.W. Norton, 1997

Brett Kahr, 'The History of Sexuality: From Ancient Polymorphous Perversity to Modern Genital Love', *Digital Archive of Psychohistory*, Vol. 26, No. 4, Spring 1999 *http://www.geocities.com/kidhistory/ja/hissex.htm*

Barbara Keesling, *How to Make Love All Night*, HarperCollins, 1994

Otto Kiefer, *Sexual Life in Ancient Rome* (1934), ninth impression, Abbey Library, 1976

Pierre Kohler, *La Dernière Mission: Mir, l'aventure humaine*, Editions Calmann-Lévy, 2000

Marvin Krims, 'A Psychoanalytic Exploration of Shakespeare's Sonnet 129', *Psychoanalytic Review*, Vol. 86, 1999, *www.clas.ufl.edu/ipsa/journal/articles/psyart2000/krims02.htm*

Irma Kurtz, *Irma Kurtz's Ultimate Problem Solver*, Avon, 1995

Thomas W. Laqueur, *Solitary Sex: A History of Masturbation*, Zone Books, 2003

Sally Lehrman, 'The Virtues of Promiscuity', *www.alternet.org/story.html?StoryID=13648*

Howard S. Levy, *Chinese Footbinding: The History of a Curious Erotic Custom*, in various editions from 1966 onwards

Hans Licht, *Sexual Life in Ancient Greece*, George Rowledge and Sons, 1932

Hanny Lightfoot-Klein, M.A., 'The Sexual Experience and Marital Adjustment of Genitally Circumcised and Infibulated Females in The Sudan', *Journal of Sex Research*, August 1989

Brenda Love, *Encyclopedia of Unusual Sexual Practices*, Abacus, 1995

L.J. Ludovici, *The Final Inequality: A Critical Assessment of Woman's Sexual Role in Society*, Frederick Muller, 1965

Rachel P. Maines, *The Technology of Orgasm: 'Hysteria', the Vibrator, and Women's Sexual Satisfaction*, John Hopkins University Press, 1999

Steven Marcus, *The Other Victorians: A Study of Sexuality and Pornography in Mid-Nineteenth-Century England*, W.W. Norton, 1985

Donald S. Marshall and Robert C. Suggs, eds. *Human Sexual Behavior: Variations in the Ethnographic Spectrum*, Basic Books, 1971

William Masters and Virginia Johnson, *Human Sexual Response*, Little, Brown, 1966

John Maynard, *Victorian Discourses on Sexuality and Religion*, Cambridge University Press, 1993

Margaret Mead, *Male and Female: A Study of the Sexes in a Changing World*, William Morrow, 1967

Johann Jakob Meyer, *Sexual Life in Ancient India: A Study in the Comparative History of Indian Culture* (1930), republished Dorset Press, 1995

Robert T. Michael and Edward O. Laumann, *Sex in America: A Definitive Survey*, Warner Books, 1995

Desmond Morris, *The Naked Ape*, Vintage, 1994; *The Human Zoo*, Vintage, 1994; *Manwatching*, Triad, 1978

Toni Morrison, *The Bluest Eye*, Vintage, 1999

Clelia Duel Mosher, *The Mosher Survey: Sexual Attitudes of 45 Victorian Women*. James Mahood and Kristine Wenburg, eds, Arno Press, 1980

Kim Murphy, 'Frigid Victorian Women?' *Citizens' Companion*, December-January 2002–2003

John T. Noonan, *Contraception*, Harvard University Press, 1966

Jeannette Parisot, *Johnny Come Lately: A Short History of the Condom*, tr. Bill McCann, Journeyman, 1987

Burgo Partridge, *A History of Orgies* (1958), republished Prion Books, 2002

Ian Penman, 'The Big "Oh!": Having them, delaying them, faking them; The troubled history of orgasms in cinema', *Independent*, November 30, 1989

Melanie Phillips, *The Ascent of Woman: A History of the Suffragette Movement*, Little, Brown, 2003

Roy Porter, *The Greatest Benefit to Mankind: A Medical History of Humanity*, Fontana Press, 1999

Gordon Rattray Taylor, *Sex in History*, Thames & Hudson, 1953

Elizabeth Roberts, *A Woman's Place: An Oral History of Working-Class Women 1890–1940*, Blackwell, 1986

Alan Rusbridger, *A Concise History of the Sex Manual*, Faber and Faber, 1986

George Ryley Scott, *Phallic Worship: A History of Sex and Sexual Rites*, Luxor Press, 1966

Mary Jane Sherfey, *The Nature and Evolution of Female Sexuality*, Random House, 1972. Also 'A Theory of Female Sexuality', a 1966 essay included in *Sisterhood Is Powerful: An Anthology of Writings from the Women's Liberation Movement*, ed. Robin Morgan, Vintage Books, 1970

Edward Shorter, *The Making of the Modern Family*, Basic Books, 1977

G.L. Simons, *The Illustrated Book of Sexual Records*, Virgin Books, 1986

Joan Smith, *Moralities: Sex, Money and Power in the 21st Century*, Penguin, 2001

Donna C. Stanton, ed., *Discourses of Sexuality: From Aristotle to AIDS*, University of Michigan Press, 1992

Gloria Steinem, *A Book of Self-Esteem: Revolution from Within*, Bloomsbury, 1992

David Stevenson, *The Beggar's Benison: Sex Clubs of Enlightenment Scotland and Their Rituals*, Tuckwell Press, 2001

Houzan Suzuki, *Zen Tantra: Total Orgasm, www.mumyouan.com*

Donald Symons, *The Evolution of Human Sexuality*, Oxford University Press, 1979

Thomas Szasz, *Sex: Facts, Frauds and Follies*, Blackwell, 1981

Simon Szreter, *Fertility, Class and Gender in Britain, 1860–1940*, Cambridge University Press, 1996

Reay Tannahill, *Sex in History*, Scarborough House, 1992

C. Tarvis and S. Sadd, *The Redbook Report on Female Sexuality*, Delacorte Press, 1977

Stanley Thomas, 'Did you have your orgasm today?' *The Week* (India), Dec. 6, 1998, *http://www.the-week.com/98dec06/ cover. htm*

Randy Thornhill and Craig T. Palmer, *A Natural History of Rape: Biological Bases of Sexual Coercion*, MIT Press, 1999

Lionel Tiger, *The Pursuit of Pleasure*, Little, Brown, 1992

E.B. Vance and N.N. Wagner, 'Written Descriptions of Orgasm: A Study of Sex Differences', *Archives of Sexual Behavior*, Vol. 5, 1976

Theodore van de Velde, *Ideal Marriage: Its Physiology and Technique*, Heinemann, 1960 (last edn.)

Robert Hans van Gulik, *Sexual Life in Ancient China*, E. J. Brill, 1961

Andre Van Lysebeth, *Tantra: The Cult of the Feminine*, Red Wheel/Weiser, 1995

Mallanaga Vatsyayana, Wendy Doniger (tr.), Sudhir Kakar (tr.), *Kamasutra*, Oxford University Press, 2002

'Walter', *My Secret Life*, Wordsworth Editions, 1996

Hope Wells, 'Omigod! The things you didn't know about sex; facts about foreplay, sex techniques and sexual behavior', *Cosmopolitan*, October 1996

Naomi Wolf, *The Beauty Myth*, Vintage, 1990; *Promiscuities: A Secret History of Female Desire*, Chatto and Windus, 1997

Alayne Yates, *Sex Without Shame: Encouraging the Child's Healthy Sexual Development*, William Morrow, 1978

Toby Young, *How to Lose friends and Alienate People*, Abacus, 2002.

Zaehner, R.C., *Mysticism Sacred and Profane: An Inquiry into Some Varieties of Preternatural Experience*, Clarendon Press, 1957; Oxford University Press, 1961.

Additional Reading

Iwan Bloch, *Anthropological Studies in the Strange Sexual Practices of all Races in All Ages*, Anthropological Press, 1933

Sally Cline, *Women, Celibacy and Passion*, Andre Deutsch, 1993
John D'Emilio and Estelle B. Freedman, *Intimate Matters: A History of Sexuality in America*, Harper and Row, 1988
Robert T. Francoeur, *The Descriptive Dictionary and Atlas of Sexology*, Greenwood Press, 1991
Michael Gordon, 'From an Unfortunate Necessity to a Cult of Mutual Orgasm: Sex in American Marital Education Literature, 1830–1940', in James M. Henslin and Edward Sagarin, eds., *The Sociology of Sex*, Schocken Books, 1978
Walter Kendrick, *The Secret Museum: Pornography in Modern Culture*, University of California Press, 1987
Edward O. Laumann, John H. Gagnon, Robert T. Michael and Stuart Michaels, *The Social Organisation of Sexuality: Sexual Practices in the United States*, University of Chicago Press, 1994
G. Legman, *Oragenitalism: Oral Techniques in Genital Excitation*, Julian Press, 1969
Hoag Levin, *American Sex Machines: The Hidden History at the US Patent Office*, Adams Media Corp., 1996
Brenda Love, *Encyclopedia of Unusual Sex Practices*, Barricade Books, 1992
R. Osborne, ed., *Classical Greece*, Oxford, 2000; chapter on 'Private Life', by James Davidson, Reader in Ancient History at Warwick University
G. L. Simons, *Simons' Book of World Sexual Records*, Pyramid Books, 1975
Dr Robin Smith, *The Encyclopedia of Sexual Trivia*, St Martin's Press, 1990
Paul Tabori, *The Humor and Technology of Sex*, Julian Press, 1969
Sallie Tisdale, *'Rockets, earthquakes, fireworks, full-excursion pelvic thrusting, the final engorgement of the late plateau phase, and the high mountaintop of love of which the poets sang: The Orgasm, what else?'*, *Esquire*, October 1994
Theodore Zeldin, *An Intimate History of Humanity*, Vintage, 1998